PLACE IN RETURN BOX to remove this checkout from your record.
TO AVOID FINES return on or before date due.
MAY BE RECALLED with earlier due date if requested.

DATE DUE	DATE DUE	DATE DUE
NOV 2 4 2000 0 3 2 3 0 1	JAN 1 0 2005 0111 05	
JUN 0 7 2001 0518 01		
OCT 1 5 2001 1 0 1 5 0 1 0 7 3 0 1 5		

Fate and Management of Turfgrass Chemicals

ACS SYMPOSIUM SERIES 743

Fate and Management of Turfgrass Chemicals

J. Marshall Clark, EDITOR
University of Massachusetts

Michael P. Kenna, EDITOR
United States Golf Association

American Chemical Society, Washington, DC

Library of Congress Cataloging-in-Publication Data

Fate and management of turfgrass chemicals / J. Marshall Clark, Michael P. Kenna.

   p.   cm.—(ACS symposium series : 743)

Includes bibliographical references and indexes.

ISBN 0–8412–3624–0

1. Turfgrasses—Diseases and pests—Control. 2. Agricultural chemicals. 3. Turf management.

I. Clark, J. Marshall (John Marshall), 1949– . II. Kenna, Michael P. III. Series.

SB608.T87F38   2000
635.9´642995—dc21   99–38608
                                                 CIP

The paper used in this publication meets the minimum requirements of American National Standard for Information Sciences—Permanence of Paper for Printer Library Materials, ANSI Z39.48-94 1984.

PRINTED IN THE UNITED STATES OF AMERICA

Foreword

THE ACS SYMPOSIUM SERIES was first published in 1974 to provide a mechanism for publishing symposia quickly in book form. The purpose of the series is to publish timely, comprehensive books developed from ACS sponsored symposia based on current scientific research. Occasionally, books are developed from symposia sponsored by other organizations when the topic is of keen interest to the chemistry audience.

Before agreeing to publish a book, the proposed table of contents is reviewed for appropriate and comprehensive coverage and for interest to the audience. Some papers may be excluded in order to better focus the book; others may be added to provide comprehensiveness. When appropriate, overview or introductory chapters are added. Drafts of chapters are peer-reviewed prior to final acceptance or rejection, and manuscripts are prepared in camera-ready format.

As a rule, only original research papers and original review papers are included in the volumes. Verbatim reproductions of previously published papers are not accepted.

ACS BOOKS DEPARTMENT

Contents

INDEXES

Preface

Turfgrass provides beautiful green areas within our urban and suburban landscapes. Turfgrasses have been maintained by humans to enhance their environment for more than ten centuries. The complexity and comprehensiveness of the environmental benefits of turfgrass that improve our quality of life are just now being documented through research.

A properly planned and maintained turfgrass facility, such as a golf course, offers a diversity of functional benefits to the overall community, in addition to the physical and mental benefits provided by outdoor exercise. Given the pest complex that damages turfgrass, a variety of cultural practices, nutrients, and pesticides may be used to promote turf health. With the dramatic growth in popularity during the past 20 years, turfgrass sports, especially golf, now face the challenge of providing quality playing surfaces while minimizing or eliminating any deleterious impact these practices may have on the environment. One of the greatest concerns is that nutrients and pesticides used to maintain turfgrass will pollute drinking water or will be responsible for other unwanted exposures that could affect human health. In addition, there is concern for preserving the natural ecology of surface waters and the wildlife habitat within or abutting areas where turfgrasses are used.

In response to such concerns, the U.S. Golf Association (USGA) sponsored research at 12 universities that examined the degradation and fate of turfgrass chemicals, the evaluation of best management strategies for the environmentally sound use of turfgrass chemicals, and the development of alternative pest control strategies using biological and biotechnological approaches. The overall goal of the Environmental Research Program of the USGA was to build a scientifically validated foundation for discussing the effects of golf course activities and turfgrass management on the environment. Since 1991, the program has evaluated the effects of turfgrass maintenance on people, wildlife, and the environment with three major objectives in mind: (1) Understand the effects of turfgrass pest management and fertilization on water quality; (2) evaluate valid alternative pest control methods; and (3) document the turfgrass and golf course benefits to humans, wildlife, and the environment. This scenario provided the scientific framework for the organization of an American Chemical Society (ACS) symposium entitled "Fate of Turf-grass Chemicals and Pest Management Approaches" upon which this book is based.

This book is divided into four sections. The first presents three overview chapters that introduce the reader to the current understanding of the environmental issues concerning the USGA, the scope, benefits, and make-up of the turfgrass industry with special emphasis on the golf courses, and a review of turfgrass management from an agrochemical viewpoint. The second section presents seven chapters that review research on the environmental fate of nutrients and pesticides applied to turfgrass. The following section evaluates seven chemical management strategies for reducing environmental and human exposures following application of nutrients and pesticides to turfgrass. The final section presents novel biotechnological and alternative (non-chemical) pest management approaches to maintain turfgrass in a less chemically-dependent fashion.

Acknowledgments

We thank all the authors for their presentations at the symposium and for their contributed chapters that encompass this volume. In particular, we extend our deepest appreciation to the many expert colleagues who provided helpful and necessary critical reviews. We thank Anne Wilson and Kelly Dennis of the ACS Books Department for all their help, suggestions, and encouragement; and Amity Lee-Bradley of the Department of Legal Studies, University of Massachusetts at Amherst for endless organizational and editorial concerns. Their efforts and the generous financial support of the USGA and the ACS Division of Agrochemicals made this book possible.

J. MARSHALL CLARK
Environmental Sciences Program
University of Massachusetts
Amherst, MA 01003

MICHAEL P. KENNA
Research Office
U.S. Golf Association Green Section
P.O. Box 2227
Stillwater, OK 74076

OVERVIEW OF THE TURFGRASS INDUSTRY AND ENVIRONMENTAL ISSUES

Chapter 1

The U.S. Golf Association Turfgrass and Environmental Research Program Overview

Michael P. Kenna[1] and James T. Snow[2]

[1]Research Office, U.S. Golf Association Green Section,
P.O. Box 2227, Stillwater, OK 74076
[2]National Office, U.S. Golf Association, P.O. Box 708, Far Hills, NJ 07931

Golf courses provide beautiful green areas within our urban and suburban landscapes. However, there is public concern about the possible effects of golf courses on the environment. In response to this concern, the USGA sponsored research at 12 universities that examined the degradation and fate of turfgrass chemicals, as well as the development of alternative pest control. The pesticide and nutrient fate research demonstrated that nitrogen and pesticide leaching generally was minimal; the turf/soil ecosystem enhances pesticide degradation; and that the current agricultural models over predict nitrogen and pesticide leaching. Alternative pest management research, including plant improvement through conventional breeding and molecular approaches, was initiated to reduce pesticide applications on golf course turfgrasses. Breeding efforts and genetic engineering have shown promise to increase resistance to abiotic and biotic stresses. The introduction of antagonistic organisms to decrease turfgrass pathogens and insects has produced variable field results. An improved understanding of pest ecology and biocontrol mechanisms is needed. These biological controls will need to be highly effective under realistic golf course conditions in order to be adopted by the turfgrass industry.

Despite a dramatic growth in popularity during the past 20 years, golf now faces one of its greatest challenges from people who believe that course maintenance practices have a deleterious impact on the environment. One of the greatest fears is that nutrients and pesticides used to maintain golf courses will pollute drinking water supplies. Many people are concerned about the potential effects of elevated pesticide and nutrient levels on human health. In addition, there is concern for the ecology of surface waters and wildlife health problems associated with the use of pesticides and nutrients on golf courses. Claims have been made that up to 100 percent of the fertilizers and pesticides applied to golf course turf end up in local water supplies, a claim with no basis in fact. However, a significant emotional reaction from a scientifically illiterate public, unaware of what happens to chemicals in the environment, has increased the urgency and hysteria surrounding the pesticide and nutrient fate issues.

People became concerned about how golf courses affect the environment during a series of widespread droughts that occurred in the late 1970's and early1980's. Severe drought in California and other western states resulted in extreme restrictions on water-use. Golf courses were among the first and most severely restricted operations in many areas, due in part to their visibility and because they were considered non-essential users of water. Similarly, increased golf course construction in the 1980's and 1990's brought golf courses under attack because of the potential impacts on natural areas and the use of pesticides on existing courses. In many cases, anti-development groups made unsubstantiated claims about the negative effects of golf courses in an effort to kill housing developments or commercial real estate development.

As an important leadership organization in golf, the USGA initiated a research program focused on environmental issues in 1989. In particular, university research studies examining the effects of fertilizers and pesticides on surface and groundwater resources were initiated. The studies were conducted on the major pathways of chemical fate in the environment, including leaching, runoff, plant uptake and utilization, microbial degradation, volatilization, and other gaseous losses. The studies were conducted at twelve universities throughout the United States, representing the major climatic zones and turfgrass types. The goal of the program was to build a solid foundation for discussing the effects of golf course activities and turfgrass management on the environment (*1*).

Since 1991, the Environmental Research Program has evaluated the effects of golf courses on people, wildlife and the environment. The three major objectives of the program were to:

1. Understand the effects of turfgrass pest management and fertilization on water quality;
2. Evaluate valid alternative pest control methods; and
3. Document the turfgrass and golf courses benefits to humans, wildlife and the environment.

Turfgrass Pest Management and Fertilization Effects on Water Quality

Golf course superintendents apply pesticides and fertilizers to the course, and depending on an array of processes, these chemicals break down into biologically inactive by products. There are several interacting processes that influence the fate of pesticides and fertilizers applied to turf. For the purposes of this section, the following seven categories that influence the fate of pesticides and nutrients will be discussed:

1. Volatilization
2. Water solubility
3. Sorption
4. Plant Uptake
5. Degradation
6. Runoff
7. Leaching

The role these processes play in the likelihood that the pesticides will reach ground or surface water will be addressed somewhat. The relative importance of each process is controlled by the chemistry of the pesticide or fertilizer and environmental variables such as temperature, water content, and soil type.

Volatilization. Volatilization is the process by which chemicals transform from a solid or liquid into a gas. The vapor pressure of a chemical is the best indicator of its potential to volatilize. Pesticide volatilization increases as the vapor pressure increases. As temperature increases, so do vapor pressures and the chance for volatilization loss. Volatilization losses are generally lower following a late afternoon or an early evening pesticide application rather than in the late morning or early afternoon when temperatures are increasing. Volatilization also will increase with air movement and can be greater from unprotected areas than from areas with windbreaks. Immediate irrigation is usually recommended for highly volatile pesticides to reduce loss.

Nitrogen Volatilization. Only a few studies have evaluated nitrogen volatilization from turfgrass. Nitrogen volatilization depends on the degree of irrigation after the application of fertilizer (*2, 3, 4*). When no irrigation was used, as much as 36 percent of the nitrogen volatilized. A water application of 1 cm reduced volatilization to 8 percent. Branham et al. (*5, 6*) could not account for 36 percent of the nitrogen applied one year after a spring application. It was suggested that volatilization and denitrification could be responsible.

In a greenhouse experiment, Starrett et al. (*7, 8*) investigated the fate of ^{15}N applied as urea from intact soil columns (Table I). Nitrogen recovery averaged 82.4 and 89.7 percent for single (one 25 mm per week) and split (four 6 mm per week) irrigation treatments. Nitrogen volatilization was higher for the split application; however, this difference was not significant.

Table I. Percentage of nitrogen recovered from intact soil cores for two irrigation treatments.

Sample	Weekly Irrigation Regime[1]	
	One 25-mm	Four 6-mm
	----- % of Total Applied -----	
Volatilization	0.9	2.3
Clippings, Verdure	14.3	37.3
Thatch-Mat	11.3	16.7
0-10 cm	13.4	12.6
10-20 cm	7.7	6.4
20-30 cm	7.2	5.6
30-40 cm	7.8	6.7
40-50 cm	7.6	2.2
Leachate**	12.3	0.4
Total	82.4	89.7

[1] Each treatment had six replications.
** Significantly different (t-test, $P < 0.001$)
SOURCE: Adapted with permission from reference 8. Copyright 1995.

Pesticide Volatilization. The volatilization studies determined the amount of several pesticides that volatilized into the air for several days after application (Table II). Reported volatile losses over a one to four-week period, expressed as a percentage of the total applied, ranged from less than one percent to 16 percent (*9-14*). Results of volatilization studies showed that maximum loss occurred when surface temperature and solar radiation were highest, and that volatile losses were directly related to the vapor pressure characteristics of the pesticide. Thus, examining the physical and chemical properties of the pesticide is a good way to determine if volatilization losses are likely to occur under particular weather and application conditions.

Table II. Summary of volatile insecticide residues recovered from putting green and fairway plots.

Pesticide	Volatile Residues % applied	Comments	ref.
Trichlorfon	11.6	Applied 9/28/91 on bentgrass fairway, no irrigation following application. Sampled for 15 days.	(10)
	9.4	Applied 7/7/93 on bentgrass fairway, 1.3 cm of irrigation after application. Sampled for 15 days.	(10)
	0.09	Applied 6/4/96 on bentgrass green. Sampled for 29 days.	(13)
Isazofos	11.4	Applied 8/22/93 on bentgrass fairway, 1.3 cm of irrigation after application. Sampled for 15 days.	(10)
	1.04	Applied 10/4/96 on bermudagrass green, 0.6 cm of irrigation followed by 3.94 cm of rainfall over 24 hrs. Sampled 48 hours during cloudy, rainy conditions.	(12)
	9.14	Applied 10/10/96 on bermudagrass green, 0.6 cm of irrigation after application. Sampled 22 hrs, no rainfall.	(12)
Chlorpyrifos	2.7	Applied 10/4/96 on bermudagrass green, 0.6 cm of irrigation followed by 3.94 cm of rainfall over 24 hrs. Sampled 48 hours during cloudy, rainy conditions.	(12)
	11.6	Applied 10/10/96 on bermudagrass green, 0.6 cm of irrigation after application. Sampled 22 hrs, no rainfall.	(12)
	15.7	Applied 6/4/96 on bentgrass green. Sampled 29 days.	(13)
Fenamiphos	0.04	Applied 10/10/96 on bermudagrass green, 0.6 cm of irrigation after application. Sampled 22 hrs, no rainfall.	(12)
	0.25	Applied 10/4/96 on bermudagrass green, 0.6 cm of irrigation followed by 3.94 cm of rainfall over 24 hrs. Sampled 48 hours during cloudy, rainy conditions.	(12)
Carbaryl	0.03	Applied 8/93 to bentgrass green and bermudagrass fairway plots. Value is average over turfgrass and soil types.	(13)
Triadimefon	7.3	Applied 8/23/91 on bentgrass fairway, 1.3 cm of irrigation after application. Sampled 15 days.	(11)
Metalaxyl	0.08	Applied 9/27/96 on bentgrass green. Sampled 8 days.	(13)
Chlorthalonil	0.02	Applied 9/27/96 on bentgrass green. Sampled 8 days.	(13)
Mecoprop	0.08	Applied 9/24/92 on bentgrass fairway, no irrigation following application. Sampled 15 days.	(11)
2,4-D	0.67	Applied 8/93 to bentgrass green and bermudagrass fairway plots. Averaged of turfgrass and soil types.	(14)

Post-application irrigation has an effect on the volatilization of trichlorfon (*10, 15*). The pesticide was applied once, followed by 13 mm of irrigation, and again separately with no post-application irrigation. The application rate for both occasions was 9 kg a.i. ha^{-1}. Without post-application irrigation, trichlorfon volatile loss totaled 13 percent compared to 9 percent when irrigated. Also, withholding post-application irrigation resulted in less conversion of trichlorfon to its more toxic breakdown product, DDVP (Figure 1). It appears that light post-application irrigation may have a small, positive effect on preventing volatile loss of pesticides. However, more research investigating light irrigation on several pesticides is needed to confirm this observed trend.

Water Solubility. The extent to which a chemical will dissolve in a liquid is referred to as solubility. Although water solubility is usually a good indicator of mobility (Figure 2), it is not necessarily the only criterion. In addition to pesticide solubility, the affinity of a pesticide to adhere to soils must be considered (*16*).

Sorption. The tendency of a pesticide to leach or runoff is strongly dependent upon the interaction of the pesticide with solids in the soil. Sorption includes the process of adsorption and absorption. Adsorption refers to the binding of a pesticide to the soil particle surface. Absorption implies that the pesticide penetrates into a soil particle or is taken up by plant leaves or roots.

This difference is important because pesticides may become increasingly absorbed with time (months to years), and desorption (or release) of the absorbed pesticide may be reduced with time. The unavailable or undetachable pesticide is often referred to as bound residue and is generally unavailable for microbial degradation or pest control.

Factors that contribute to sorption of pesticides on soil materials include a) chemical and physical characteristics of the pesticide, b) soil composition, and c) nature of the soil solution. In general, sandy soils offer little in the way of sorptive surfaces. Soils containing higher amounts of silt, clay and organic matter provide a rich sorptive environment for pesticides. Research conducted during the past eight years (*17, 18, 19*) indicates that turfgrass leaves and thatch adsorb a significant amount of pesticide (See Figure 3).

Adsorption of pesticides is affected by the partition coefficient, which is reported as K_d or more accurately as K_{oc}. A K_{oc} less than 300 to 500 is considered low. The strength of adsorption is inversely related to the pesticide's solubility in water and directly related to its partition coefficient. For example, chlorinated hydrocarbons are strongly adsorbed, while phenoxy herbicides like 2,4-D are much more weakly adsorbed.

Plant Uptake. Plants can directly absorb pesticides or influence pesticide fate by altering the flow of water in the root zone. Turfgrasses with higher rates of transpiration can reduce the leaching of water-soluble pesticides. In situations where the turf is not actively growing or root systems are not well developed, pesticides are more likely to migrate deeper into the soil profile with percolating water.

Degradation. Degradation occurs because of the presence of soil microorganisms and chemical processes in the turfgrass-soil system. Pesticides are broken down in a series of steps that eventually lead to the production of CO_2 (carbon dioxide), H_2O (water) and some inorganic products (i.e., nitrogen, phosphorus, sulfur, etc.).

Microbial Degradation. Microbial degradation is a biological process whereby microorganisms transform the original compound into one or more new compounds. Each of these new compounds has different chemical and physical

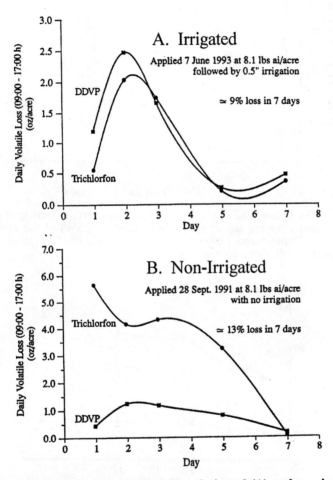

Figure 1. Trichlorfon volatilization from irrigated (A) and non-irrigated (B) bentgrass fairway plots. SOURCE: Reproduced with permission from ref. 9. Copyright 1995 United States Golf Association.

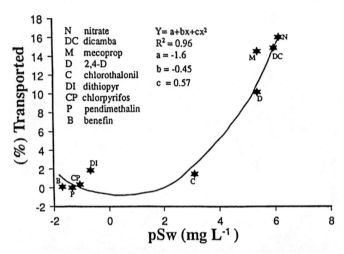

Figure 2. Fraction of the applied pesticides transported from simulated fairway plotted for the log of the water solubility (pSW) of the analyte. SOURCE: : Reproduced with permission from ref. *16*. Copyright 1998 United States Golf Association.

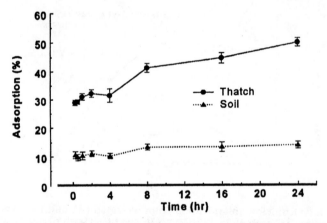

Figure 3. Adsorption kinetics for 2,4-D in thatch and soil. SOURCE: Reproduced with permission from ref. *17*. Copyright 1998 United States Golf Association.

properties that make them behave differently in the environment. Microbial degradation may be either direct or indirect. Some pesticides are directly utilized as a food source by microorganisms. In most cases, though, indirect microbial degradation of pesticides occurs though passive consumption along with other food sources in the soil.

Chemical Degradation. Chemical degradation is similar to microbial degradation except that pesticide break down is not achieved by microbial activity. The major chemical reactions such as hydrolysis, oxidation, and reduction occur in both chemical and microbial degradation. Photochemical degradation is an entirely different break down process driven by solar radiation. It is the *combined pesticide degradation* that results from chemical, microbial, and photochemical processes under field conditions that was of the most interest in the USGA sponsored studies.

Degradation rates are also influenced by factors like pesticide concentration, temperature, soil water content, pH, oxygen status, prior pesticide use, soil fertility, and microbial population. These factors change dramatically with soil depth and greatly reduce microbial degradation as pesticides migrate below the soil surface. An interesting result occurred at the University of Florida study when fenamiphos was applied twice at a monthly interval. Although leaching from the first application amounted to about 18 percent, leaching from the second application was just 4 percent (*20, 21, 22, 23*). These results suggest that microbial degradation was enhanced due to microbial buildup after the first application, thereby reducing the amount of material available for leaching after the second application.

In the case of degradation rates, the average DT_{90} (days to 90% degradation) in turf soils generally is significantly less than established values based upon agricultural systems (See Figure 4). Thus, leaching potential for most pesticides is less in turfgrass systems because turfgrass thatch plays an important role in adsorbing and degrading applied pesticides (*24, 25*).

Persistence of a pesticide, expressed as half-life (DT_{50}), is the time required for 50 percent of the original pesticide to degrade. Half-life measurements are commonly made in the laboratory under uniform conditions. On the golf course, soil temperature, organic carbon and moisture content change constantly. These factors dramatically influence the rate of degradation. Consequently, half-life values should be considered as guidelines rather than absolute values.

Leaching. The downward movement of nutrients and pesticides through the turfgrass-soil system by water is called leaching. Compared to some agricultural crops, the USGA-sponsored research demonstrates that leaching is reduced in turfgrass systems. This occurs because of the increase in adsorption on leaves, thatch, and soil organic matter; a high level of microbial and chemical degradation; and reduced percolation due to an extensive root system, greater plant uptake, and high transpiration rates. Separate discussions on nitrogen and pesticide leaching follow.

Nitrogen Leaching. Golf courses use a significant amount of nitrogen (N) fertilizer and there is concern that nitrogen leaching is affecting groundwater supplies. Seven different universities investigated nitrogen leaching, most using bucket lysimeters to measure leaching potential. In general, very little nitrogen leaching occurred when nitrogen was applied properly, i.e., according to the needs of the turf and in consideration of soil types, irrigation regimes and anticipated rainfall. Properly maintained turf grown in a loam soil allowed less than one percent of the nitrogen applied to leach to a depth of 1.2 m (*5, 6*). Sandy soils are more prone to leaching losses than loam soils. Results averaged over seven

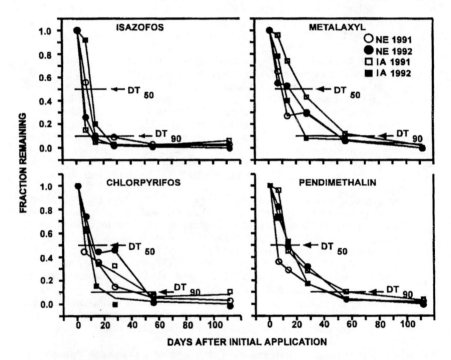

Figure 4. Total pesticide residue in verdure, thatch and soil as a function of sampling time. Intercepts with 0.5 and 0.1 estimate time to 50 (DT_{50}) and 90 (DT_{90}) percent dissipation, respectively. SOURCE: Reproduced with permission from ref. *25*. Copyright 1996 G. L. Horst, P. J. Shea, N. E. Christians, R. D. Miller, C. Stuefer-Powell, and S. K. Starrett.

leaching projects during the establishment year indicate that nitrogen leaching ranged from 11 percent of applied for pure sand rootzones to one percent or less for root zones containing more silt and clay (Figure 5). When more nitrogen is applied than is needed, both the amount and the percentage of nitrogen lost increases. Nitrogen leaching losses can be greatly reduced by irrigating lightly and frequently, rather than heavily and less frequently. Applying nitrogen in smaller amounts on a more frequent basis also reduced leaching losses.

Braun et al. (26) found that nitrogen leaching was significant when applied at heavy rates to newly established turfgrass on pure sand rootzones. For example, they found that 7.6 percent of an annual application rate of 585 kg N ha^{-1} yr^{-1} (12 lb. N 1000 ft^{-2} yr^{-1}) applied to immature turf grown on a pure sand rootzone leached through the profile (Table III). Leaching was significantly less when peat was added to the sand (USGA recommended mix), occurring at a level of about three percent. On pure sand, nitrogen concentrations exceeded federal drinking water standards (10 ppm NO$_3$-nitrogen) several times at the 585 kg rate during the first year, whereas nitrogen concentrations in leachate never exceeded federal standards from the sand/peat mix. Significantly less leaching also occurred when less nitrogen was applied (<380 kg N ha^{-1} yr^{-1}) and when application frequency was increased (22 vs. 11 times annually). During years two and three, on mature turf, much less nitrogen leaching occurred for all treatments. In putting green construction, mixing peat moss with sand significantly reduced nitrogen leaching compared to pure sand rootzones during the year of establishment. Light applications of slow-release nitrogen sources on a frequent interval provided excellent protection from nitrate leaching.

Table III. Percent of Total Applied Nitrogen Leached as Nitrate.

Rootzone Medium	Nitrogen kg N ha^{-1}yr^{-1}	NO$_3$-N, % of Total N		
		Year 1	Year 2	Year 3
Sand	190	5.37	0.06	2.71
	390	6.31	0.04	3.17
	585	7.55	0.70	4.28
Amended Sand	190	5.37	0.40	0.16
	390	0.91	0.02	0.17
	585	3.37	1.26	2.31

SOURCE: Adapted from ref. 26.

Branham et al. (5, 6, 27) reported that less than one percent of the applied nitrogen leached through a 1.2-m (4-foot) deep profile of undisturbed loam-soil during a 2.5 year period (Figure 6). Most of the nitrogen was recovered in clippings, thatch, and soil. They suggested that the remaining amount volatilized or was lost through denitrification. Starrett et al. (7, 8) observed similar results when nitrogen was applied at moderate rates and lightly irrigated (one 25-mm vs. four 6-mm applications). However, up to 30 times more nitrogen (Table I) was leached after the single 25 mm irrigation application, perhaps in part due to macropore flow caused by earthworm activity. Yates (14) reported that nitrogen leaching from a USGA profile sand-based green was generally less than one percent when nitrogen was applied lightly and frequently.

Irrigating bermudagrass and tall fescue turf with adequate amounts (no drought stress) of moderately saline water did not increase the concentration or amount of

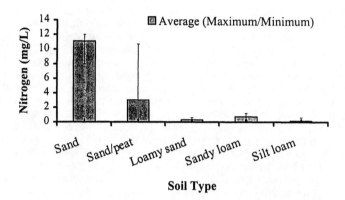

Figure 5. Summary of the nitrogen leaching results (mg L⁻¹) from five soil types reported from USGA-sponsored research studies.

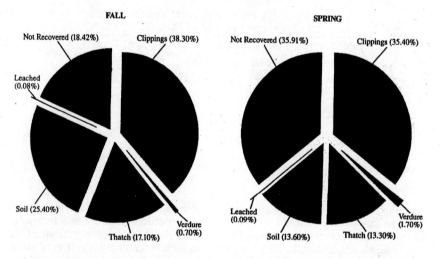

Figure 6. Percent of Total Nitrogen applied recovered two years after spring and fall applications. SOURCE: Reproduced with permission from ref. 27. Copyright 1994 United States Golf Association.

nitrate leached (2). Higher amounts of salinity in the root zone, drought, or the combination of these two stresses caused high concentrations and larger amounts of nitrate to leach from both a tall fescue and bermudagrass turf. This suggests that drought, high salinity, or both impair the capacity of the root system of the turf, and that management modification may be needed to prevent nitrate leaching.

Pesticide Leaching. Pesticide leaching studies were conducted at universities throughout the United States. Treatments were made to a variety of soils and turf species, and plots received varying irrigation regimes or rainfall events. During the first year of the studies, most turf areas were relatively immature. Results showed that very little pesticide leaching occurred with most products, generally less than one percent of the total applied. However, significant leaching occurred with certain products and under certain circumstances (Table IV).

Table IV. Summary of Total Pesticide Mass and Percent of Total Applied Recovered in Water Effluent from Putting Green and Fairway Lysimeters.

Common Name	Trade Name	n	Total Recovered Mean	(Range)	Percent Recovered Mean	(Range)
			-------- $\mu g\ m^{-2}$ --------		---------- % ----------	
Fairway Plots						
Chlorothalonil	Daconil	1	0	(0)	0.00	(0.00)
Fenarimol	Rubigan	1	0	(0)	0.00	(0.00)
Metalaxyl	Subdue	1	0	(0)	0.00	(0.00)
Propiconazole	Banner	1	0	(0)	0.00	(0.00)
Triadimefon	Bayleton	8	2,312	(27 - 11,160)	0.51	(0.01 - 2.44)
2,4-D	2,4-D	9	155	(0 - 329)	0.28	(0.00 - 0.60)
Dicamba	Banvel	2	4,750	(3,700 - 5,800)	39.58	(30.83 - 48.33)
MCPP	Mecoprop	6	44,236	(1,006 - 142,062)	19.34	(0.44 - 62.12)
Carbaryl	Sevin	8	132	(24 - 375)	0.01	(0.00 - 0.02)
Isazofos	Triumph	6	5,590	(5,590)	2.44	(0.00 - 10.40)
Trichlorfon	Proxol	6	21,527	(5,763 - 40,341)	2.35	(0.63 - 4.41)
Putting Green Plots						
Chlorothalonil	Daconil	4	2,961	(749 - 5,486)	0.08	(0.02 - 0.14)
2,4-D	2,4-D	7	871	(347 - 1,808)	2.25	(1.12 - 3.79)
2,4-D amine	2,4-D	6	46	(0 - 133)	0.12	(0.00 - 0.48)
Dicamba	Banvel	7	201	(0 - 1,173)	3.07	(0.00 - 19.55)
MCPP	Mecoprop	4	109	(0 - 329)	0.08	(0.00 - 0.25)
Carbaryl	Sevin	4	372	(205 - 642)	0.04	(0.02 - 0.07)
Chlorpyrifos	Dursban	4	92	(0 - 193)	0.04	(0.00 - 0.08)
Dithiopyr	Dimension	4	139	(101 - 196)	0.24	(0.18 - 0.35)
Ethoprop	Mocap	1	41,138	(1,138)	0.05	(0.05)
Fenamiphos	Nemacur	4	53,121	(419 - 199,038)	4.70	(0.04 - 17.61)
Fonofos	Dyfonate	2	54	(4 - 103)	0.01	(0.00 - 0.02)
Isazofos	Triumph	2	123	(41 - 204)	0.05	(0.02 - 0.09)
Isofenphos	Oftanol	2	43	(33 - 53)	0.02	(0.01 - 0.02)

The physical and chemical properties of the pesticides proved to be good indicators of the potential for leaching, runoff and volatilization (27). Products that exhibit high water solubility, low soil adsorption potential, and greater persistence are more likely to leach and run off. For example, fenamiphos, a commonly used nematicide, is highly water soluble and has low adsorption potential, and its toxic breakdown metabolite tends to persist in the soil. As expected, losses of fenamiphos and its metabolite due to leaching were as high as 18 percent from a sand-based green in Florida (20, 21, 22, 23), though when all studies are considered, the average loss was about five percent. Soil type and precipitation/irrigation amount also were important factors in leaching losses. Table V shows the effects of soil type and precipitation on leaching of MCPP and triadimefon, two pesticides whose chemical and physical properties indicate a relatively high potential for leaching. Results show significant leaching from coarse sand profiles, especially under high precipitation, and much less leaching from sandy loam and silt loam soils (28).

Table V. Effect of Soil Type and Precipitation on the Leaching Loss of Two Pesticides from an Immature Turf, Expressed as Percent of Total Applied.

Pesticide	Precipitation	Sand	Sandy Loam	Silt Loam
		--- % Recovered of Total Applied ---		
MCPP	Moderate	51.0	0.8	0.4
	High	62.1	0.5	1.2
Triadimefon	Moderate	1.0	0.06	0.2
	High	2.4	0.01	0.3

SOURCE: Adapted from ref. 28.

There are several simulation models currently used to predict the downward movement of pesticides through soil. A good review on pesticide transport models was conducted by Cohen et al. (29). The USGA-sponsored studies indicate that many of these models will need adjustments that take into account the role of a dense turf canopy and thatch layer. For example, the actual leaching loss of 2,4-D was less than the predicted value using the GLEAMS (i.e., Groundwater Loading Effects of Agricultural Management Systems) computer model. Without proper parameter adjustments, the model over-predicted the actual amount leached by 10 to 100 times, or more, for five of the seven pesticides screened by the computer (30, 31).

In summary, the pesticide leaching studies indicate that dense turf cover reduced the potential for leaching losses of pesticides; conversely, more leaching occurred from newly planted turf stands. Generally, sandy soils are more prone to leaching losses than clayey soils. The physical and chemical properties of the pesticides were good indicators of leaching potential. Finally, current pesticide models over-predict the leaching loss of most pesticides applied to turf if valid adjustments are not made to account for the role the turf canopy and thatch.

Runoff. An important finding from the USGA-sponsored research was that pesticide and nutrient runoff pose a greater threat to water quality than leaching. Runoff refers to the portion of precipitation (rainfall) that is discharged from the area through stream channels. The water lost without entering the soil is called surface runoff, and that which enters the soil before reaching the stream is called groundwater runoff or seepage flow from ground water. Pesticides and nutrients applied to golf course turf, under some circumstances, can be transported off site in surface runoff.

Nitrogen Runoff. Three USGA-sponsored studies examined runoff from fairway research plots. In Pennsylvania, runoff experiments were conducted on plots characterized by slopes of 9 to 13 percent, good quality loam soil, and turf cover consisting of either creeping bentgrass or perennial ryegrass cut at a 12.5 mm (1/2-inch) fairway height (*32, 33, 34*). Typical of that part of the country, the fairway-type plots received 195 kg N ha^{-1} yr^{-1} (4 lb. N per 1000 ft^2 per year). The irrigation water used to simulate rainfall contained a relatively high level of nitrate-nitrogen, ranging from 2 to 10 ppm. They reported that nitrate concentrations in the runoff or leaching samples did not differ significantly from the nitrate concentration in the irrigation water (Figure 7). The study was conducted on excellent quality turf and on soil with a high infiltration rate.

Nitrogen runoff also was measured as part of the studies in Georgia (*35, 36*). Nitrogen applications were made and 24 hours after a simulated storm event (25 mm applied at a rate of 50 mm hr^{-1}) as much as 40 to 70 percent of the rainfall left the plots as runoff. A total of 16 percent (12.5 mg L^{-1}) of the nitrate nitrogen

Table VI. The Percentage of Applied and Concentration of Product Transported from Runoff Plots during Rainfall 24 hours after Application on Heavy Textured Soils with High Antecedent Moisture.

Pesticide or Nitrogen Treatment	Application Rate	Percentage Transported	Conc. 24 h after Application
	-- kg ha^{-1}--	---- % -----	----- µg L^{-1} -----
NO$_3$ active growth	24.4	16.4	12,500
NO$_3$ dormant	24.4	64.2	24,812
Dicamba	0.56	14.6	360
Dicamba (dormant)	0.56	37.3	752
Mecoprop	1.68	14.4	810
Mecoprop (dormant)	1.68	23.5	1,369
2,4-D DMA	2.24	9.6	800
2,4-D DMA (dormant)	2.24	26.0	1,959
2,4-D DMA (injected)	2.24	1.3	158
2,4-D DMA (buffer)	2.24	7.6	495
2,4-D LVE	2.24	9.1	812
Trichlorfon[1]	9.15	32.5	13,960
Trichorfon [1](injected)	9.15	6.2	2,660
Chlorothalonil[2]	9.50	0.8	290
Chlorpyrifos[2]	1.12	0.1	19
Dithiopyr	0.56	2.3	39
Dithiopyr (granule)	0.56	1.0	26
Benefin	1.70	0.01	3
Benefin (granule)	1.70	0.01	6
Pendimethalin	1.70	0.01	9
Pendimethalin (granule)	1.70	0.01	2

[1] Trichlorfon + dichlorvos metabolite.
[2] Total for active ingredient and metabolites.
SOURCE: Adapted from ref. *36*.

applied at 24 kg ha^{-1} applied to actively growing bermudagrass was found in surface runoff water (Table IV). However, 64 percent (24.8 mg L^{-1}) of the nitrate nitrogen applied at 24 kg ha^{-1} applied to dormant bermudagrass was found in surface runoff water.

In Oklahoma, the effects of buffer strips and cultivation practices on pesticide and nitrogen runoff were investigated. It was concluded that antecedent soil moisture was the major factor influencing runoff (Table VII). During the first simulated rainfall event in July, soil moisture conditions were low to moderate. After a 50 mm (2 inches) rainfall event, less than one percent of the applied nitrogen was collected in the runoff (*37, 38*). In August, when the simulated rainfall occurred after 150 mm (6 inches) of actual rainfall the previous week (i.e., high soil moisture), the amount of nitrogen collected after the simulated rainfall averaged more than 8 percent. When soil moisture was moderate to low in the Oklahoma study, the presence of a 2.4 to 4.9 m (8 to 16 ft.) untreated buffer strip significantly reduced nitrogen runoff, whereas when soil moisture was high, the buffer strips made no difference. In both cases, less runoff occurred when sulfur-coated urea was applied compared to straight urea.

Table VII. Effect of Low/moderate versus High Antecedent Soil Moisture Levels on Pesticide and Nutrient Runoff Losses from Bermudagrass maintained as Fairway Turf, expressed as a Percent of Total Applied.

Soil Moisture	NO$_3$	PO$_4$	NH$_4$	Dicamba	2,4-D	MCPP	Chlorpyrifos	ref.
Low/mod.	0.09	0.2	0.2	0.35	0.79	0.81	0.04	(*38*)
	-	-	-	3.1	2.6	1.3		(*36*)
High	3.1	7.7	5.1	5.4	8.7	9.3	0.025	(*38*)
	-	-	-	9.7	7.3	9.5		(*36*)

SOURCE: Adapted from ref. *36, 38*.

In summary, results from USGA-sponsored runoff studies showed that dense turf cover reduces the potential for runoff losses of nitrogen, and significant runoff losses are more likely to occur on compacted soils. Much greater nitrogen runoff occurred when soil moisture levels were high, as compared to moderate or low. Buffer strips reduced nitrogen runoff when soil moisture was low to moderate at the time of the runoff event. However, buffer strips were ineffective when soil moisture levels were high. Nitrogen runoff was significantly less when a slow release product (sulfur-coated urea) was used compared to a more soluble product (urea).

Pesticide Runoff. In Georgia, studies were conducted on plots with a 5 percent slope and a sandy clay soil typical of that region (*15, 30, 31, 35*). Pesticides were applied and 25-mm (1-inch) simulated rainfall events occurred 24 and 48 hours afterward. At a rainfall rate of 50 mm (2 inches) per hour, as much as 40 to 70 percent of the rainfall left the plots as runoff during simulated storm events. The collected surface water contained moderately high concentrations of treatment pesticides having high water solubility (Table VI). For example, under these conditions, only very small amounts (<1%) of chlorthalonil and chlorpyrifos could be detected in the runoff. However, between 10 to 13 percent of the 2,4-D, MCPP and dicamba was transported off the plots over an 11-day period. About 80 percent of this transported total moved off the plots with the first rainfall event 24 hours after pesticide application. Also, the amount of the trichlorfon that ran off the plots was 5.2 times greater when broadcast as a granular product compared to being pressure injected. Finally, the runoff loss of several herbicides was greater when applied to dormant turf as compared to an actively growing turf.

Antecedent soil moisture was a significant factor in determining how much pesticide ran off the plot areas (36, 37, 39). Where soil moisture was low to moderate, buffer zones were effective in reducing pesticide runoff; when soil moisture was high they were not effective except for the insecticide chlorpyrifos (Table VII).

In both Oklahoma and Georgia, best management practice studies investigated how cutting heights and buffers of varying lengths could minimize fertilizer and pesticide runoff. The effect of soil cultivation (core aerification) on runoff potential also was studied. In Oklahoma, a 4.9-m buffer cut at 5 cm (3 inches) significantly decreased the amount of 2,4-D found in runoff water from a 4.9-m treated bermudagrass fairway (Figure 8). However, the results in Georgia that used smaller buffer strips indicated no reduction in the amount of pesticide transported in the surface-water solution (39).

Among the conclusions or trends observed from the pesticide runoff studies were the following: 1) dense turf cover reduces the potential for runoff losses of pesticides; 2) the physical and chemical properties of pesticides are good indicators of potential runoff losses; 3) heavy textured, compacted soils are much more prone to runoff losses than sandy soils; 4) moist soils are more prone to runoff losses than drier soils; 5) buffer strips at higher cutting heights tend to reduce runoff of pesticides when soil moisture is low to moderate prior to rainfall events; and, 6) the application of soluble herbicides on dormant turf can produce high levels of runoff losses.

Alternative Pest Management

Alternative pest management is intended to reduce the amount of pesticide needed to maintain golf course turfgrasses. The USGA provided funding for the development and evaluation of alternative methods of pest control. Research projects focused on the following areas: biological control, cultural and mechanical practices, allelopathy, selection and breeding for pest resistance, and application of integrated pest management strategies.

Biological Control. The microorganisms that inhabit the turfgrass root zone are just starting to be characterized in a way that will lead to positive developments in biological control. It is important that the mechanisms of biological control be understood thoroughly before products are commercialized. The USGA has sponsored several projects in the area of biological control and we are just starting to understand why some of these organisms fail when used in field situations.

Biological control of turfgrass pests is generally accomplished with a living organism that either lowers the population density of the pest problem or reduces its ability to cause injury to the turf. In this section, a list of potential problems that any organic or biological control agent must overcome to be a commercial success is first covered. After this discussion, biocontrol approaches investigated to solve some of our common turfgrass disease and insect problems are reviewed.

Potential Problems of Biocontrol. Once laboratory or greenhouse evidence demonstrates that a specific predator or microbial antagonist controls a turfgrass pest problem, additional field research must be performed to determine whether it is a functional biocontrol agent acceptable for commercial use (40). Specifically, sound scientific research must:

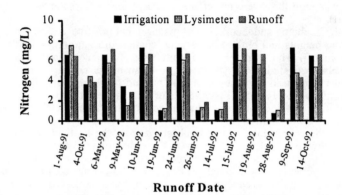

Figure 7. Concentration of nitrate-nitrogen in irrigation, leachate, and runoff water collected during the 1993 growing season. SOURCE: Adapted from ref. *32.*

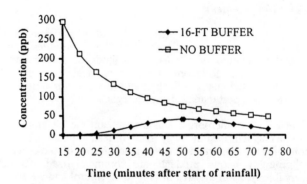

Figure 8. Plot of the predicted concentration of 2,4-D in surface runoff versus time in the 1996 buffer length experiment. SOURCE: Reproduced with permission from ref. *39.* Copyright 1998 United States Golf Association.

1. Determine if the biocontrol agent readily establishes in the turf or surrounding areas.
2. Evaluate the effects of pesticides on the growth and development of the antagonist.
4. Estimate the likelihood of resistance to the effects of the biological agent.
5. Reduce the need for conventional pesticides by providing an adequate level of control.
6. Provide evidence that the biological agent is safe for people, wildlife and the environment.
7. Develop methods for producing commercial quantities with an acceptable shelf life and cost.

The antagonist must be readily established in the turf in order for biocontrol to be effective. This means that the antagonist must become quickly established in the turf and is able to maintain the population level needed to reduce the pest problem. At the present time, repeat applications of large quantities of preparations of the antagonistic organism are usually required in order to achieve and maintain levels of pest suppression that approximate those provided by pesticides. The use of an antagonist must take into consideration how much of the material is required to provide a significant reduction of the pest problem, how long the individual applications will last, and whether the method is economically feasible.

Antagonist growth and development must not be affected by pesticides. Generally, biocontrol procedures alone will not provide satisfactory levels of pest reduction in intensely managed golf course turf. Biocontrol products will probably be used in conjunction with a pesticide program. There are some situations where the pesticides used on the golf course may be toxic to the antagonist or may restrict the growth and development of indigenous species that are necessary for the biocontrol species to be effective. Therefore, the impact of pesticides on the growth and development of the antagonist is an important consideration in the selection of biocontrol agents.

Is Resistance likely? Many of the pest problems that occur in turfgrass are known to have biotypes that differ in their vulnerability to pesticides, their ability to infest the turf area, or in the environmental conditions needed for optimum growth and development. This means that the biocontrol will need to have a wide range of effectiveness on the various pest biotypes. For this reason, before being sold commercially, an antagonist must be field tested within the prospective marketing regions to determine its level of effectiveness against the target pests in these localities.

The pest reduction level must permit a decrease in pesticide use. A biocontrol system based on the use of a single predator or microbial species is highly specific. Unlike most pesticides, it will not control a wide range of disease or insect problems. The level of pest reduction provided by the antagonist should at least offset pesticide consumption for control of the target pest problem.

The antagonist must be safe for people, wildlife and the environment. Protection of the environment is one of the primary reasons that the concept of biological control has gained much support. Before a biocontrol agent is placed into general use, however, it must be clearly demonstrated that it does not jeopardize the stability of the biological environment or the health of people or wildlife. Ideally, the selected agents should be similar to indigenous microflora so that the danger of introducing potentially damaging organisms is minimal.

Methods for production and formulation must be developed. A commercial product must be easily produced and have a reasonable shelf life in order to be accepted by turfgrass managers. Several effective biocontrol agents cannot be produced in commercial quantities. These biological agents either need a live host to reproduce or the current laboratory production methods available would make the product too expensive to produce.

Disease Biocontrol. There are two general approaches to the biocontrol of turfgrass diseases. First, preparations, containing known microbial species detrimental to the growth and development of a specific pathogen, are used to prevent or control the disease. Second, the use of organic materials, colonized by a complex mixture of microbial species that restrict the pathogen growth and development, are applied to a turfgrass area for disease control. Varying degrees of success have been reported (*41*) from field and greenhouse-based tests for the control of turfgrass diseases by applying a known microbial agent or complex mixture of bacteria and fungi to the turf area (Table VIII).

Table VIII. Examples for biological control of turfgrass diseases.

Disease Pathogen	Antagonists	Location
Brown Patch	*Rhizoctonia* spp.	Ontario, Canada
Rhizoctonia solani	*Laetisaria* spp.	North Carolina
	Complex mixtures	New York, Maryland
Dollar Spot	Enterobacter cloacae	New York
Sclerotinia homoeocarpa	*Fusarium heterosporum*	Canada
	Trichoderma harzianum	New York
	Gliocladium virens	South Carolina
	Actinomycetes spp.	New York
	Complex mixtures	New York
Pythium Blight	*Pseudomonas* spp.	Illinois, Ohio
Pythium aphanidermatum	*Trichoderma* spp.	Ohio
	Trichoderma hamatum	Colorado
	Enterobacter cloacae	New York
	Various bacteria	New York
	Complex mixtures	Pennsylvania
Pythium Root Rot	*Enterobacter cloacae*	New York
Pythium graminicola	Complex mixtures	New York
Red Thread	Complex mixtures	New York
Laetisaria fuciformis		
Southern Blight	*Trichoderma harzianum*	North Carolina
Sclerotium rolfsii		
Take-All Patch	*Pseudomonas* spp.	Colorado
Gaeumannomyces graminis	*Gaeumannomyces* spp.	Australia
var. *avenae*	*Phialophora radicicola*	Australia
	Complex mixtures	Australia
Typhula Blight	*Typhula phacorrhiza*	Ontario, Canada
Typhula spp.	*Trichoderma* spp.	Massachusetts
	Complex mixtures	New York
Summer Patch		
Magnaporthae poae	*Stenotrophomonas maltophilia*	New Jersey
Spring Dead Spot	*Pseudomonas* spp.	California
Leptosphaeria korrae		
Sting Nematodes	*Pasteuria* spp.	Florida
Belonolaimus longicaudatus		

SOURCE: Adapted from ref. *41*.

Use of Natural Organic Products. Many natural organic products such as composts, sewage sludge, organic fertilizers and manure-based preparations are colonized by complex mixtures of microorganisms. When some of these materials are applied to turf, the microbial species they carry restrict the growth and development of the resident pathogens. Among the most suppressive composts tested in laboratory and field experiments, brewery and sewage sludge composts have shown the most promise. Organic fertilizers derived from poultry litter also have provided disease suppression in field experiments.

Field tests indicate highly variable results for the reduction of dollar spot incidence in creeping bentgrass golf greens (*40, 41*). Products included sewage sludge materials, brewery and manure composts, and natural organic fertilizers. The natural organic fertilizers contained hydrolyzed poultry feather meal, wheat germ, soybean meal, brewers yeast, bone meal, blood meal, sulfate of potash, and supplemented with species of *Bacillus* and the fungus *Trichoderma viride*. Ammonium nitrate and sulfur coated urea often provided an equal level of dollar-spot control (*41*).

It is important to know something about how the material was produced. Compost that has demonstrated some disease suppressive characteristics are produced under ideal physical and environmental conditions (*41*). The correct starting material composted in this manner allows for the colonization of both mesophilic (moderate temperature) and thermophilic (high temperature) microflora (Figure 9). This complex mixture of microorganisms makes an important contribution to the disease suppressive characteristics of the composted material. Also, composts that have cured for a long period will have a more diverse mesophilic microflora and have the potential to perform better for disease suppression.

Unfortunately, at this time there is no reliable way to predict the disease suppressive properties of composts or organic fertilizers since the nature of these colonizing microbial antagonists is left to chance. It is very important to proceed with caution when using these materials. Even though some research has documented up to 75 percent control of common turfgrass diseases, there generally is a great deal of variability in their performance. Also, at application rates of greater than 480 kg ha^{-1} (10 lbs. 1000 ft^{-2}) there exists the potential problem of creating an unwanted organic layer above a well-drained putting green or fairway root zone.

Use of Preparations Containing Known Microbial Antagonists. The use of a specific microbial antagonist to reduce the incidence and severity of diseases has been slightly more promising and consistent. Take-all patch of bentgrass has been controlled by infesting the soil with the nonpathogenic fungus *Phialophora graminicola*. Steam-treated preparations of the nonpathogenic fungus *Typhula phacorrhiza* applied as topdressing to creeping bentgrass provided up to 74 percent control of Typhula blight. The incidence and severity of Rhizoctonia blight of creeping bentgrass golf greens was reduced by inoculating the turf with nonpathogenic *Rhizoctonia* species.

The USGA provided funding for three projects that used specific microbial antagonists to suppress turfgrass pathogens. *Trichoderma harzianum* and *Enterobacter cloacae* have demonstrated promising disease suppression for dollar spot and different *Pythium* species. *T. harzianum* is a fungus that parasitizes *Pythium* blight and other turfgrass pathogens. The current study demonstrates that commonly used fungicides have little effect on the growth and performance of this biocontrol (*42*). *Enterobacter cloacae* has shown promising results for controlling *Pythium* species by removing turfgrass root exudates (waste products) found in the soil. These compounds stimulate the germination of disease sporangium, and once they are removed, pathogenic species of *Pythium* are less likely to germinate and infest turfgrass roots (*43*). *Stenotrophomonas maltophilia* (formerly the genus

22

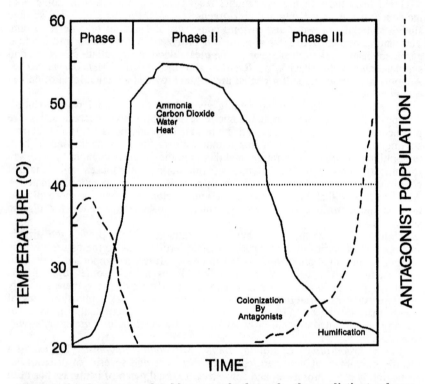

Figure 9. Composts should proceed through three distinct phases involving: Phase I, a rapid rise in temperature; Phase II, a prolonged high-temperature decomposition phase; and Phase III, a curing process where temperatures and decomposition rate decrease. SOURCE: Reproduced with permission from ref. *41*. Copyright 1992 United States Golf Association.

Xanthomonas) has provided excellent summer patch control in growth chamber and greenhouse trials. The field research confirms the importance of understanding the population dynamics of the biological control. Populations of the biocontrol agent were maintained above 10^5 cfu g^{-1} of soil; however, as many as 10^7 cfu g^{-1} of soil may be needed to control summer patch on field plots (*44*).

The USGA also supported a project that is developing a bacterial parasite (*Pasteuria* sp.) of sting nematodes (*Belonolaimus longicaudatus*). The bacterium is very effective in greenhouse trials; however, live nematodes are needed in order for the bacteria to reproduce. Additional field studies were conducted to evaluate natural populations of *Pasteuria* sp. under a variety of turf management levels. Results indicate nematode populations are significantly reduced in turf areas where the bacteria were present (*45*).

Insect Biocontrol. Biological controls for insect problems in agricultural crops have been used successfully in the United States. Research into the use of biological controls for turfgrass insects recently gained attention because of the push for natural, less toxic, pest control products. Rather than introducing a complex mixture of microorganisms with applications of organic materials, the research effort has focused on increasing our understanding of natural predators and parasites that occur in turfgrass areas. Similar to the preparations of known microbial antagonists to disease, specific organisms such as entomopathogenic nematodes or bacteria are introduced to reduce insect problems.

Natural Predators and Parasites in Turf. The fact that severe insect outbreaks are relatively uncommon in low-maintenance turf suggests that many pests are normally held in check by natural buffers. Environmental stresses such as drought can take a heavy toll on some insects. Natural insect pathogens, including bacteria, fungi, parasitic nematodes, and other disease causing agents also help to reduce pest populations.

Field surveys have shown that predatory insects are often very diverse and abundant in turf (*46*). Dozens of different species of ants, ground beetles, spiders and rove beetles feed on the eggs or larval (immature) stages of plant-eating insects. The common tiger beetle was observed to kill as many as 20 fall armyworm caterpillars in a single hour. There is some evidence to suggest that pest outbreaks can sometimes be attributed to inadvertent elimination of natural enemies when using broad-spectrum insecticides. A surface application of chlorpyrifos or isofenphos in June reduced predator populations such as spiders and rove beetles by as much as 60 percent. Similar work has documented that during a 48-hour period, predators consumed or carried off as many as 75 percent of sod webworm eggs laid on untreated turf. Chlorpyrifos treated plots had lower numbers of predators and subsequently resulted in significantly reduced consumption of the sod webworm eggs for three weeks (Figure 10).

A better understanding of the interactions between commonly used turfgrass pesticides and beneficial organisms that prey on insect pests is needed. Insecticides and other pesticides are powerful tools that often provide the only method to prevent severe damage from unexpected or heavy pest infestations. Broad-spectrum insecticides kill beneficial insects as well as pests, but fortunately, populations of predators and parasites seem to recuperate relatively quickly following individual applications. Turf managers need to apply pesticides at the proper time and rate, and only as needed to reduce the impact on beneficial organisms.

Use of Parasitic Agents for Control of Turfgrass Insects. Several interesting biological controls have performed well under ideal conditions (*47*). Research programs are field testing a variety of organisms to determine which products will establish quickly and have predictable control. In laboratory and greenhouse studies, entomopathogenic nematodes have shown the most promise for

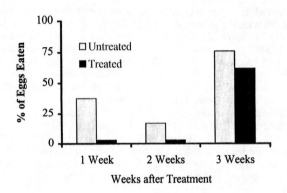

Figure 10. Reduced predation on sod webworm eggs placed in Kentucky bluegrass at one or three weeks after the turf was treated with chlorpyrifos. Predation levels recovered five weeks after application of the pesticide. The difference between treated and untreated plots was statistically significant. SOURCE: Reproduced with permission from ref. *46*. Copyright 1992 United States Golf Association.

controlling grub and caterpillar species. The nematode effectiveness is a function of the species selected and the conditions under which the organism is applied to the turf area. *Steinernema glaseri* and *Heterorhabditis bacteriophora* are effective on white grubs while *S. carpocapsae* is most effective against billbugs, cutworms, webworms and armyworms. Critical conditions include sunlight, irrigation and storage. Applications made in the early morning or evening followed by irrigation were most effective.

Bacteria such as *Bacillus thuringiensis* (Bt) and milky spore disease have been around for many years. The current commercial Bt products are slow to act, have a short residual time in the soil, and are ineffective on larger larvae. In recent years, the U.S. Department of Agriculture has identified a better strain of Bt (*B. japonesis* var. 'buibui') which is effective against the Oriental and Japanese beetle grubs. Milky spore disease against Japanese beetle grubs has been around for decades. Unfortunately, the disease has not been effective for turfgrass areas with high economic value.

Two fungi for controlling turfgrass pests include *Beauveria balliana* and *Metarhizium anisopliae*. Fungi may be more effective because they do not need to be ingested by the target insect. The fungal spore attaches itself to the insect, penetrates the circulatory system, and uses the host as an incubator to reproduce more spores. Research studies on the commercial production, cost and reliability of this type of biological control is just underway.

Cultural and Mechanical Practices. The USGA sponsored a few projects that evaluated the impact of a cultural or mechanical practices on reducing pest infestations. One of the best examples is the research conducted at University of Kentucky (*48*). Very little was known about the behavior of black cutworms in turf. The study determined where the moths laid their eggs and nightly behavior of cutworms on putting greens. It was found that black cutworm moths laid single eggs on the tips of the putting green leaf blades. More importantly, daily mowing removed 87 percent of the eggs.

Black cutworms thrived on perennial ryegrass and tall fescue surrounds (*48*). One source of infestation could be the eggs from putting green clippings strewn in adjacent areas. During nightly monitoring, it was observed black cutworms crawled considerable distances within a single night. Of the caterpillar tracks measured, about half originated from high grass off the putting green and terminated on the putting surface. The average length of the tracks was about 9 meters (30 feet).

These results suggest that black cutworm infestations on putting greens may originate, at least in part, from peripheral areas such as collars and roughs. Since eggs removed from the green can survive in the clippings, it is important to dispose of the clippings well away from the green. In addition, control measures should include a 9-meter buffer zone around the putting green to reduce reservoir populations in the surrounds. Because black cutworms are nocturnal, treatments are best applied at dusk.

Allelopathy. Chemical substances produced by some plants prevent or inhibit the growth of neighboring plants. For example, English walnut (*Juglens* sp.) are known to produce allelopathic substances. In Arkansas, twelve perennial ryegrasses that ranged from moderate to high stand density and zero to 95 percent endophyte infection were evaluated for their ability to decrease crabgrass populations. Bermudagrass fairway plots were overseeded with new seed lots of the 12 cultivars in the fall of 1994 and 1995. Half of each plot was then overseeded with crabgrass each spring and evaluated for crabgrass suppression. No differences in crabgrass stand could be attributed to any of the 12 cultivars (*49*).

A basic laboratory evaluation for allelopathy using *Lemna minor* L. (duckweed) measured allelopathic effects of plant tissue extracts on the growth rate of duckweed fronds. Extracts from shoots were applied to duckweed cell plates at three concentrations. The amount of allelopathic inhibition (or stimulation) of duckweed varied with shoot tissue sample season and extract concentration. All cultivars affected duckweed growth, but inconsistently. Perhaps eventually, selection of ryegrass cultivars for crabgrass inhibition may become an important part of IPM programs.

Selection and Breeding for Pest Resistance. Since 1982, the USGA has sponsored turfgrass breeding programs directed toward significantly reducing water use and maintenance costs. Among the maintenance costs, the amount of money spent on mowing, fertilization, and pesticide use is ranked very high. Turfgrass improvement efforts are very important because an integral part of IPM strategies is to use varieties that are well adapted to the local environmental and climatic conditions. Thus, new varieties with better abiotic and biotic stress resistance will have a positive role in IPM practices.

Conventional Breeding. One strategy to accomplish this goal was to improve specific characteristics within each species, such as water use or heat tolerance. This is a long-term approach, however, and can be expected to produce results in small increments over a period of many years. As an example, improved creeping bentgrasses began with vegetative cultivars in the 1920's to 'Penncross' (a seeded polycross) in 1954, to the recently developed seeded varieties.

Another strategy, and one that has the potential to produce a bigger return in a shorter period of time, is to replace stress-susceptible cool season grasses with stress-tolerant grasses that have improved turf characteristics. A good example of this strategy is buffalograss (*Buchloe dactyloides*), which in a short period has been significantly improved for turf quality. Buffalograss now is being used to replace perennial ryegrass (*Lolium perenne*) and Kentucky bluegrass (*Poa pratensis*) roughs in the central United States, resulting in water use savings of 50 percent or more and a large reduction in maintenance costs. Soon there will be buffalograsses for fairways, producing corresponding savings on those areas. Other non-traditional grasses being improved for turf characteristics are alkaligrass (*Puccinellia* spp.), blue grama (*Bouteloua gracilis*), fairway crested wheatgrass (*Agropyron cristatum*), inland saltgrass (*Distichlis* spp.), and curly mesquitegrass (*Hilaria belangeri*).

Yet another breeding strategy employed has been to expand the range of adaptation or use of existing stress-tolerant turfgrasses, with the hope of replacing less tolerant species. For example, if warm season grasses such as bermudagrass (*Cynodon dactylon*) can be improved for cold tolerance, it could be used to replace bentgrass (*Agrostis* spp.) and perennial ryegrass fairways in the transition zone areas of the United States. This would result in a tremendous reduction in water and pesticide use in these areas.

The last strategy is to develop improved, seeded, warm season grasses, such as bermudagrass, zoysiagrass (*Zoysia* spp.) and buffalograss, which could replace existing vegetatively propagated cultivars and save significant establishment costs. Availability of seeded types also would encourage courses in the transition zone to convert from high maintenance cool-season grasses to stress-tolerant warm-season species.

The USGA has financially supported turfgrass breeding programs in Georgia, New Jersey, Rhode Island, and Pennsylvania for many decades, with significant results. The new programs, initiated in the 1980's, have produced a number of improved cultivars or selections commercially available or released to seed companies for development (Table IX). All exhibit improved turf characteristics, stress tolerance or pest resistance. During the next decade, the number of new introductions will increase significantly, and the golf industry will be in a much better position to conserve and protect our natural resources.

Table IX. Summary of USGA Turfgrass Breeding Projects.

Turfgrass	University	Varieties and Germplasm Released
Creeping Bentgrass *Agrostis palustris*	Texas A&M University	CRENSHAW (Syn3-88), CATO (Syn4-88) and MARINER (Syn1-88), CENTURY (Syn92-1), IMPERIAL (Syn92-5), BACKSPIN (92-2)
	University of Rhode Island	PROVIDENCE
	Pennsylvania State University	PENNLINKS (PSU-126)
Colonial Bentgrass *Agrostis tenuis*	DSIR-New Zealand and University of Rhode Island	BR-1518 developed
Bermudagrass *Cynodon dactylon*	New Mexico State University	NuMex SAHARA, SONESTA, PRIMAVERA and other seed propagated varieties were developed
	Oklahoma State University	OKS 91-11 and OKS 91-1
C. transvaalensis	Oklahoma State University	A release of germplasm for university and industry use is under consideration.
C. dactylon X C. transvaalensis	University of Georgia	TIFEAGLE (TW-72), TIFTON 10 and TIFSPORT (TIFTON 94 or MI-40)
Buffalograss *Buchloe dactyloides*	University of Nebraska	Vegetative varieties 609, 315, and 378 Seeded varieties CODY and TATANKA Three new vegetative selections, NE 86-61, NE 86-120 and NE 91-118, are currently being processed for release
Alkaligrass *Puccinellia* spp.	Colorado State University	Ten improved families developed.
Blue grama *Bouteloua gracilis*	Colorado State University	ELITE, NICE, PLUS and NARROW.
Fairway Crested Wheat-grass *Agropyron cristatum*	Colorado State University	Narrow leafed and rhizomatous populations were developed
Curly Mesquitegrass *Hilaria belangeri*	University of Arizona	A 'fine' and 'roadside' population was developed.
Annual bluegrass *Poa annua* var *reptans*	University of Minnesota	DW-184 (MN #184) is commercially available. Selections MN #42, #117, #208, and #234 were released.
Zoysiagrass *Zoysia japonica* and *Z. matrella*	Texas A&M University	DIAMOND (DALZ8502), CAVALIER (DALZ8507), CROWNE (DALZ8512) and PALISADES (DALZ8514)
Seashore Paspalum *Paspalum vaginatum*	University of Georgia	One green type (AP 10) and one fairway type (PI 509018-1) released

Molecular Genetics. Over the last 75 years, a tremendous amount of effort has been exerted toward improving turfgrass species with conventional plant breeding methods. During the past decade, turfgrass improvement using cell and molecular techniques has been initiated. These new techniques are still in the early stages of development but hold great promise for the future. Biotechnology will enhance breeding efforts to develop turfgrasses that will meet the challenges of the next century in the following ways:

1. Molecular marker analysis
2. Genes with potential for turfgrass improvement
3. *In vitro* culture
4. Genetic engineering

Molecular Marker Analysis. Identifying the differences among varieties or cultivars of turfgrass species using molecular genetic analysis has made substantial progress in recent years (*50*). The new techniques are faster and more accurate than techniques used just a short time ago. Molecular markers linked to important genes controlling a desirable trait should make breeding programs more effective.

Genetic characterization of open-pollinated turfgrass species will present a different set of problems compared to the characterization of asexually propagated cultivars. We need to know something about the breeding behavior and genetics of turfgrass species to establish which genetic characterization techniques should be applied. Is the turfgrass species apomictic? Is it a homogeneous or heterogeneous population? In addition, the statistical analysis techniques to help interpret the tremendous amount of data that can be produced from molecular analysis need further development and need implementation in order to interpret molecular analysis results (*51*).

Genes with Potential for Turfgrass Improvement. Molecular and biochemical techniques allow researchers to examine how turfgrass plants respond to abiotic and biotic induced plant stress. Turfgrasses, like all plants, respond to plant stresses by up- or down-regulating genes. Some examples include the production of chitinase in response to disease pressure and cold temperatures (*52, 53*), the increased presence of desaturases that improve the properties of the cell wall (*54*), or the production of heat shock proteins in response to high temperatures (*55, 56*).

With regard to biotic stresses, such as disease problems, research with the chitinase gene demonstrates that genes for disease resistance may be introduced into turfgrass plants (*53*). However, a great deal more needs to be understood about how these disease-resistant genes will work in an entirely different species. Also, since turfgrasses are perennial, there are concerns that the introduction of a single gene will not produce long-lasting disease resistance.

In vitro Culture. Tissue or *in vitro* culture already has proven to be a very useful tool in turfgrass breeding programs. The presence of somaclonal variation in many of the commercially important turfgrass species has been documented (*55, 57*). The presence of this genetic variation has allowed turfgrass scientists and breeders an additional means to improve turfgrass quality through selection of abiotic and biotic stress tolerance.

In vitro culture was used to produce interesting somaclonal variants of seashore paspalum (*57*). The application of this technique on seashore paspalum is particularly useful because this open-pollinated species produces less than 5 percent viable seed and is self-incompatible. Conventional breeding methods are limited, and *in vitro* culture produced more than 4,000 regenerants that varied in genetic color, growth rate, density and winter hardiness. More than 100 selections with improved turfgrass traits were selected for further evaluations.

Extensive selection and breeding of regenerated or transformed plants for agronomic characteristics is still required prior to their release as new cultivars.

Using parental clones of existing cultivars as the explant material may provide the quickest and most rewarding approach. Work with the six parental clones of 'Crenshaw' creeping bentgrass and five new zoysiagrass cultivars is evaluating this approach (58).

Several forage and turfgrass species do not produce economically viable quantities of seed for commercial production. With the development of artificial seed technology through *in vitro* culture, it may be feasible to use this system to propagate parents for commercial production and release of hybrid varieties (59). In addition, if the parents of a commercial variety can be manipulated *in vitro,* the artificial seed method could facilitate the creation of varieties with value-added traits introduced by genetic transformation.

For turfgrass species, *in vitro* culture has been the primary method used to produce somaclonal variants that are screened for an important characteristic under greenhouse or field evaluations. There is hope that *in vitro* culture can be used to actually select for superior somaclonal variants at the cellular level. Embryogenic bentgrass callus was screened *in vitro* for heat tolerance and disease resistance (55). The results of this effort led to the development of HPIS (i.e., host plant interaction system) that allows the disease pathogen and turfgrass species to be cultured together *in vitro.*

Genetic Engineering of Turfgrasses. This research effort has been possible due to the advances made with important food and fiber crops. However, a pleasant surprise is how easily the success with agricultural crops can be applied to turfgrass species. In less than five years, early efforts in genetic transformation have already produced turfgrass parental clones of commercial interest. Herbicide-resistant genes have been successfully introduced into creeping bentgrass (52, 60). Through traditional breeding methods, it now may be possible to incorporate these herbicide resistance genes into putting green bentgrasses. This will allow golf course superintendents to keep annual bluegrass (*Poa annua)* and other weeds out of their greens and fairways with fewer pesticide applications. Potentially, this could reduce water and pesticide use on greens.

It is hopeful that genetic engineering will expedite the development of improved disease and insect resistance in turfgrass varieties for golf courses. Two projects were initiated to introduce chitinase production genes into bentgrass, providing an internal mechanism to help control turfgrass diseases and thereby reduce pesticide use. In 1996, the USGA sponsored a symposium on turfgrass biotechnology, and in 1997 published the proceedings in book form titled, *Turfgrass Biotechnology: Molecular Approaches to Turfgrass Improvement (61).*

Application of Integrated Pest Management (IPM). This approach relies on a combination of preventative and corrective measures to keep pest densities below levels that would cause unacceptable turf damage. Its goal is to manage pests effectively, economically, and with minimal risks to people and the environment. The process of IPM involves the following steps:

1. Sampling and monitoring turf areas to detect pests and evaluate how well control tactics have worked.
2. Pest identification in order to adequately understand the habits and life cycle of the organism and how it can be managed.
3. Decision making guided by action thresholds based on pest damage that justifies treatment or intervention.
4. Appropriate intervention that determines why a pest outbreak occurred, and if cultural practice adjustments can reduce damage or future outbreaks.
5. Follow-up that includes record-keeping to help predict future pest problems and to plan accordingly. This process also helps to evaluate which management practices work effectively.

6. Education to develop ongoing employee training focusing on pest recognition and agronomic factor that help them to make sound management decisions. Client education on management action to reduce pest problems (i.e., soil tests, aerification, dethatching, fertilization, and tree care) while providing a safe, environmentally responsible golf course.

Although USGA-sponsored research has not specifically addressed investigations on IPM as defined above, most of the research results will help with one or more of the steps outlined above. For example, in North Carolina, with cooperation with Cornell University, extensive research was conducted on the mole cricket behavior (62, 63). The results indicate that mole crickets are capable of avoiding pesticides applied to control them by burrowing deep into the soil until the active ingredient has degraded. A better understanding of mole cricket response to pesticide applications will help determine proper placement and timing of products that enhance contact with the insect.

Research conducted on white grubs (scarab beetle larvae) in Kentucky examined the role of several cultural practices on insect outbreaks (64). Field studies showed that withholding irrigation during peak flight of beetles, raising cutting height, and a light application of aluminum sulfate in the spring helped to reduce severity of subsequent infestations of Japanese beetle and masked chafer grubs. Grub densities were not affected by spring applications of lime or urea, but use of organic fertilizers (composted cow manure or activated sewage sludge) increased problems with green June beetle grubs. Use of a heavy roller was not effective for curative grub control. Soil moisture was the overriding factor determining distributions of root-feeding grubs in turf.

People and Wildlife

Pesticides and People. People are concerned about exposure they receive during a round of golf on turf that has been treated with pesticides. Exposure in this situation is caused by pesticide residues on the turf surface that rub off onto people or their equipment during a round of golf. A preliminary risk assessment was conducted for golfers who are exposed to a putting green treated with insecticides. Under the assumptions of this study, the golfer would have received about 1/3 of the lifetime reference dose considered safe by the USEPA (65). This is not to say that golfers or workers could not receive unsafe exposure to pesticides on golf courses. Under the conditions of this study, though, the golfer would not have been at significant risk.

Other studies that have determined dislodgeable residue levels indicate that less than one percent of the pesticides could be rubbed off immediately after application when the turf was still wet. In addition, the one percent could be reduced significantly by irrigating after the pesticide was applied. After the pesticides dried on the turf, only minimal amounts could be rubbed off. Volatilization studies report that organophosphate insecticides that possess high toxicity and volatility might result in exposure situations that cannot be deemed completely safe as judged by the US EPA Hazard Quotient determination (Table X). Additional biomonitoring studies will be needed to determine the extent of the risk, if any (9, 10, 11, 15).

Golf Course Benefits. Most media attention concerning golf courses tends to focus on the potential negative impacts of turfgrass and golf course management. Although people in our industry know intuitively that there are many benefits associated with turfgrasses and golf courses, the scientific basis for many of these benefits had not been documented. In 1994, through a USGA grant, Drs. James Beard and Robert Green conducted an exhaustive literature search and

Table X. Inhalation Hazard Quotients (IHQs) for Turfgrass Pesticides in the High, Intermediate and Low Vapor Pressure Group.

Group	Pesticide	Day 1	Day 2	Day 3
		----------- HQ[1] -----------		
Group 1: high vapor pressure	DDVP	0.06	0.04	0.02
(i.e., vapor pressures > 1.0×10^{-5} mm Hg)	ethoprop	50.0	26	1.2
	diazinon	3.3	2.4	1.2
	isazofos	8.6	6.7	3.4
	chlorpyrifos	0.09	0.1	0.04
Group 2: intermediate vapor pressure	trichlorfon	0.02	0.004	0.004
(i.e., 10^{-5} mm Hg > vapor pressures > 10^{-7} mm Hg)	bendiocarb	0.02	0.002	0.002
	isofenphos	n/d[2]	0.02	n/d
	chlorthalonil	0.001	0.001	0.0003
	propiconazole	n/d	n/d	n/d
	carbaryl	0.0005	0.0001	0.00004
Group 3: low vapor pressure	thiophanate-methyl	n/d	n/d	n/d
(i.e., vapor pressure < 10^{-7} mm Hg)	iprodione	n/d	n/d	n/d
	cyfluthrin	n/d	n/d	n/d

[1]The HQs reported are the maximum daily IHQ's measured, all of which occurred during the 11:00 a.m. to 3:00 p.m. sampling period.
[2] n/d = non-detect
SOURCE: Adapted from ref. *15*.

documented many of these benefits. The results of their work were published in the *Journal of Environmental Quality* (*66*), and were summarized by and are available from the United States Golf Association.

Wildlife and Golf Courses. Eight years ago, the USGA and Audubon International, Inc. developed the Audubon Cooperative Sanctuary Program (ACSP) to educate everyone involved with the game of golf about the environment. The goal was not to list the things that golf courses did wrong, but teach people what could be done to make the golf course better. The program encourages golf courses to become certified in the following six areas: 1) environmental planning, 2) wildlife and habitat management, 3) water conservation, 4) water quality management, 5) integrated pest management, and 6) member/public involvement.

This certification program has implemented findings from the USGA Turfgrass and Environmental Research Program. It is great to produce research information, but it is more important to put it into practice. Pesticide and nutrient fate, water conservation, and integrated pest management are areas where a significant amount of research has been completed. In order for the golf course superintendent to successfully implement suggested practices from the research studies, the golfer must first be educated about the environment. The ACSP is an important asset in this education effort!

Conclusion

The university research investigating pesticide and nutrient fate was the first extensive self-examination of golf's impact on water quality and the environment. What has the environmental research program told us? First, the university research shows that most pesticides used on golf courses have a negligible effect on the environment. The word "negligible" is used because we did find pesticides and fertilizers in runoff or leachate collected from research plots. However, under most conditions, the small amount collected (parts per billion) were found at levels well below the health and safety standards established by the Environmental Protection Agency (EPA). The studies demonstrated that the turfgrass canopy, thatch, and root system were an effective filter or sponge. As one would expect, the results documented that heavy textured soils adsorbed pesticides and fertilizers better than light textured or sandy soils.

The environmental research program has had a positive impact on golf. The program was run in an unbiased fashion, results have been published in peer-reviewed scientific journals, and the message, *be careful and responsible* is getting out to golf course superintendents around the country. Future efforts will focus on scaling up the size of research plots to simulate entire fairways or greens. Some of these studies will be conducted on golf courses. The new projects will document the impact of properly constructed and maintained golf courses on water quality.

The USGA has made progress in understanding the impact that golf courses have on the environment. More is being done to make golf courses both a recreational and environmental asset to the community. An excellent foundation of environmental information, based on scientific investigations rather than emotional rhetoric, has been established. The USGA, and the game of golf, need to keep asking questions and looking for new ways to maintain golf course grasses. More important, efforts should be increased to educate the golfer about environmental issues.

Literature Cited

1. Snow, J. T. *USGA Green Section Record.* **1995**, *33(3)*, 3-6.
2. Bowman, D. C.; Devitt, D. A.; Miller, W. W. *USGA Green Section Record.* **1995**, *33(1)*, 45-49.
3. Joo, Y. K.; Christians, N. E.; Blackmer, A. M. *J. of Fert. Issues.* **1987**, *4*, 98-102.
4. Joo, Y. K.; Christians, N. E.; Blackmer, A. M. *J. of Soil Sci.* **1991**, *55*, 528-530.
5. Branham, B. E.; Miltner, E. D.; Rieke. P. E. *USGA Green Section Record.* **1995**, *33(1)*, 33-37.
6. Miltner, E. D.; Branham, B.E.; Paul, E. A.; Rieke, P. E. *Crop Sci.* **1996**, *36*, 1427-1433.
7. Starrett, S. K.; Christians. N. E. *USGA Green Section Record.* **1995**, *33(1)*, 23-25.
8. Starrett, S. K.; Christians, N. E.; Austin, T. A. *J. of Irrg. Drain. Eng.* **1995**, *121*, 390-395.
9. Cooper, R. J.; Clark; J. M.; Murphy. K. C. *USGA Green Section Record.* **1995**, *33(1)*, 19-22.
10. Murphy, K.C.; Cooper, R. J.; Clark, J. M. *Crop Sci.* **1996**, *36*, 1446-1454.
11. Murphy, K.C.; Cooper, R. J.; Clark, J. M. *Crop Sci.* **1996**, *36*, 1455-1461.
12. Snyder, G. H.; Cisar, J. L. In *USGA Environmental Research Program: Pesticide and Nutrient Fate 1996 Annual Project Reports;* Kenna, M. P., Ed.; USGA Green Section Research: Stillwater, OK. **1996**, pp 109-140.

13. Yates, M. V.; Green, R. L.; Gan, J. In *USGA Environmental Research Program: Pesticide and Nutrient Fate 1996 Annual Project Reports*; Kenna, M. P., Ed.; USGA Green Section Research: Stillwater, OK. **1996**, pp 80-94.

14. Yates, M. V. *USGA Green Section Record.* **1995**, *33(1)*, 10-12.

15. Clark, J. M. In *USGA 1996 Turfgrass and Environmental Research Summary;* Kenna, M. P.; Snow, J. T., Eds.; United States Golf Association: Far Hills, NJ. **1997**, pp 60-63.

16. Smith, A. E.; Bridges, D. C. In *USGA 1997 Turfgrass and Environmental Research Summary;* Kenna, M. P.; Snow, J. T., Eds. ; United States Golf Association: Far Hills, NJ. **1998**, pp 74-75.

17. Carroll, M. J.; Hill, R. L. In *USGA 1997 Turfgrass and Environmental Research Summary;* Kenna, M. P.; Snow, J. T., Eds.; United States Golf Association: Far Hills, NJ. **1998**, pp 70-71.

18. Turco, R. In *USGA 1997 Turfgrass and Environmental Research Summary;* Kenna, M. P.; Snow, J. T., Eds.; United States Golf Association: Far Hills, NJ. **1998**, pp 79-82.

19. Lickfeldt, D. W.; Branham, B. E. *J. of Environ. Qual.* **1995**, *24*, 980-985.

20. Snyder, G. H.; Cisar, J. L. *Int. Turfgrass Soc. Res. J.* **1993**, *7*, 978-983.

21. Snyder, G. H.; Cisar, J. L. *USGA Green Section Record.* **1995**, *33(1)*, 15-18.

22. Cisar, J. L.; Snyder, G. H. *Int. Turfgrass Soc. Res. J.* **1993**, *7*, 971-977.

23. Cisar, J. L.; Snyder, G. H. *Crop Sci.* **1996**, *36*, 1433-1438.

24. Horst, G. L.; Shea, P. J.; Christians, N. E. *USGA Green Section Record.* **1995**, *33(1)*, 26-28.

25. Horst, G. L.; Shea, P. J.; Christians, N. E.; Miller, D. R.; Stuefer-Powell, C.; Starrett, S. K. *Crop Sci.* **1996**, *36*, 362-370.

26. Brauen, S. E.; Stahnke, G. *USGA Green Section Record.* **1995**, *33(1)*, 29-32.

27. Kenna, M. P. *USGA Green Section Record.* **1994**, *32(4)*, 12-15.

28. Petrovic, A. M. *USGA Green Section Record.* **1995**, *33(1)*, 38-41.

29. Cohen, S. Z.; Wauchope, R. D.; Klein, A. W.; Eadsforth, C. V.; Graney, R. *Pure & Applied Chem.* **1995**, *67*, 2109-2148.

30. Smith, A. E.; Tilloston, T. R. In *Pesticides In Urban Environments;* Racke, K. D.; Leslie, A. R., Eds.; ACS Symposium Series 522; American Chemical Society: Washington, D.C. **1993**, pp. 168-181.

31. Smith, A. E. *USGA Green Section Record.* **1995**, *33(1)*, 13-14.

32. Linde, D. T. *Surface Runoff and Nutrient Transport Assessment on Creeping Bentgrass and Perennial Ryegrass Turf.* M. S. Thesis, Pennsylvania State University, The Graduate School College of Agricultural Sciences, **1993**, p 84.

33. Linde, D. T.; Watschke, T. L.; Borger, J. A. *USGA Green Section Record.* **1995**, *33(1)*, 42-44.

34. Linde, D. T.; Watschke, T. L.; Jarrett A. R.; Borger, J. A. *Agron. J.* **1995**, *87*, 176-182.

35. Smith, A. E.; Bridges, D. C. In *USGA 1995 Turfgrass and Environmental Research Summary*; Kenna, M. P., Ed.; United States Golf Association: Far Hills, NJ. **1996**, pp 76-77.

36. Smith, A. E.; Bridges, D. C. In *USGA 1996 Turfgrass and Environmental Research Summary;* Kenna, M. P.; Snow, J. T., Eds.; United States Golf Association: Far Hills, NJ. **1997**, pp 72-73.

37 Baird, J. H. In *USGA 1995 Turfgrass and Environmental Research Summary*; Kenna, M. P., Ed.; United States Golf Association: Far Hills, NJ. **1996**, pp 64-65.

38 Cole, J. H.; Baird, J. H; Basta, N. T.; Hunke, R. L.; Storm, D. E.; Johnson, G. V.; Payton, M. E.; Smolen, M. D.; Martin, D. L.; Cole, J. C. *J. of Environ. Qual.* **1997**, *26*, 1589-1598.

39. Baird, J. H. In *USGA 1997 Turfgrass and Environmental Research Summary*; Kenna, M. P.; Snow, J. T., Eds.; United States Golf Association: Far Hills, NJ. **1998**, pp 56-59.

40. Couch, H. B. In *Diseases of Turfgrasses;* Krieger Publishing Company: Malabar, FL. **1995**, pp 249-257.

41. Nelson, E. B. *USGA Green Section Record.* **1992**, *30(2)*, 11-14.

42. Harmon, G. E. In *USGA 1996 Turfgrass and Environmental Research Summary;* Kenna, M. P.; Snow, J. T., Eds.; United States Golf Association: Far Hills, NJ. **1997**, pp 92-93.

43. Nelson, E. B. In *Turfgrass Biotechnology: Cell and Molecular Genetic Approaches to Turfgrass Improvement;* Sticklen, M. B.; Kenna, M. P., Eds.; Ann Arbor Press: Chelsea, MI. **1997**, pp 55-92.

44. Kobayashi, D. In *1995 Turfgrass and Environmental Research Summary;* Kenna, M. P., Ed.; United States Golf Association: Far Hills, NJ. **1996**, pp 50-51.

45. Giblin-Davis, R. M. In *1996 Turfgrass and Environmental Research Summary;* Kenna, M. P.; Snow, J. T., Eds.; United States Golf Association: Far Hills, NJ. **1997**, pp 54-56.

46. Potter, D. A. *USGA Green Section Record.* **1992**, *30(6)*, 6-10.

47. Schrimpf, P. *Lawn & Landscape.* **1997**, *18 (3)*, 66-69, 122.

48. Williamson, R. C.; Potter, D. A. *USGA Green Section Record.* **1998**, *36(1)*, 6-8.

49. King, J. W. In *1997 Turfgrass and Environmental Research Summary;* Kenna, M. P.; Snow, J. T., Eds.; United States Golf Association: Far Hills, NJ. **1998**, pp 53-54.

50. Kenna, M. P. In *Turfgrass Biotechnology: Cell and Molecular Genetic Approaches to Turfgrass Improvement;* Sticklen, M. B.; Kenna, M. P., Eds.; Ann Arbor Press: Chelsea, MI. **1997**, pp vii-xi.

51. Huff, D. R. In *Turfgrass Biotechnology: Cell and Molecular Genetic Approaches to Turfgrass Improvement;* Sticklen, M. B.; Kenna, M. P., Eds.; Ann Arbor Press: Chelsea, MI. **1997**, pp 19-30.

52. Anderson, M. P.; Taliaferro, C. M.; Gatschet, M.; de los Reyes, B.; Assefa, S. In *Turfgrass Biotechnology: Cell and Molecular Genetic Approaches to Turfgrass Improvement;* Sticklen, M. B.; Kenna, M. P., Eds.; Ann Arbor Press: Chelsea, MI. **1997**, pp 115-134.

53. Zhong, H.; Liu, C. A.; Vargas, J.; Penner, D.; Sticklen, M. B. In *Turfgrass Biotechnology: Cell and Molecular Genetic Approaches to Turfgrass Improvement;* Sticklen, M. B.; Kenna, M. P., Eds.; Ann Arbor Press: Chelsea, MI. **1997**, pp 203-210.

54. Baird, W. V.; Samala, S.; Powell, G. L.; Riley, M. B.; Yan, J.; Wells, J. In *Turfgrass Biotechnology: Cell and Molecular Genetic Approaches to Turfgrass Improvement;* Sticklen, M. B.; Kenna, M. P., Eds.; Ann Arbor Press: Chelsea, MI. **1997**, pp 135-142.

55. Krans, J. V.; Park, S. L.; Tomaso-Peterson, M.; Luthe, D. S. In *Turfgrass Biotechnology: Cell and Molecular Genetic Approaches to Turfgrass Improvement;* Sticklen, M. B.; Kenna, M. P., Eds.; Ann Arbor Press: Chelsea, MI. **1997**, pp 211-222.

56. Sweeney, P. M.; Danneberger, T. K.; DiMascio, J. A.; Kamalay, J. C. In *Turfgrass Biotechnology: Cell and Molecular Genetic Approaches to Turfgrass Improvement;* Sticklen, M. B.; Kenna, M. P., Eds.; Ann Arbor Press: Chelsea, MI. **1997**, pp 143-152.

57. Cardona, C. A.; Duncan, R. R. In *Turfgrass Biotechnology: Cell and Molecular Genetic Approaches to Turfgrass Improvement;* Sticklen, M. B.; Kenna, M. P., Eds.; Ann Arbor Press: Chelsea, MI. **1997**, pp 229-238.

58. Yamamoto, I.; Engelke, M. C. In *Turfgrass Biotechnology: Cell and Molecular Genetic Approaches to Turfgrass Improvement;* Sticklen, M. B.; Kenna, M. P., Eds.; Ann Arbor Press: Chelsea, MI. **1997**, pp 165-172.

59. Brittan-Loucas, H.; Tar'an, B.; Bowley, S. R.; McKersie, B. D.; Kasha, K. J. In *Turfgrass Biotechnology: Cell and Molecular Genetic Approaches to Turfgrass Improvement;* Sticklen, M. B.; Kenna, M. P., Eds.; Ann Arbor Press: Chelsea, MI. **1997**, pp 173-182.

60. Lee, L.; Day, P. In *Turfgrass Biotechnology: Cell and Molecular Genetic Approaches to Turfgrass Improvement;* Sticklen, M. B.; Kenna, M. P., Eds.; Ann Arbor Press: Chelsea, MI. **1997**, pp 195-202.

61. *Turfgrass Biotechnology: Cell and Molecular Genetic Approaches to Turfgrass Improvement;* Sticklen, M. B.; Kenna, M. P., Eds.; Ann Arbor Press: Chelsea, MI. **1997**, 256 p.

62. Brandenburg, R. L.; Villani, M. G. In *USGA 1997 Turfgrass and Environmental Research Summary*; Kenna, M. P.; Snow, J. T., Eds.; United States Golf Association: Far Hills, NJ. **1998**, pp 43-44.

63. Brandenburg, R. L. *TurfGrass TRENDS.* **1997**, *6(1),* 1-8.

64. Potter, D. In *USGA 1997 Turfgrass and Environmental Research Summary*; Kenna, M. P.; Snow, J. T., Eds.; United States Golf Association: Far Hills, NJ. **1998**, pp 39-42.

65. Borgert, C. J.; et. al. *USGA Green Section Record.* **1994**, *33(2),* 11-14.

66. Beard, J. B.; Green, R. L. *J. of Environ. Qual.* **1994**, *23,* 452-460.

Chapter 2

Turfgrass Benefits and the Golf Environment

J. B. Beard[1]

International Sports Turf Institute, College Station, TX 77840

There is a need to recognize and quantify the beneficial contributions of turfgrasses when properly maintained as part of a positive golf course environment. Specific functional benefits include: excellent soil erosion control and dust stabilization; enhanced runoff water entrapment and soil water infiltration; improved biodegradation of organic compounds; soil improvement and restoration of disturbed lands; substantial urban heat dissipation; reduced noise and glare; decreased noxious pests, allergy-related pollens, and human diseases; and a favorable wildlife habitat. The recreational-health benefits include enhanced physical health of participants and a unique low-cost cushion against personal injuries; plus improved mental health with a positive therapeutic impact and an overall better quality-of-life, especially in densely populated urban areas. This paper updates a 1992 publication concerning our current state of knowledge as to the benefits of golf course turfs as documented by scientific data.

An 18-hole golf course facility in the United States typically is comprised of (a) 0.8 to 1.2 ha of putting green area, (b) 0.6 to 1.2 ha of teeing area, and (c) 10 to 20 ha of fairway area (Table I). In other words, only 20 to 30% of the area on a golf course is used and maintained to specific criteria as part of the playing requirements of the game. Except for the various types of physical structures that may be constructed on site, a majority of the golf course property is devoted to a low maintenance, natural landscape. A properly planned and maintained golf course facility offers a diversity of functional benefits to the overall community, in addition to the physical and mental health benefits provided from the game itself. These benefits are significant when compared to alternate uses such as industry, business, and residential housing,

[1]Current address: ISTI, 1812 Shadowood Drive, College Station, TX 77840.

especially in the case of urban areas where a majority of the golf courses are located. In addition, greater benefits are derived from golf course facilities when compared to agricultural production operations in rural areas.

Table I. Comparative turf use by area for a representative 18-hole golf course.

Type of use	Area (ha)	Percent of Area
Rough/water/woodland	52.65	72.2
Fairways	16.2	22.2
Buildings/parking lots	2.11	2.9
Putting greens	1.01	1.4
Tees	0.93	1.3
Total	72.9	100.0

Turfgrasses have been used by humans to enhance their environment for over ten centuries. The complexity and comprehensiveness of the environmental benefits that improve our quality-of-life are just now being quantitatively documented through research. Benefits from golf course turfs may be divided into (a) functional and (b) recreational-health components.

Functional Benefits

Soil Erosion Control And Dust Stabilization - Vital Soil Resource Protection. Turfgrasses serve as an inexpensive, durable ground cover that protects our valuable, non-renewable soil resources. Properly maintained turfgrass stands are not a significant source of sediment entering bodies of water. It is generally recognized that a few large storms each year are responsible for most soil erosion losses. Dense, high-cut turfgrass stands modify the overland flow process such that runoff is insignificant in all but the most intense rainfall events.

The erosion control effectiveness of turfgrass is the combined result of a high shoot density and root mass for soil surface stabilization, plus a high canopy biomass matrix that provides resistance to lateral surface water flow, thus slowing potentially erosive water velocities. A key characteristic of mowed turfgrasses that contributes to very effective erosion control is a dense, low-growing ground cover, with a high shoot density ranging from 75 million to more than 20 billion shoots ha^{-1}. Regular mowing, as practiced in turf culture, increases the shoot density substantially because of enhanced tillering when compared to natural grasslands (2). Therefore, perennial turfgrasses offer one of the most cost-efficient methods to control wind and water erosion of soil. Such control is very important in eliminating dust and mud problems. When this major erosion control benefit is combined with the ground water recharge and organic chemical decomposition discussed in the next two sections, the resultant relatively stable turfgrass ecosystem is quite effective in both soil and water conservation and quality protection.

Enhanced Surface Water Capture and Ground Water Recharge. A key mechanism by which turfgrasses conserve water is their superior capability to essentially trap and hold potential water runoff. This facilitates more water infiltrating and filtering downward through the soil-turfgrass ecosystem. A mowed turfgrass possesses a leaf and stem biomass ranging from 1,000 to 30,000 kg ha^{-1}, depending on the grass species, season, and cultural regime. This biomass is composed of a matrix of relatively fine-textured stems and narrow leaves with numerous, random open spaces. The matrix is of a porous nature which enhances surface water retention and subsequent water infiltration. The reduced runoff volume, due to a turfgrass cover, can decrease the costly storm-water structures and management requirements for urban development. Turfgrass ecosystems often support abundant populations of earthworms (*Lumbricidae*) ranging from 200 to 300 earthworms m^{-2}. Earthworm activity increases the amount of macropore space within the soil, which results in higher soil water infiltration rates and water-retention capacity *(13)*. Numerous studies have shown the ability of a turfgrass cover to reduce runoff water and enhance soil water infiltration, thereby furthering ground water recharge.

Surface Water Protection. The excellent ability of high density, low-growing turfgrasses to capture surface water and the dense root biomass provide a superior ability to control the loosening and overland movement of soil particles in surface water which may eventually enter streams, ponds, and lakes. These soil particles are typically silts and clays, with the clays tending to have phosphates attached. The phosphates contribute significantly to the eutrophication of ponds and lakes. Thus, high cut turfgrass covers contribute substantially in protecting the quality of surface waters.

Improved Soil Biodegradation Of Organic Chemicals. A diverse, large population of soil microflora and microfauna are supported by the decomposition of turfgrass roots and rhizomes. These same organisms offer one of the most active biological systems for the degradation of organic chemicals and pesticides trapped by the turf. Thus, this turf ecosystem is important in the protection of ground water quality.

The runoff water and sediment that occurs from impervious surfaces in urban areas carries many pollutants: including metals such as lead, cadmium, copper, and zinc; hydrocarbon compounds from oil, grease and fuels; and household/industrial hazardous wastes such as waste oils, paint thinners, organic preservatives, and solvents. Turfgrass areas can be designed for the catchment and filtration of these polluted runoff waters.

It is significant that large populations of diverse soil microflora and microfauna are supported by this same soil-turfgrass ecosystem. Microflora constitute the largest proportion of the decomposer biomass of most soils. The bacterial biomass component ranges from 30 to 300 g m^{-2}, and for fungi from 50 to 500 g m^{-2}, with actinomycetes probably in a similar range *(12)*. The soil invertebrate decomposer biomass ranges from 1 to 200 g m^{-2}, with higher values occurring in soils dominated by earthworms. The earthworm component is especially active in the soil incorporation and decomposition of the turf thatch and clippings, which contributes to the soil organic matter level.

The bacterial population in the moist litter, grass clippings, and thatch of a turf commonly is in the order of 10^9 organisms cm^{-2} of litter surface. These organisms offer one of the most active biological systems for the degradation of trapped organic chemicals and pesticides. The average microbial biomass pool is reported to be 700, 850, and 1,090 kg of carbon ha^{-1} for arable, forest, and grassland systems, respectively. A microbial biomass of 1,200 kg C ha^{-1} has been reported for grasslands in the United States *(15)*. Microbial biomass values of mowed turfgrasses are not yet available, but are probably even higher due to the high carbon biomass contained in the senescent leaves and grass clippings that accumulate near the soil surface and to a more favorable soil moisture regime due to irrigation.

The turfgrass ecosystem also supports a diverse community of non-pest invertebrates. For example, a *Poa pratensis - Festuca rubra* polystand in New Jersey supported 83 different taxa of invertebrates including insects, mites, nematodes, annelids, gastropods, and other groups. Similarly, dozens of species of *Staphylinidae* (rove beetles), *Carabidae* (ground beetles), *Formicidae* (ants), *Araneae* (spiders), and other groups of invertebrates have been recovered from turfgrass sites *(13)*

Finally, there also is the gaseous dimension of atmospheric pollution control. Carbon monoxide (CO) concentrations greater than 50 μl often occur in urban environments, especially near roadsides. Turfgrasses, such as *Festuca arundinacea*, have been shown to be useful as an absorber of CO from the urban environment.

Ground Water Protection. Turfgrasses are characterized by a combination of enhanced surface water capture including associated pollutants, soil water infiltration, and enhanced decomposition of organic compounds including pesticides. These attributes result in turfgrasses serving a vital role in ground water protection.

Soil Improvement And Restoration. An extremely important function of turfgrasses is soil improvement through organic matter additions derived from the decomposition of roots and other plant tissues that are synthesized in part from atmospheric carbon dioxide via photosynthesis (Table II). A high proportion of the world's most fertile soils has been developed under a vegetative cover of grass *(8)*. The root depth potential of turfgrasses ranges from 0.5 to 3 m, depending on the species, extent of defoliation, and soil/environmental conditions. Generally, C$_4$ warm-season turfgrasses produce a deeper, more extensive root system than the C$_3$ cool-season species *(2)*. More work has been reported on the rooting characteristics of *Poa pratensis* turf than any other grass species. Its root system biomass is in the range of 11,000 to 16,100 kg ha^{-1}, with approximately 122,000 roots and 6.1 x 10^7 root hairs per liter in the upper 150 mm of soil, plus a combined length of over 74 km and a surface area of about 2.6 m^2.

Table II. The 9-year contribution of a *Cynodon* sod production operation[1], involving two sod harvests annually, on the underlying soil organic matter content.

Year	Organic Matter Content (%)
1985	0.50
1987	0.98
1988	0.70
1989	1.20
1990	1.60
1992	2.46
1993	2.80

[1]Sod production was initiated in 1985 by C.C.Willis of Western Sod near Casa Grande, Arizona on a clay loam soil. The land had previously been in a two-crop per year rotation of cotton and wheat.

The annual root system turnover rate has been estimated at 42% for a turf. Using Falk's estimate, 6,761 kg of root biomass per hectare would be turned over into the soil each year. This estimate is low because it did not account for root secretions, death and decay of fine roots and root hairs, and consumption of roots by soil insects and animals. The amount of root biomass annually produced and turned over into the soil, or root net productivity, for a defoliated grassland is higher than the amount reported for ungrazed prairie ecosystem. Similarly, the net effect of regular mowing on prostrate growing turfgrasses would be to concentrate energies into increased vegetative growth, as opposed to reproductive processes, and to form a canopy of numerous dense, short, rapid growing plants with a fibrous root system. Also, many prairie lands in the United States show decreased productivity under regular defoliation, as by mowing, since most native grass species found in these ecosystems form meristematic crowns that are elevated higher above the soil where removal is more likely when compared to turfgrass species.

Land Restoration. The turfgrass benefits of enhanced soil erosion control, surface water capture, biodegradation of organic chemicals, and soil improvement via organic matter additions from an extensive grass root biomass jointly provide a favorable environment for land restoration. Accelerated soil restoration of environmentally damaged areas by planting perennial grasses is employed effectively on highly eroded rural landscapes, burned-over lands, garbage dumps, mining operations, and timber harvest areas. Some of these previously unwanted areas may then be developed as functional golf courses, recreational, and wildlife areas.

Enhanced Heat Dissipation - Temperature Moderation. Turfgrasses dissipate high levels of radiant heat in urban areas through the cooling process of transpiration. The overall temperature of urban areas may be as much as 5 to 7°C warmer than that of nearby rural areas. Maximum daily canopy temperatures of a green *Cynodon* turf was found to be 21°C cooler than a brown dormant turf and 39°C cooler than a synthetic surface (Table III). The transpirational cooling effect of green turfs and landscapes can save energy by reductions in the energy input required for interior mechanically cooling of adjacent homes and buildings.

Table III. Comparative temperatures of four surfaces assessed
in August in College Station, Texas *(9)*.

Type of Surface	Maximum Temperature (°C)	Percent Temperature Increase Compared to a Green Turf
Green growing turf	31.1	—
Dry, bare soil	38.9	16
Brown, dormant turf	52.2	43
Synthetic turf	70.0	80

Noise Abatement and Glare Reduction. The surface characteristics of turfgrasses function in significant noise abatement of from 20 to 40%, as well as in multi-directional light reflection that reduces glare. Both result in less stress to humans. Studies have shown that turfgrass surfaces absorb harsh sounds significantly better than hard surfaces such as pavement, gravel, or bare ground. These benefits are maximized by an integrated landscape of turfgrasses, trees, and shrubs. Unfortunately, the proper use of turfgrasses, trees, and shrubs in concert to maximize noise abatement has received little attention within the scientific community.

Decreased Noxious Pests, Allergy-Related Pollens, And Human Diseases.

Noxious Pests. Closely mowed turfs reduce the numbers of nuisance pests such as snakes (*Serpentes* suborder), rodents (*Rodentia* order), mosquitoes (*Culicidae* family), scorpions (*Scorpionida* order), ticks (*Ixodoidea* superfamily), and chiggers (*Trombiculidae* family) due to a less favorable habitat *(13)*.

Allergies. Allergy-related pollens can cause human discomfort and potentially serious health concerns to susceptible individuals. Dense turfs typically are void of the many weedy species that often produce allergy-related pollens. In addition, most turfgrasses that are mowed regularly at a low height tend to remain vegetative with minimal floral development *(2)*, and thus have reduced pollen production.

Human Diseases. Exposure to a number of serious human diseases is facilitated by key insect vectors, such as mosquitos and ticks. Of current concern is Lyme disease, which is spread by a tick commonly found in unmowed tall grass and woodland-shrub habitats. A closely mowed turf offers a less favorable habitat for unwanted nuisance insects and disease vectors

Provides A Favorable Wildlife Habitat. More than 70% of the golf course allocated to rough and non-play area encompasses turfgrasses, trees, and water in the primary rough and turfgrasses, flowers, shrubs, trees, and water in the secondary rough and perimeter areas. A diverse animal and plant wildlife population can be achieved by an integrated golf course landscape. A study of golf courses and parks in Cincinnati, Ohio has shown that passerine birds benefit from golf courses, when compared to adjacent urban and agricultural land uses, even to the extent that golf courses may be described as bird sanctuaries. A comparison of bird communities on a golf course and on a natural area both in Kansas revealed a higher density of birds on the golf course, a comparable number of bird species, and species characterizations that varied *(16)*.

Ponds, lakes, and wetlands are very desirable features as used in golf courses because they create aquatic habitats, as well as diversity in visual landscape aesthetics. Properly designed urban landscape "green" areas such as golf courses can maintain and even promote plant and animal diversity, natural habitats, and wetlands, especially when compared to intensive agriculture and urban residential usage. Thus, golf courses are important naturalized open spaces, especially in areas of urban development.

Substantial Contribution To The National Economy. From a monetary standpoint, the golf industry contributes in excess of US $18 billion annually to the United States economy.

Recreational Health Benefits

Enhanced Physical Health Of Golf Participants. The enjoyment and benefits of improved physical/mental health derived from golfing activities on turfgrasses are vital to a contemporary, fast-paced industrialized society, especially in densely populated urban areas. There are more than 16,000 golf courses in the United States offering 24.7 million golfers more than 2.4 billion hours of healthy, outdoor recreation.

Playing golf involves the expenditure of energy that is reflected in weight loss *(14)*. The caloric cost of golf play has been reported to be 3.7 kcal min^{-1} *(7)*. Research has shown that walking an 18-hole golf course 3 times per week resulted in a lowered total cholesterol (TC) level, a lowered low-density lipoprotein cholesterol (LDL-C) level, and an improved risk ratio of LDL-C to TC *(11)*.

The relative hardness of a surface affects the likelihood that foot, knee, and back injuries will occur to those walking or running of the respective surfaces. The turfed surfaces of golf courses have a relatively soft cushioning characteristic (Table IV).

Table IV. Comparisons of the hardness of nine representative surfaces in the College Station, Texas area, expressed as means of multiple observations of Clegg Impact Value (CIV) using the fourth 300-mm drop reading *(13)*.

Representative Types of Recreational/Sport Surfaces	Clegg Impact Value (g) (2.25-kg hammer)
Cement floor	1426
Asphalt road	1442
Tennis court–outdoor composition	1422
Running track-outdoor composition	1432
Basketball court–permanent wood	640
Baseball infield–bare clay	504
Football field-outdoor, 4-year old artificial surface	175
Football field-indoor, 1-year old artificial surface	141
Baseball field–natural turf of irrigated *Cynodon*	100

Positive Therapeutic Impact. Most city dwellers attach considerable importance to urban green areas, with views of grass, trees, and open space. Cities can be very dismal without green turfgrasses, with the result being a loss of human productivity, and more susceptibility to anxieties and mental disease. When the visual content response of a golf course landscape was compared to a structured urban and a forest setting, the former lowered the average blood pressure level and the skin conductance level, with both returning to baseline levels more rapidly and the viewer subsequently performing mental arithmetic tasks more rapidly *(12)*. There is an increased sense of residential neighborhood satisfaction and of general well-being when there was a nearby nature landscape, such as parks, golf courses, woodland areas, and large landscape sites

Turfgrasses provide beauty and attractiveness that enhance the quality-of-life for human activities. The clean, cool, natural green of turfgrasses provides a pleasant, serene environment in which to live, work, and play. Such aesthetic values are of increasing importance to the human spirit of citizens because of rapid-paced lifestyles and increasing urbanization.

Acknowledgments

This paper represents an update of an original paper coauthored with R.L. Green in the Journal of Environmental Quality in 1994 (6). For citations of the research documenting the non-referenced statements presented herein, please refer to the original 1994 paper. This series of review papers was partially supported by a grant from the United States Golf Association.

Literature Cited

1. Alexander, M. *Introduction to Soil Microbiology, 2nd ed.* Wiley, New York, N.Y. **1977.**

2. Beard, J.B. *Turfgrass: Science and Culture.* Prentice-Hall, Inc., Englewood Cliffs, N.J. **1973.** 658 pp.

3. Beard, J.B. *Turf Management for Golf Courses.* Macmillan Company, New York, N.Y. **1982.** 642 pp.

4. Beard, J.B. *The role of Gramineae in enhancing man's quality of life.* Symp. Proc. Nat. Comm. Agric. Sci., Japanese Sci. Council. **1989.** pp 1-9.

5. Beard, J.B. *Environmental protection and beneficial contributions of golf course turfs.* Science and Golf II: Proceedings of the World Scientific Congress of Golf. A.J. Cochran and M.R. Farrally, Ed. E & FN Spon, London, England. **1994.** pp 399-408.

6. Beard, J. B and R.L. Green. *The role of turfgrasses in environmental protection and their benefits to humans.* J. Environmental Quality. **1994.** *Vol. 23, no. 3.* pp 452-460.

7. Gretchell, L.H. *Energy cost of playing golf.* Arch. Phys. Med. Rehab. **1968.** *Vol. 49,* pp 31-35.

8. Gould, F.W. *Grass Systematics.* McGraw-Hill Book Company, New York, N.Y. **1968.** 382 pp.

9 Johns, D., and J.B Beard. *A quantitative assessment of the benefits from irrigated turf on environmental cooling and energy savings in urban areas.* In Texas Turfgrass Research--1985. Texas Agric. Exp. Stn. PR-4330. **1985.** pp 134-142.

10. Kaplan, R. and S. Kaplan. *The Experience of Nature.* Cambridge University Press, New York, N.Y. **1989.**

11. Palank, E.A. and E.H. Hargreaves, Jr. *The benefits of walking the golf course - Effects on lipoprotein levels and risk ratios.* The Physician and Sports Medicine. **1990.** *Vol. 18, no. 10,* pp 77-80.

12. Parsons, R., L.G. Tassinary, R.S. Ulrich, M.R. Hebl, and M. Brossman-Alexander. *The view from the road: Implications for stress recovery and immunization.* Journal of Environmental Psychology. **1998.** *Vol. 17, no. 3.*

13. Potter, D.A. *Destructive Turfgrass Insects: Biology, Diagnosis, and Control.* Ann Arbor Press, Chelsea, MI. 344 pp.

14. S.I. Sifers and J.B Beard. *Enhancing participant safety in natural turfgrass surfaces including use of interlocking mesh element matrices.* In Safety in American Football, E.F. Hoerner, Ed. Am. Soc. for Testing and Materials, STP 1305. **1996.** pp 156-163.

15. Smith, J.L., and E.A. Paul. *The significance of soil microbial biomass estimations.* In Soil Biochemistry. J.M. Bollag and G. Stotzky, Ed. Vol. 6. Marcel Dekker Inc., New York, N.Y. **1990.** pp 357-396.

16. Terman, M.R. *Natural links: naturalistic golf courses as wildlife habitat.* Landscape and Urban Planning. **1997.** *Vol. 38,* pp 183-197.

Chapter 3

Pesticides for Turfgrass Pest Management: Uses and Environmental Issues

Kenneth D. Racke

Global Health, Environmental Science and Regulatory, Building 308/2B, Dow AgroSciences, Indianapolis, IN 46268

Given the pest complex which may damage turfgrass, a variety of pesticides may be used to promote turf health. Insecticides such as acephate, chlorpyrifos, fipronil, halofenozide, imidacloprid, and spinosad may be applied to control soil inhabiting or surface feeding insects. Broadleaf and annual grass weeds may be controlled with such herbicides as 2,4-D, pendimethalin, triclopyr, and oxadiazon. Fungicides such as azoxystrobin, chlorothalonil, fenarimol, or triadimefon may be employed for fungal pests. Aspects of the chemical control paradigm for turf pest management will be examined. Pesticides for turfgrass pest management are selectively applied to achieve turf protection yet minimize potential environmental impacts. Environmental assessments of existing and new products for use on turf require generation of data related to persistence, mobility (leaching, surface runoff, volatility), bioavailability, and ecological impact. Research programs and approaches for environmental assessment of turfgrass pesticides will be reviewed. Finally, case study examples regarding research that may be associated with an existing turf pesticide product as well as research required for development of a new pesticide for the turf market will be examined.

Although much focus has been placed on the use of pesticides in agriculture, the urban environment also represents an important arena for the use of pest control chemicals [1]. This chapter will examine aspects of the chemical control paradigm for turfgrass pest management, with particular emphasis on golf course uses.

Turfgrass Pesticide Market

Pesticide sales in the urban consumer and professional markets for 1991 were estimated at approximately 2.2 billion dollars (2). EPA estimates for 1993 place the volume of pesticides for agriculture, industrial/commercial/government (including golf courses), and home and garden use at 839, 193, and 73 million pounds of active ingredient, respectively (3). In the urban environment, turfgrass pest control scenarios represent significant avenues of pesticide application and use. These include home lawns, golf courses, parks, athletic fields, commercial properties, and sod farms. A survey of home and garden pesticide use in the United States reported that 22% of home lawns were treated with pesticides by homeowners, and of those with private lawns, 12% utilized the services of a commercial lawn-care company (4). For golf courses, a recent survey reported only 20% utilized contract pesticide application services, with most opting for application by trained, in-house staff (5). Professional lawncare and golf course uses represent the major non-retail markets for turf pesticide products (Figure 1). Pesticides are heavily relied upon for pest management programs at the nation's 14,000+ golf courses, which represented upwards of 1.6 million acres in 1996 (Kline and Company, unpublished data).

Approximately 300,000 acres of golf courses received at least one insecticide treatment during 1996. Insecticide products are most likely to be applied to fairways, tees, and greens, with proportionately less applied to roughs. Pre- and postemergence herbicide application represents the largest use on an acreage basis as compared with all pesticide products. Between 1.4 and 3.2 million herbicide acre-treatments per year were made between 1992 and 1996 (Kline and Company, unpublished data). The greatest quantities of herbicide use by volume during this period were applied to fairways and roughs. Compared to other turf markets, fungicides are more heavily relied upon in the golf course segment than for other segments. Approximately 160,000 and 189,000 acres of turf were treated with fungicides during 1994 and 1996, respectively, with greens being treated at a relatively higher percentage (73-93%) than other golf course areas (Kline and Company, unpublished data). It should be pointed at that as far as quantity of active ingredients applied, herbicides still predominate due to the large overall acreage and also use of older products with relatively higher application rates. For example, a survey of Hawaiian golf courses found a statewide average of 94,025 pounds of herbicide applied yearly, versus 19,051 pounds fungicide and 4,463 pounds of insecticide (6).

Individual golf course data provided by the Golf Course Superintendents Association of America substantiate the importance of fungicides in golf course pest control programs, with comparable reliance upon herbicides and insecticides from an expense standpoint (Figure 2). An interesting observation from this survey is that spending on chemical fertilizers was roughly equivalent to that for all pesticide products combined for the average golf course (5).

Pests and Products

A plethora of pest insects, weeds, and diseases attack turfgrass areas of golf courses and lead to the need for management programs. Expectations of golfers are high with respect to the quality of turf on tees, fairways, and greens. At present, use of

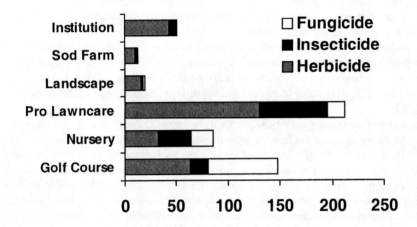

Figure 1. Turfgrass pesticide market for 1995/1996 with sales shown in millions of dollars (Kline and Company, unpublished data).

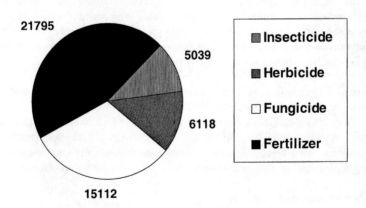

Figure 2. Average annual expenditures (dollars) on turfgrass chemicals for individual golf courses during 1997 (5).

pesticides represents a major means of control of turfgrass pests of golf courses, and the chemical control paradigm has been widely adopted.

Insects and Insecticides. Insect pests of turf include surface feeders which actively damage foliage through chewing or sucking, and also subsurface feeding insects which may devour roots and leave plants vulnerable to drought and desiccation. In addition, the burrowing habit of the mole cricket results in unsightly disruption of the turf/ thatch/soil surface. There are significant regional differences between the types of insect pests encountered on golf courses. For example, mole crickets and fire ants are typical pests of Southeast U.S. turf whereas *Hyperodes* weevils are common pests of the Northeast U.S. Examples of surface and subsurface feeding insect pests are listed in Table I.

For surface feeding insects, well-established products such as chlorpyrifos, carbaryl, and acephate are the major compounds of use (Kline and Company, unpublished data; 5). Newly introduced insecticides for surface feeder control include bifenthrin and spinosad. For subsurface grub control, although established products such as chlorpyrifos and trichlorfon continue to be widely used, there have been recent and highly successful introductions of the new products imidacloprid and halofenozide (7). For mole crickets, established products such as acephate and isazophos have been joined by the recent introduction of fipronil.

Weeds and Herbicides. Weeds infesting turf include both grasses and broadleaves. Examples of common grass and broadleaf weed pests of turf are listed in Table I. In addition to terrestrial weeds, pond areas of golf courses are also subject to problems with nuisance aquatic weeds and algae.

Grass weed control largely involves preemergent applications of such products as pendimethalin, trifluralin, benefin, dithiopyr, prodiamine, and oxadiazon. Broadleaf weed control predominantly relies upon postemergent applications of such products as 2,4-D, MCPA, and dicamba. Combination products containing one or more of these active ingredients are commonly employed to achieve an acceptable spectrum of activity across weed pest varieties. Several of the active ingredients employed for weed control on golf courses have been employed for many years. An example would be 2,4-D, which has seen over 50 years of use.

Diseases and Fungicides. A large variety of plant pathogens may infect and damage turfgrass. Examples of common turf diseases are listed in Table I.

Products for turf disease control include those which have contact or systemic activity such as chlorothalonil and flutolanil, respectively. In addition, products demonstrating broad-spectrum disease control both systemically and by contact have been highly successful, as exemplified by the recent introduction of azoxystrobin.

Turf Pest Management Trends

Although the chemical control paradigm is still largely employed in turf pest management systems, there have been significant changes during the past several years. These changes include the increasing importance of integrated pest

Table I. Common Turfgrass Pests and Pesticides.

Insect/Nematode Control

Insects	Surface Feeders Chinch bugs Sod webworms Cutworms Armyworms	Subsurface Feeders Grubs Mole crickets Nematodes
Insecticides	Chlorpyrifos (Dursban*) Carbaryl (Sevin) Acephate (Orthene) Bifenthrin (Talstar) Spinosad (Conserve*)	Imidacloprid (Merit) Chlorpyrifos (Dursban) Trichlorfon (Dylox) Halofenozide (Mach II) Fipronil (Chipco Choice) Acephate (Orthene) Isazophos (Triumph)

Weed Control

Weeds	Grasses Crabgrass Goosegrass Annual bluegrass	Broadleaf White clover Dandelion Plantain
Herbicides	Dithiopyr (Dimension) Prodiamine (Barricade) Oxadiazon (Ronstar) Pendimethalin Simazine Benefin + Trifluralin (Team*)	2,4-D MCPA, MCPP Dicamba Dichlorprop 2,4-D + MCPP + Dicamba (Trimec) Triclopyr + Clopyralid (Confront*)

Disease Control

Diseases	Brown patch Pink snow mold Dollar spot	Pythium blight Gray snow mold Anthracnose
Fungicides	Contact or Systemic Mancozeb (C) Chlorothalonil (C) Flutolanil (Prostar) (S) Metalaxyl (Subdue)(S) Fosetyl al (Aliette) (S)	Contact/Systemic Azoxystrobin (Heritage) Iprodione Propiconazole (Banner) Triadimefon (Bayleton) Fenarimol (Rubigan*)

*Trademark of Dow AgroSciences

management and the continued introduction of both conventional and biological pesticides into the turfgrass manager's chemical toolbox. In the future, the role of biotechnology may also play a larger role with respect to the incorporation of favorable traits (e.g., insect resistance) into turf varieties.

Integrated Pest Management. The concept of Integrated Pest Management (IPM) arose originally within the agricultural pest control arena. The philosophy of IPM places less emphasis on chemical control measures and more emphasis on the use of all available control measures (e.g., chemical, cultural, biological) in an integrated fashion for effective management of insect, weed, and disease pests. Use of chemicals has a clear place in the IPM approach, but rather than rely on prophylactic or calendar-based spray application schedules, the use of chemicals at specific times and sites identified by scouting, monitoring, or modeling, and justified economically, is encouraged.

There has been a significant degree of recent research on IPM for turfgrass pest management, and comprehensive summaries are available (8,9). Despite the volume of research methodologies and academic approaches, IPM has been only slowly adopted in the turfgrass pest management arena. With respect to the largest turf market segments (home lawncare, professional lawncare, and golf course), the golf course environment has offered the greatest potential for adoption of IPM approaches because golf course superintendents already monitor their landscapes and use cultural tactics to minimize disease and insect problems. A survey of golf course superintendents revealed that an average of 86% reported at least some involvement with an IPM program, which resulted in a reported average reduction in pesticide use of 21% (5). Few comprehensive IPM programs have been developed for turfgrass, however, and Potter (10) summarized major impediments to broader adoption of IPM in the golf course environment. These included high expectations of golfers for near-perfect turf quality, cost-ineffectiveness of available pest monitoring and sampling methods, lack of reliable and cost effective alternatives to pesticides, and a relatively weak research and extension base. It is anticipated that continued efforts will be devoted to meeting these challenges given the cost of chemical control measures and issues associated with their use.

Continued Introduction of Conventional Products. Use of some turf pesticide products has endured for many years. Such herbicide products as 2,4-D and atrazine, and insecticide products as chlorpyrifos and carbaryl have been employed for 25-plus years for turf pest control. There have continued to be new product introductions for all classes of pesticide products for turf use through the years. The past several seasons have seen the introduction of several innovative and noteworthy products, especially those targeted at insect control. These include imidacloprid (1995) and halofenozide (1997) for grub control, fipronil (1996) for mole cricket control, and spinosad (1997) for surface feeder control. The recent introduction to the turfgrass market of azoxystrobin (1997), a broad-spectrum fungicide, has greatly impacted professional turf disease management. Although several of these products have been approved by EPA as "reduced risk" alternatives to existing chemical products, it is noteworthy that characteristics of some new products may actually discourage IPM approaches. For example, due to residuality of control and desirability of targeting early life stages of grub pests, both imidacloprid and halofenozide may be applied weeks or months prior to pest egg hatch. This practice might tend to discourage soil sampling and use of selective, curative control measures (*7, 11*).

Availability of Biologicals. Biological products, especially those targeted at insect control, are also available to the turfgrass pest manager. These include the recently approved *Beauvaria bassiana* (Naturalis-T), a fungi which affects grasshoppers, mole crickets, and sod webworms. Other options available include various *Bacillus thuringiensis* (Bt) products (XenTari, Condor, Dipel) which produce bacterial toxins active against sod webworms, and *Heterorhabditis bacteriophora* (Cruiser) a nematode that attacks white grub larvae. Nearly all of the biological products, with the exception of *B. bassiana*, are narrow in terms of spectrum of activity, often targeting just one specific pest or at most a specific taxonomic group. The extent to which these biologicals are being employed within the golf course environment is somewhat unclear, but limited adoption may be due to perceived lower consistency, reliability, and versatility and higher cost than conventional pesticides (*9, 10*). A recent survey of golf course superintendents did not identify any of the biological products as primary treatments, although a marketing assessment revealed that during its first year of introduction *B. bassiana* accounted for approximately 1% of total insecticide active ingredient use by golf courses (Kline and Company, unpublished data; *5*).

Increased Regulatory Scrutiny. The turfgrass care industry in general is accustomed to a significant degree of regulation, as exemplified by a recent series of articles in *Grounds Maintenance* magazine titled "Are regulations killing our industry?" (*12, 13*). Pesticides represent one of the most highly regulated product groups, and an incredible set of requirements must be met in order to introduce and market a product for use in the agricultural or urban environment. The costs of required pesticide registration studies have been steadily rising, and one important factor behind the increasing costs is increasing regulatory scrutiny. For example, the Food Quality Protection Act of 1996 amended basic U.S. pesticide regulation to require assessment of potential human exposure to a given pesticide from all potential sources (dietary, occupational, drinking water, residential reentry,...), simultaneous consideration of pesticides with common mechanisms of toxicity, and additional testing of potential endocrine effects of pesticides. One outcome of the high cost of product development is that it is difficult for basic manufacturers to economically justify pursuit of new products exclusively for the turfgrass market. Instead, new active ingredients are often simultaneously developed for introduction both for urban and agricultural uses. Increased regulatory scrutiny also favors the development of new pesticide products with improved safety profiles. For example, the U.S. EPA has established a "reduced risk" pesticide program whereby new active ingredients with lower toxicity or less potential for environmental impact are given preferential treatment and accelerated registration review and approval. Such was the case with the recently introduced fungicide azoxystrobin and insecticide spinosad, both of which were approved as reduced risk products by the EPA during early 1997 (U.S. EPA Press Release, February 28, 1997).

Environmental Considerations

There are a number of environmental considerations that arise from the use of pesticides in the turfgrass environment. These include potential impacts on surface and groundwater quality, persistence in soil, effects on nontarget terrestrial and aquatic organisms, and human exposure. These topics have been summarized in recent volumes by Racke and Leslie (*1*) and Balogh and Walker (*14*). A brief discussion of environmental considerations related to turf pesticide use and some relevant examples are described in the following section.

Surface Water Quality. Pesticides present in surface water may be of concern from the standpoint of potential impacts on nontarget aquatic organisms (fish, invertebrates, plants) as well as with respect to potential drinking water exposure of humans. Pesticides and fertilizers applied to turfgrass areas may reach water via spray drift or surface runoff. Research on potential surface runoff of pesticides has resulted in the general finding that limited transport occurs from treated turfgrass areas under most conditions (*15-18*). For example, Harrison et al. (*15*) reported no detectable concentrations of the highly sorbed pesticides chlorpyrifos and pendimethalin and small quantities (0.8-1.6% of applied) of several soluble herbicides (2,4-D, dicamba) in runoff water from treated turf plots subject to a >100-year return frequency simulated rainstorm. The magnitude of pesticide transport observed in turfgrass environments is much lower as compared with transport observed from bare soil or row crop situations under similar treatment for several reasons. These include the lack of significant sediment erosion that might carry sorbed pesticides from turf areas, increased sorption by the thatch as compared with soil, and increased infiltration of water and decreased runoff water volumes in turfgrass. The increased retention of pesticides in turfgrass areas is exemplified by the employment of untreated grassed buffer/filter strips as a mitigation measure for reducing edge-of-field losses of pesticides from both agricultural fields and turfgrass areas (*18-20*).

Surface water monitoring of golf course and other urban watersheds reveals detectable but trace quantities of turfgrass pesticides. Cohen et al. (*21*; Cohen personal communication) summarized the results of a number of North American golf course surface water monitoring studies, and overall results indicated that widespread and/or repeated water quality impacts by golf courses was not occurring at the study sites. None of the authors of the individual studies concluded that toxicologically significant impacts were observed, although scattered exceedences of some water quality criteria were observed. Similar results have been reported from golf course surface water monitoring studies conducted in Japan (*22-24*).

Under certain conditions, such as when high rainfall occurs shortly after application to steep slopes, surface transport with water can move greater quantities of less highly sorbed pesticides. For example, Smith and Bridges (*25*) reported that between 10 and 14% of several soluble herbicides (dicamba, mecoprop, 2,4-D) were transported with runoff water from simulated fairways subjected to 4 intense rainfall events within 192 hours after treatment. In rare instances, such large runoff events coming immediately after application may lead to some impacts on aquatic organisms, as exemplified by a reported fish kill in a New Jersey lake which followed a fenamiphos application to a nearby golf course (*26*). Clearly, specific local conditions conducive to surface runoff impacts may justify both prudent application procedures and water management practices (*18*).

Groundwater Quality. Pesticides and nutrients (e.g., nitrate) present in groundwater are primarily of concern from a human drinking water exposure standpoint. Pesticides and fertilizers applied to turfgrass areas may leach to subsurface water via chromatographic movement through the network of soil micropores and/or via macropore flow. Research on potential leaching mobility of pesticides has resulted in the general finding that limited transport occurs from treated turfgrass areas under most conditions (*15-18, 27-29*). For example, Smith and Bridges (*25*) reported only trace quantities (0.1-0.9% of applied) of several water-soluble herbicides (2,4-D, dicamba, mecoprop) leached from simulated bentgrass and bermudagrass greens during a 70 day period. The organic thatch layer, which has a significant ability to sorb and retain pesticide residues, has been credited with much of the "filter" effect which turfgrass confers with respect to soil leaching potential (*27, 30-32*).

Groundwater monitoring of golf course areas reveals detectable but trace quantities of turfgrass pesticides and fertilizer components may reach groundwater under some conditions, but the frequency of detection and magnitude of residues are generally lower than observed during similar surface water monitoring surveys. For example, Cohen et al. (*21*; Cohen personal communication) summarized the results of a number of North American golf course groundwater monitoring studies, and overall results indicated that widespread and/or repeated water quality impacts by golf courses was not occurring at the study sites. The frequency of exceedence of Health Advisory Limits (HALs) by individual pesticides in the groundwater samples analyzed in the studies was approximately 0.07%. Specific instances in which pesticide residues are detected in groundwater beneath turfgrass areas at concentrations of concern are limited, but may be associated with unusual conditions or management practices. For example, a survey of groundwater under Cape Cod golf courses resulted in detection of chlordane and heptachlor residues above HAL limits, and the presence of these discontinued pesticides may have been attributable to repeated, heavy applications coupled with preferential flow of bound particulate phase (*33*). Similarly, an EPA study reported that the presence of trace residues of DCPA (chlorthal-dimethyl) in water samples taken from individual wells was directly correlated with its heavy use on golf courses and commercially maintained landscaping in urban areas (*34*).

Nontarget Wildlife Impacts. Considerations regarding turf pesticides and nontarget wildlife include potential impacts on birds, wild mammals and amphibians, and soil invertebrates. Golf courses appear to represent significant wildlife habitat in the urban environment, as documented by the numbers and variety of bird species that have been observed during surveys (*14, 35*). It is also worth noting that wildlife habitat areas on and surrounding golf courses may not be routinely treated with pesticides; the most heavily treated areas tend to be those of limited value for wildlife (e.g., tees, greens).

Research on earthworm and soil arthropod populations in turfgrass has demonstrated that pesticides differ greatly in their potential for causing adverse effects. For example, use of most herbicides and some insecticides (e.g., 2,4-D, dicamba, isofenphos) had little or no impact on earthworm populations, whereas some insecticides and fungicides (e.g., carbaryl, bendiocarb, benomyl) caused severe or very severe toxicity (51-99% population reduction) under some conditions examined (*36-38*).

With respect to turf pesticides and avian species, detailed accounts of bird impacts are not abundant in the open literature, and available data generally indicates that bird populations in golf course areas are not adversely affected by the use of pesticides. In the specific case of granular diazinon, incident reports of bird mortalities following use of this product on turfgrass led the U.S. EPA in 1990 to prohibit use of the product on golf courses and sod farms (*39*). The move to ban versus tighten labeling restrictions was a controversial one in that several field monitoring studies were completed which demonstrated a lack of bird impacts with a proposed label limit of 2 applications of 2 lb a.i./acre per year (*39*). EPA later completed a comparative analysis of acute avian risk resulting from use of granular insecticides and nematicides on turfgrass (*40*). Although the screening level approach employed in the risk analysis in reality represented a hazard ranking (i.e., avian LD50/square foot of applied granules), a number of manufacturers responded to the analysis with voluntary labeling modifications that reduced potential avian exposure (e.g., application rate reductions, post-application irrigation requirements) (*41*). Despite theoretical concerns that may be raised based on such calculations, evidence from large-scale monitoring studies of bird populations on treated and untreated golf courses has demonstrated a lack of significant impacts for currently used products under most conditions (*35, 42, 43*).

Persistence and Degradation. Pesticides present in the turfgrass ecosystem are degraded through a variety of mechanisms, both abiotic (e.g., photolysis, hydrolysis, volatilization) and biological (e.g., plant metabolism, microbial degradation). A full discussion of the variety of factors that may affect pesticide persistence in turf can be found in available review articles (*1, 14*). It should be recognized that observed pesticide dissipation within an environmental compartment (e.g., soil, plan) can be the result of both degradation and transport processes (e.g., volatilization).

Some of the earliest synthetic turf pest products were chlorinated hydrocarbon insecticides (e.g., chlordane, dieldrin, aldrin, heptachlor) which were eventually removed from the marketplace due to undesirably long persistence and potential bioaccumulation within ecosystems. These products were replaced by much less residual products, often with degradation half-lives of a few weeks or months. For at least some insecticide products, biodegradability can lead to the problem of enhanced rates of microbial degradation. Enhanced degradation is the accelerated rate of pesticide degradation, by adapted soil microorganisms, observed in soil following previous application of the pesticide (*44*). Isofenphos is one chemical that is susceptible to this phenomenon, and although a first soil application of the product results in effective residuality, subsequent applications are rapidly degraded by adapted microorganisms. The decreased persistence and associated soil insect pest control failures led to the withdrawal of this product from the corn rootworm market during the early-1980's (*45*), and enhanced degradation was subsequently observed in turfgrass thatch and soil with a prior history of isofenphos use (*46*). Not all pesticides are able to induce enhanced rates of microbial degradation, and the phenomenon is fairly specific with respect to chemical structure. The inherent biodegradability of a pesticide, its degree of bioavailability in soil and thatch, and nutritive value of one or more metabolites to support microbial metabolism and growth influence whether microbes can proliferate and cause rapid rates of pesticide degradation for subsequent applications. At least several additional turfgrass pesticides, including 2,4-D, carbaryl, and fenamiphos have been shown to undergo enhanced rates of degradation in soil or turfgrass thatch under some conditions (*47, 48*), whereas other products are apparently resistant to the phenomenon (*49*).

Human Exposure Considerations. Although a detailed discussion is beyond the scope of this chapter, there are also human exposure considerations related to the use of pesticides on turfgrass. In addition to potential applicator exposure, there is the potential for exposure to those reentering treated turfgrass areas. The primary route of potential reentry exposure involves dermal uptake from dislodgeable residues on turfgrass foliage, although inhalation exposure from the presence of volatilized pesticide residues in air can also occur. Studies of the fate of total and dislodgeable (i.e., removed via contact and abrasion) residues of pesticides on turfgrass foliage have demonstrated that a very low percentage (5-10% at most) of the total residue present immediately after application is in dislodgeable form and concentrations decrease rapidly with time (*50, 51*). Some researchers have completed biomonitoring studies of the uptake and excretion of pesticide residues in individuals reentering treated turf areas, and results demonstrate adequate safety margins (*52, 53*). Additional research on both application and reentry exposure in turfgrass is being completed by the Occupational and Residential Exposure Task Force, an industry association currently working cooperatively to meet an EPA data call-in (*54*). Regarding potential inhalation exposure, monitoring of air during and following application to turfgrass confirms the very low levels that may be present, indicating that this is at most a secondary route of potential exposure (*55*).

Turf Pesticide Research and Testing

The safe and effective use of pesticides for turfgrass pest management requires the availability of a significant body of research information related to the behavior and potential effects of these chemicals. This information is employed by users, regulators, manufacturers, researchers, and consultants in making informed pest management recommendations and risk management decisions. Research on turfgrass pesticides begins at the basic manufacturer level in association with new product development and testing to meet regulatory requirements. Industry research related to product stewardship as well as independent research to address new areas of science and emerging issues follow the introduction of a pesticide product into the turf pest management market. All of these research avenues are important for increasing the knowledge base on turf pesticides and for providing information related to effective pest management as well as environmental and human health issues that may arise from turf pesticide use. The following paragraphs will briefly describe the nature of these complementary research areas in relation to turfgrass pesticides, and case study examples of an older, established product and a very recently introduced product will be examined.

Development and Registration. Research and development related to discovery of new pest control products is an integral part of a number of major agrochemical manufacturers and suppliers including Novartis, Monsanto, Zeneca, Bayer, DuPont, Dow AgroSciences, and AgrEvo. Screening biology and chemistry programs to identify promising new chemistries for further testing are important components of this effort and, as noted earlier, pursuit of turfgrass pest products is usually incorporated into broader discovery targets related to major pest groups (e.g., broadleaf weeds, Lepidoptera). Significant efforts are undertaken to characterize biological efficacy under real-world field conditions, and importance is placed on understanding factors potentially affecting or enhancing performance. Generation of human safety and environmental impact data for regulatory review and approval is another major undertaking of the new product development process (see Table II). Over 120 studies for an individual pesticide candidate may be completed during a development period of some 7 to 9 years to generate the necessary data submission package for agencies such as the Office of Pesticide Programs at U.S. EPA or the Department of Pesticide Regulation at California EPA. Although the majority of regulatory testing is generic in nature and related to the molecule rather than specific environments or use patterns, there are studies directly related to use on turfgrass that are either routinely required or triggered by results of preliminary assessments. Examples of these would include generation of turf foliar dislodgeable residue data (for assessment of potential human exposure) and monitoring of avian impacts under field use conditions.

Table II. Major Categories and Selected Examples of New Product Registration Testing Requirements.

Product Chemistry	Physical/Chemical Characteristics
Description of manufacturing process	Solubility
Technical impurity identification	Vapor pressure
Analytical methods	Octanol water partition coefficient
Mammalian Toxicology	Ecotoxicology
Acute oral toxicity	Acute avian toxicity
90-day feeding study	Avian dietary toxicity
Chronic feeding study	Acute fish toxicity
Teratogenicity study	Fish life cycle study
2-Generation reproduction study	Acute freshwater invertebrate toxicity
Residue Chemistry	Environmental Fate
Nature of residues in plants & animals	Hydrolysis
Analytical methods for residue analysis	Photodegradation in water
Magnitude of residues in field crops	Aerobic soil metabolism
	Leaching and adsorption/desorption
Worker and Reentry Exposure	Soil field dissipation
Foliar residue dissipation	Accumulation in fish
Dermal and inhalation dosimetry	Groundwater monitoring

In addition to external review and approval, R&D companies involved in new product development utilize a series of decision-making or advancement gates at various points in the process so that assessments based on results of safety testing, efficacy trials, and marketing evaluations can occur. These assessments can result in advancement or in delay or in project cancellation based on the accumulated information, and current indications are that no more than 1 or 2 products reach commercial status of more than 50,000-100,000 that may have been through the screening process. The new product development process is a costly one, and it may take from 10-15 years or more for the costs incurred during the development process (between 50 and 100 million dollars) to be recouped through commercial product sales (56). A final note is that through the reregistration and data call-in processes, basic manufacturers are obliged to maintain their databases of safety information on existing products so as to meet current study guidelines and new standards of performance.

Product Stewardship. Once a turfgrass product has been introduced into the marketplace, there is a need for ongoing generation of new research information related to product performance and safety. This product stewardship-based research is not driven by regulatory requirements, but rather by the desire to uphold the highest standards of ethics and environmental responsibility by addressing new areas of science and emerging issues. Such research may be related to product quality, performance, and efficacy. It is difficult within new product development programs to evaluate every possible use scenario, and additional testing may be required. For example, the suitability of a tank mix partner and potential for turfgrass phytotoxicity may need to be evaluated when another new product enters the marketplace. Alternately, product stewardship research may address a specific question or technical issue related to environmental behavior, ecological impact, or human safety. For example, although EPA registration requirements for the field soil dissipation guideline may be met by completion of a bare-soil study at maximum application rate, a manufacturer with a turfgrass product may wish to examine the fate of a more typical application rate of the product in turf foliage, thatch, and underlying soil to provide data for more realistic assessments. Product stewardship research represents continuing investment in product support by the basic manufacturer to ensure that products are long-lived and continue to be used efficaciously and safely with appropriate management practices. Product stewardship research may be conducted in a proprietary nature by scientists on staff with the product manufacturer or funds may be provided to an outside contractor or university for investigation support.

Independent Research. Independent research is another important avenue by which the knowledge base on turfgrass pesticides is expanded. Such turf pesticide research might be completed at a university, a federal or state research laboratory, or by a private research organization. Funding for such research may come from multiple sources including government agencies, foundations, user groups, or industry. Independently funded research often carries great credibility with the general public, user groups, and advocacy groups, and serves as an important complement to work that might be completed from a more proprietary standpoint. The recent environmental turfgrass research program sponsored by the U.S. Golf Association is an excellent example of this type of research. Results of this program, summarized in other chapters within this volume, have added considerably to the general knowledge base on turfgrass pesticide behavior, particularly with respect to the fate of pesticides in the golf course environment.

Case Study: Chlorpyrifos (Dursban). Chlorpyrifos provides an example of turfgrass research efforts that may be associated with a well-established product. The original environmental safety and health data for this product, first discovered by the Agricultural Products Department of The Dow Chemical Company (now Dow AgroSciences) in the early 1960's, supported the registration of chlorpyrifos for turfgrass use as Dursban 2E in 1965. Since that time, from a registration research standpoint, additional data call-ins have occurred periodically to upgrade and modernize the database. For example, in 1991 the U.S. EPA requested generation of new data for chlorpyrifos fate on turfgrass under field conditions. As a result, studies to meet current guideline requirements were completed on Kentucky bluegrass in Indiana and St. Augustinegrass in Florida (57). Another example of regulatory research was occasioned by a data call-in issued by U.S. EPA in 1993 for foliar residue dissipation, applicator exposure, and reentry exposure (dermal and inhalation) for all turfgrass pesticide products. In support of chlorpyrifos, Dow AgroSciences and a number of other basic manufacturers of turf pesticide products formed an Occupation and Residential Task Force to jointly develop generic data (e.g., transfer factors, spray deposition on clothing) to meet the general aspects of the data call-in requirement (54). Each company is also generating foliar dislodgeable residue data on representative formulations for each active ingredient to provide chemical-specific data.

Given the economic importance of urban uses of chlorpyrifos, Dow AgroSciences has supported a number of product stewardship-based research initiatives on the product. One example is related to a company research project which addressed the potential for migration of chlorpyrifos from treated turfgrass areas to sensitive surface water habitats with runoff water. To this end, steeply-sloped bluegrass and Bermudagrass plots on a golf course in Kentucky were marked off after application, subject to simulated rainfall, and runoff water analyzed for residues of transported chlorpyrifos (58). Another example is provided by company research efforts related to development of appropriate methodology for accurate estimation of potential reentry exposure to chlorpyrifos or other turfgrass pesticides. To this end, indirect methods for estimation of exposure (Dow "drag-sled" method to determine dislodgeable foliar residues and potential dermal exposure and air monitoring to determine potential inhalation exposure) were compared to a biomonitoring method (urine of test subjects reentering treated turfgrass area monitored for appearance and level of TCP, the major metabolite of chlorpyrifos). Results confirmed the usefulness of the more flexible and less costly indirect method developed by Dow AgroSciences (53). A final example may be provided by reference to research on potential avian effects of turfgrass pesticide use sponsored by Dow AgroSciences with a Canadian research group. This group examined the potential impact resulting from the use of several turfgrass pesticides on population levels and nesting success of the American robin in suburban turfgrass environments (59).

The widespread and historic use of chlorpyrifos in turfgrass markets (golf course, professional lawncare, home lawncare) has also made it a common subject of independent research efforts. Many of these efforts have involved highly focused and unique approaches to determination of the potential environmental fate, ecological impact, and human safety assessment for chlorpyrifos in the turfgrass system. Independent research on chlorpyrifos has examined such diverse topics as dissipation under golf course fairway conditions (60), leaching mobility on golf course greens (61), sorption characteristics of turfgrass thatch (31), fate of residues during composting of grass clippings (62), and dissipation of foliar dislodgeable residues (50, 51).

Case Study: Spinosad (Conserve). Spinosad provides an example of research efforts that may be associated with development of a new turfgrass pesticide product. Screening of fermentation broths by Dow AgroSciences during the mid-1980's demonstrated activity against several key target insect pest groups in a broth derived from a soil sample. This activity was attributed to metabolites produced by a specific soil bacterium during fermentation, and these were characterized as "spinosyns". The structures of spinosyn A and spinosyn D, the active ingredients in spinosad, were eventually determined (63). Key milestones in the characterization and development of spinosad as a new insecticide product are summarized in Table III.

Regulatory testing in support of spinosad development was conducted during the early 1990's. Studies ranged from determining analytical methods for detection of spinosad in environmental samples to evaluating the rate and pathway of degradation in soil, water, plants, and animals (64, 65). In addition to the standard package of studies required for registration, special testing and research was also completed. For example, the fate of spinosad following application to turfgrass and ornamentals was assessed (Dow AgroSciences unpublished report). Research on the mode of action of spinosad, which turned out to involve effects at a novel target site in the nicotinic acetylcholine receptor of the insect nervous system, was also completed and published (66). To determine the fate of spinosad in water following an accidental overspray, a sediment/water pond microcosm study was conducted (65). Results of the testing program for spinosad yielded not only a complete regulatory database, but revealed a very positive safety and environmental profile. In fact, spinosad's properties were so favorable that EPA granted the product "reduced risk" status (as compared with existing insecticide products) and approved it under an expedited regulatory review (U.S. EPA Press Release, February 28, 1997). The basis for EPA's decision was that spinosad possessed low mammalian toxicity, low nontarget organism toxicity, and excellent compatibility with IPM programs.

In addition to regulatory testing, a significant field research program to characterize performance under field conditions was undertaken. To assist with characterization of the efficacy of spinosad against turf and ornamental insect pests, an Experimental Use Permit (EUP) program for 1996 and 1997 was conducted. This EUP included real-world testing at a number of golf courses. Dow AgroSciences worked with these golf courses to collect data on effectiveness of label use rates and use directions, including facilities in Florida, California, North Carolina, Indiana, and New Jersey. Results of the EUP program indicated excellent efficacy against key Lepidoptera target pests including sod webworms, black cutworms, and fall armyworms. Testing also indicated that the Conserve formulation of spinosad proved to be non-phytotoxic to turfgrasses, which was an important finding. In addition,

Table III. Major Milestones in the Development of Spinosad

1982	Soil sample collected from rum distillery in the Virgin Islands
1985	Screening of fermentation broth from soil sample demonstrates biological activity towards mosquito and southern armyworm larvae
1985	Newly discovered bacterium isolated from fermentation broth, *Saccharopolyspora spinosa*, found to produce active substances known as "spinosyns"
1989	Structure of spinosyn A determined

spinosyn A: R = H   MW = 732
spinosyn D: R = CH3 MW = 746

1989	Field efficacy trials initiated
1991	Predevelopment regulatory research program initiated
1994	Submission of registration data package to U.S. EPA
1994	Product commercialization and development program initiated
1996	Initiation of Experimental Use Permit program for turfgrass pest control
1996, 1997	Patents issued for production of spinosyns
1997	Registrations approved for cotton (Tracer*) and turfgrass (Conserve)
1997	Commercial launch of Tracer and Conserve

*Trademark of Dow AgroSciences

helpful insight was gained into how best to make spinosad packaging, delivery and labeling as user-friendly as possible.

In addition to the research programs related to safety and efficacy testing, additional efforts with respect to development of suitable formulations and packaging, construction of a manufacturing facility, design of a distribution network, and development of marketing and sales programs also required considerable effort. The end result of the 12 year research program, which involved an R&D investment in excess of $60 million, was introduction of a new pesticide product for the turfgrass and agricultural insect pest management markets.

Summary and Conclusions

Golf course turf pest control is strongly reliant on the use of pesticides, especially fungicides. Pesticide products continue to be developed for the turf pest market; several successful launches of outstanding new products have recently occurred. Environmental considerations related to turf pesticide use include surface and groundwater quality, wildlife impacts, and degradation processes. Development of a new turf pesticide involves a major commitment to generation of data related to both efficacy and health and environmental safety. Both product stewardship-based and independent research continue to expand the knowledge-base with regard to potential impacts and management practices for use of pesticides in the turfgrass environment.

Literature Cited

1. *Pesticides in Urban Environments: Fate and Significance*; Racke, K. D.; Leslie, A. R., Eds.; ACS Symposium Series No. 522; American Chemical Society, Washington, DC, 1993.
2. Hodge, J. E. In: *Pesticides in Urban Environments: Fate and Significance*; Racke, K. D.; Leslie, A. R., Eds.; ACS Symposium Series No. 522; American Chemical Society, Washington, DC, 1993; pp 10-17.
3. Aspelin, A. L. *Pesticide Industry Sales and Usage: 1992 and 1993 Market Estimates*; U.S. Environmental Protection Agency, 733-K-94-001, June 1994, Washington, DC.
4. Whitmore, R. W.; Kelly, J. E.; Reading, P. L.; Brandt, E.; Harris, T. In: *Pesticides in Urban Environments: Fate and Significance*; Racke, K. D.; Leslie, A. R., Eds.; ACS Symposium Series No. 522; American Chemical Society, Washington, DC, 1993; pp 18-36.
5. *1998 Plant Protection and Fertilizer Usage Report*; Golf Course Superintendents Association of America, Lawrence, KS, 1998.
6. Brennan, B. M.; Higashi, A. K.; Murdoch, C. L. *Estimated Pesticide Use on Golf Courses in Hawaii*; Research Extension Series 137, College of Tropical Agriculture and Human Resources, University of Hawaii, Honolulu, HI, 1992.
7. Potter, D. A. *Prosource*; **1998**, *May-June*, 14-17.
8. *Handbook of Integrated Pest Management for Turf and Ornamentals*; Leslie, A. R., Ed.; Lewis Publishers, Boca Raton, FL, 1994.
9. *IPM Handbook for Golf Courses*; Schumann, G. L.; Vittum, P. J.; Elliott, M. L.; Cobb, P. P., Eds.; Ann Arbor Press, Chelsea, MI, 1997.
10. Potter, D. A. In: *International Turfgrass Society Research Journal 7.* Carrow, R. N.; Christians, N. E.; Shearman, R. C., Eds.; Intertec Publishing Corporation, Overland Park, KS, 1993.
11. Vittum, P. J. *Grounds Maintenance*; **1997**, *September*, 6-8.
12. Hogan, G. K. *Grounds Maintenance*; **1998**, *March*, 15-18, 103.
13. Liskey, E. *Grounds Maintenance*; **1998**, *March*, 20, 24, 28, 32, 115.

14. Balogh, J. C.; Gibeault, V. A.; Walker, W. J.; Kenna, M. P.; Snow, J. T. In: *Golf Course Management and Construction: Environmental Issues*; Balogh, J. C.; Walker, W. J., Eds.; Lewis Publishers, Boca Raton, Florida, 1992; pp 1-37.
15. Harrison, S. A.; Watschke, T. L.; Mumma, R. O.; Jarrett, A. R.; Hamilton, G. W. In: *Pesticides in Urban Environments: Fate and Significance*; Racke, K. D.; Leslie, A. R., Eds.; ACS Symposium Series No. 522; American Chemical Society, Washington, DC, 1993; pp 191-207.
16. Shuman, L. M.; Smith, A. E.; Bridges, D. C. In: Clark, J. M.; Kenna, M. P., Eds.; *Fate of Turfgrass Chemicals and Pest Management Approaches*; ACS Symposium Series, American Chemical Society, Washington, DC, 1999.
17. Watschke, T. L.; Mumma, R. O.; Linde, D.; Borger, J.; Harrison, S. In: Clark, J. M.; Kenna, M. P., Eds.; *Fate of Turfgrass Chemicals and Pest Management Approaches*; ACS Symposium Series, American Chemical Society, Washington, DC, 1999.
18. Baird, J. H.; Basta, N. T.; Huhnke, R. L.; Johnson, G. V.; Payton, M. E.; Storm, D. E.; Smolen, M. D. In: Clark, J. M.; Kenna, M. P., Eds.; *Fate of Turfgrass Chemicals and Pest Management Approaches*; ACS Symposium Series, American Chemical Society, Washington, DC, 1999.
19. Patty, L.; Real, B.; Grill, J. J. *Pestic. Sci.* **1997**, *49*, 243-251.
20. *Reducing Dormant Spray Runoff from Orchards*; Ross, L. J.; Bennett, K. P.; Kim, K. D.; Hefner, K.; Hernandez, J. State of California, Environmental Protection Agency, Department of Pesticide Regulation, Report EH 97-03, Sacramento, CA; 1997.
21. Cohen, S. Z.; Svrjcek, A.; Durborow, T. E.; Barnes, L. N. In: *Conference Proceedings of the 68th International Golf Course Conference*. Golf Course Superintendents Association of America, Lawrence, KS, 1997; pp. 19-20.
22. Morioka, T.; Cho, H. S. *Water Sci. Technol.* **1992**, *25*, 77-84.
23. Hori, S.; Kato, M.; Tsukabayashi, H.; Hashiba, H. *J. Environ. Chem.* **1992**, *2*, 65-70.
24. Tomimori, S.; Nagaya, Y.; Taniyama, T. *Japan. J. Crop Sci.* **1994**, *63*, 442-451.
25. Smith, A. E.; Bridges, D. C. *Crop Sci.* **1996**, *36*, 1439-1445.
26. Meyer, L. W.; Russell, D.; Louis, J. B.; Jowa, L.; Post, G.; Sanders, P. In: *Pesticides in the Next Decade: The Challenges Ahead*; Virginia Water Resources Research Center, Virginia Polytechnic Institute and State University, Blacksburg, VA, 1990.
27. Horst, G. L.; Shea, P. J.; Powers, W. L.; Christians, N. In: Clark, J. M.; Kenna, M. P., Eds.; *Fate of Turfgrass Chemicals and Pest Management Approaches*; ACS Symposium Series, American Chemical Society, Washington, DC, 1999.
28. Branham, B. E.; Gardner, D. S.; Miltner, E. W.; Zabik, M. J. In: Clark, J. M.; Kenna, M. P., Eds.; *Fate of Turfgrass Chemicals and Pest Management Approaches*; ACS Symposium Series, American Chemical Society, Washington, DC, 1999.
29. Petrovic, A. M.; Lisk, D. J.; Larsson-Kovach, I. In: Clark, J. M.; Kenna, M. P., Eds.; *Fate of Turfgrass Chemicals and Pest Management Approaches*; ACS Symposium Series, American Chemical Society, Washington, DC, 1999.
30. Sears, M. K.; Chapman, R. A. *J. Econ. Entomol.* **1979**, *72*, 272-274.

31. Spieszalski, W. W.; Niemczyk, H. D.; Shetlar, D. J. *J. Environ. Sci. Health*; **1994**, *B29*, 1117-1136.

32. Cisar, J. L.; Snyder, G. H. In: Clark, J. M.; Kenna, M. P., Eds.; *Fate of Turfgrass Chemicals and Pest Management Approaches*; ACS Symposium Series, American Chemical Society, Washington, DC, 1999.

33. Cohen, S. Z.; Nickerson, S.; Maxey, R.; Dupuy, A.; Senita, J. A. *Ground Water Monit. Rev.* **1990**, *10*, 160-173.

34. *National Pesticide Survey: Update and Summary of Phase II Results*; U.S. Environmental Protection Agency, 1992.

35. Potter, D. A.; Buxton, M. C.; Redmond, C. T.; Patterson, C. G.; Powell, A. J. *J. Econ. Entomol.* **1990**, *83*, 2362-2369.

36. Potter, D. A. In: *Pesticides in Urban Environments: Fate and Significance*; Racke, K. D.; Leslie, A. R., Eds.; ACS Symposium Series No. 522; American Chemical Society, Washington, DC, 1993; pp 331-343.

37. Potter, D. A.; Spicer, P. G.; Redmond, C. T.; Powell, A. J. *Bull. Environ. Contam. Toxicol.* **1994**, *52*, 176-181.

38. U.S. Environmental Protection Agency, *U. S. Federal Register*; **1990**, *55*, 31138-31146.

39. *Comparative Analysis of Acute Avian Risk from Granular Pesticides*; U. S. Environmental Protection Agency, Washington, DC, 1992.

40. Anonymous. *Pestic. Toxic Chem. News* **1992**, *November 4*, 41-44.

41. Brewer, L. W.; Hummell, R. A.; Kendall, R. J. In: *Pesticides in Urban Environments: Fate and Significance*; Racke, K. D.; Leslie, A. R., Eds.; ACS Symposium Series No. 522; American Chemical Society, Washington, DC, 1993; pp 320-330.

42. Rainwater, T. R.; Leopold, V. A.; Hooper, M. J.; Kendall, R. J. *Environ. Toxicol. Chem.* **1995**, *14*, 2155-2161.

43. Barron, M. G.; Woodburn, K. B. *Rev. Environ. Contam. Toxicol.* **1995**, *144*, 1-93.

44. Racke, K. D. In: *Enhanced Biodegradation of Pesticides in the Environment*; Racke, K. D.; Coats, J. R., Eds.; ACS Symposium Series No. 426; American Chemical Society, Washington, DC, 1990; pp 269-282.

45. Racke, K. D.; Coats, J. R. *J. Agric. Food Chem.* **1987**, *35*, 94-99.

46. Niemczyk, H. D.; Chapman, R. A. *J. Econ. Entomol.* **1987**, *80*, 880-882.

47. Felsot, A. S. *Ann. Rev. Entomol.* **1989**, *34*, 453-476.

48. Beehag, G. W. *Proceedings of the 21ˢᵗ Australian Turfgrass Research Institute Turf Research Conference*; **1995**, *2*, 61-65.

49. Racke, K. D.; Laskowski, D. A.; Schultz, M. R. *J. Agric. Food Chem.* **1990**, *38*, 1430-1436.

50. Sears, M. K.; Bowhey, C.; Braun, H.; Stephenson, G. R. *Pestic. Sci.* **1987**, *20*, 223-231.

51. Hurto, K. A.; Prinster, M. G. In: *Pesticides in Urban Environments: Fate and Significance*; Racke, K. D.; Leslie, A. R., Eds.; ACS Symposium Series No. 522 American Chemical Society, Washington, DC, 1993; pp 86-99.

52. Solomon, K. R.; Harris, S. A.; Stephenson, G. R. In: *Pesticides in Urban Environments: Fate and Significance*; Racke, K. D.; Leslie, A. R., Eds.; ACS Symposium Series No. 522; American Chemical Society, Washington, DC, 1993; pp 262-274.
53. Vaccaro, J. R.; Nolan, R. J.; Murphy, P. G.; Berbrich, D. B. In: *Characterizing Sources of Indoor Air Pollution and Related Sink Effects*; ASTM STP 1287, Tichenor, B. A., Ed.; American Society for Testing and Materials, 1996, pp. 166-183.
54. U.S. Environmental Protection Agency, Pesticide Regulation Notice 94-9, 1994.
55. Yeary, R. A.; Leonard, J. A. In: *Pesticides in Urban Environments: Fate and Significance*; Racke, K. D.; Leslie, A. R., Eds.; ACS Symposium Series No. 522; American Chemical Society, Washington, DC, 1993; pp. 275-281.
56. Leng, M. L. In: *Regulation of Agrochemicals: A Driving Force in Their Evolution*; Marco, G. J.; Hollingworth, R. M.; Plimmer, J. R.; American Chemical Society, Washington, DC, 1991; pp 27-44.
57. Racke, K. D.; Lubinski, R. N.; Fontaine, D. D.; Miller, J. R.; McCall, P. J.; Oliver, G. R. In: *Pesticides in Urban Environments: Fate and Significance*; Racke, K. D.; Leslie, A. R., Eds.; ACS Symposium Series No. 522; American Chemical Society, Washington, DC, 1993; pp. 70-85.
58. Racke, K. D.; Robb, C. K.; Coody, P. N. *Society of Environmental Toxicology and Chemistry: Second World Congress*; Vancouver, BC, 1995, Abstract PT302.
59. Decarie, R.; DesGranges, J. L.; Lepine, C.; Morneau, F. *Environ. Pollut.* **1993**, *80*, 231-238.
60. Horst, G. L.; Shea, P. J.; Christians, N.; Miller, D. R.; Stuefer-Powell, C.; Starrett, S. K. *Crop Sci.* **1996**, *36*, 362-370.
61. Cisar, J. L.; Snyder, G. H. *Crop Sci.* **1996**, *36*, 1433-1438.
62. Vandervoort, C.; Zabik, M. J.; Branham, B.; Lickfeldt, D. W. *Bull. Environ. Contam. Toxicol.* **1997**, *58*, 38-45.
63. Thompson, G. D.; Michel, K. H.; Yao, R. C.; Mynderse, J. S.; Mosburg, C. T.; Worden, T. V.; Chio, E. H.; Sparks, T. C.; Hutchins, S. H. *Down to Earth*, **1997**, *52*, 1-5.
64. Bret, B. L.; Larson, L. L.; Schoonover, J. R.; Sparks, T. C.; Thompson, G. D. *Down to Earth*, **1997**, *52*, 6-13.
65. Saunders, D. G.; Bret, B. L. *Down to Earth*, **1997**, *52*, 14-20.
66. Salgado, V. L. *Pestic. Biochem. Physiol.* **1998**, *60*, 91-102.

PESTICIDE AND NUTRIENT FATE

Chapter 4

Groundwater Contamination Potential of Pesticides and Fertilizers Used on Golf Courses

B. E. Branham[1], E. D. Miltner[2], P. E. Rieke[3], M. J. Zabik[4], and B. G. Ellis[3]

[1]Department of Natural Resources and Environmental Science,
University of Illinois, Urbana, IL 61801
[2]Department of Crop and Soil Science, Washington State University,
Puyallup, WA 98371
Departments of [3]Crop and Soil Science and [4]Entomology,
Michigan State University, East Lansing, MI 48824

Leaching losses from seven pesticides applied to 1.2 m deep soil monolith lysimeters with a turf cover were negligible while applications of phosphorus (P) to turf indicated that soil levels of P can increase rapidly after only one to two years of annual applications. Lysimeter percolate was continuously monitored and only one of seven pesticides were detected during the 2.5 y monitoring period. Triadimefon was detected at a concentration of 31 μg L^{-1} 55 days following application with subsequent detections at 10 μg L^{-1} or less. Phosphorus applications of 200 kg ha^{-1} to a P-deficient turf growing on a sandy soil increased soil P test values in the 0-7.5 cm soil layer from 3 mg kg^{-1} to 9.6 mg kg^{-1} in 1992. Soil P rose to 25.3 mg kg^{-1} in 1993 despite applying only 50 kg ha^{-1} in 1993.

Turfgrass management is a unique agricultural ecosystem characterized by continuous plant cover; a layer of partially decomposed organic matter, termed thatch or mat; and a very dense, but shallow, root system. While turfgrasses are characterized by high levels of surface organic matter, intensively maintained turfs often receive frequent applications of pesticides and fertilizers. The frequency of pesticide and nutrient application combined with the preference for growing turf on coarse-textured, rapidly draining soils leads to concerns about nutrient and pesticide leaching from highly managed turfgrass systems.

Early research suggested that excessive nutrient leaching might be a problem. Brown et al. (1) found very high levels of nitrate leaching from bermudagrass maintained at greens height. However, irrigation and fertilization rates were much higher than current recommended management practices. For example, an application of 163 Kg N ha^{-1} as NH_4NO_3 was followed with irrigation of 1.0 to 1.2 cm per day, yielding maximum concentrations of 91.7 mg L^{-1} of NO_3-N. In contrast, irrigation rates of 0.6 to 0.8 cm per day resulted in maximum concentrations of 2.5 mg L^{-1} of NO_3-N. Current fertilizer recommendations suggest no more than 50 kg N ha^{-1} per application, less than 2/3 of the application rates that, when combined with daily overwatering, resulted in significant N leaching.

In contrast to the research of Brown et al., Starr and DeRoo (2) used ^{15}N labeled $(NH_4)_2SO_4$ to determine the fate of nitrogen applied to turfgrass. They detected ^{15}N in leachate only once during the one year of their study and concluded that nitrate leaching from turf was minimal.

Pesticide leaching in turf has been studied to a lesser degree than nitrate leaching. However, several factors unique to a turfgrass system work to mitigate pesticide leaching from turf. Several authors have noted that thatch results in reduced pesticide mobility (3-4) and more rapid pesticide degradation (5). Turf root density is extremely high in the top 15 cm of the soil; however, root density drops dramatically below 30 cm for cool season grasses. The high root length density in the top 15 cm makes this soil zone high in organic carbon and microbial activity, factors which tend to mitigate pesticide movement.

Despite the positive effects of turf, pesticide leaching does occur. Cisar and Synder (6-7) studied the leaching of organophosphorus pesticides through a root zone constructed of sand with percolate samples collected at a depth of 20 to 26 cm. Under these conditions, the maximum concentration of fenamiphos (ethyl 3-methyl-4-(methylthio)phenyl (1-methylethyl)phosphoramidate) collected in the percolate from an application of 11.25 kg ha^{-1} was 8 μg L^{-1}. However, concentrations of the sulfone and sulfoxide metabolites of fenamiphos reached concentrations of 2300 μg L^{-1}. A total of 17.7 % of the applied fenamiphos was recovered in percolate as the sulfone + sulfoxide metabolites after the first application. A subsequent application of fenamiphos yielded only 1.1 % recovery of the metabolites in the percolate. The reduced recovery of fenamiphos metabolites in the leachate was attributed to enhanced degradation of fenamiphos and its metabolites.

Phosphorus (P) is an important nutrient in maintaining highly manicured turfgrass. In many native soils, the need to apply P fertilizer is often minimal due to adequate P content of the soil. However, in sand root zones, as are often found in golf course greens and athletic fields, naturally occurring soil P is usually minimal, necessitating regular, supplemental P applications. The capacity of these artificial rooting media to bind P is generally limited. In combination with a high degree of macroporosity, this creates an environment where P may be more vertically mobile than in a native soil. Golf greens and similar sites usually have tile drainage systems, and any P that reaches the drain tile could be quickly transported to surface waters, creating an increased risk for P pollution. This experiment was designed to examine the vertical movement of P through a sand putting green following various rates and methods of application.

The objectives of these studies were to assess the leaching of pesticides and phosphorus (P) applied to turfgrasses under realistic application rates and conditions. Secondly, we determined the fate of nitrogen in turf when applied either in the early spring or late fall. The nitrogen fate findings have been previously published (8) and will not be covered in this chapter.

Materials and Methods.

Pesticide Leaching. Pesticide leaching was studied using large, intact soil monolith lysimeters. The construction and design of the lysimeters has been detailed previously (8). Four lysimeters were constructed of stainless steel and were 1.14 m in diameter (1 m^2 in surface area) and 1.2 m in depth. During construction of the lysimeters, the bottom 3 cm of soil was removed and replaced with 1 - 2 cm pea gravel. The bottom of each lysimeter had a 3% slope so leachate flowed towards a stainless steel drain tube. The lysimeters and surrounding area were sodded in Sep 1990 with a Kentucky bluegrass sod blend consisting of equal percentages of 'Adelphi', 'Nassau', and 'Nugget' Kentucky bluegrasses. Soil type in the lysimeters and surrounding area was a Marlette fine sandy loam (62:22:16/ % sand:silt:clay) with a pH of 7.3 and an organic matter content of 2.5%. The lysimeters were mowed by hand at a height of 6.4 cm. Nitrogen was applied at an annual rate of 196 kg N ha⁻¹ in five applications of 39 kg ha⁻¹. Irrigation was applied only when the turf showed signs of drought stress.

Water draining from the lysimeters was collected continuously into a 19 L glass carboy. The carboy was periodically drained (usually every 7-14 days), volume measured, and a 1 L sample was collected in a glass jar with a foil-lined lid and stored at 2 C until analysis. Samples were typically extracted and analyzed within 3 weeks of sample collection. Analysis for pesticide residues began on 3 Sep 1991 and was discontinued on 20 Oct 1993.

Pesticides were applied using a hand-held back pack sprayer that applied 560 L ha⁻¹ at 0.28 MPa. The following pesticides were applied during the study: isazofos (0-[5-chloro-1-(1-methylethyl)-1H-1,2,4-triazol-3-yl] 0,0-diethylphosphorothioate) at 2.24 kg ha⁻¹ on 12 Aug 1991, chlorothalonil (2,4,5,6-tetrachloro-1,3-benzenedicarbonitrile) at 9.6 kg ha⁻¹ on 21 Aug 1991, 2,4-D ((2,4-dichlorophenoxy) acetic acid) at 1.14 kg ha⁻¹ on 17 Sep 1991, fenarimol ((±)-α-(2-chlorophenyl)-α-(4-chlorophenyl)-5-pyrimidinemethanol) at 0.76 kg ha⁻¹ on 3 May 1992, propiconazole (1-[[2-(2,4-dichlorophenyl)-4-propyl-1,3-dioxolan-2-yl]methyl]-1H-1,2,4-triazole) at 0.84 kg ha⁻¹ on 18 Jun 1992, triadimefon (1-(4-chlorophenoxy)-3,3,-dimethyl-1-(1H-1,2,4-triazol-1-yl)-2-butanone) at 1.53 kg ha⁻¹ on 21 Jul 1992, and metalaxyl (methyl N-(2,6-dimethylphenyl)-N-(methoxyacetyl)-DL-alaninate) at 1.53 kg ha⁻¹ on 5 Aug 1992. Each pesticide was applied to two lysimeters.

A standard extraction procedure was used for all pesticides in this study. A 100 mL water sample was partitioned three times with hexane. The hexane fractions were combined, dried over Na_2SO_4, and evaporated to dryness. The residue was taken up in 1 mL of benzene and analyzed by gas chromotagraphy on a Varian GC equipped with an autosampler. Pesticide detection limits were as follows: isazofos 16 ug L⁻¹, chlorothalonil 1 ug L⁻¹, fenarimol 4 ug L⁻¹, propiconazole 4 ug L⁻¹, triadimefon 2 ug L⁻¹, metalaxyl 2 ug L⁻¹, 2,4-D 1 ug L⁻¹.

Phosphorus Mobility. The site was a sand putting green at the Hancock Turfgrass Research Center at Michigan State University (MSU) in East Lansing, MI. The green was constructed and seeded with creeping bentgrass (*Agrostis palustris* L. cv. Penncross) in 1980. No phosphorus fertilizers had been used in this plot area for approximately three years prior to the initiation of this study. The site showed visible signs of P deficiency, including the characteristic red to purple coloration of the foliage. Initial soil samples yielded an average soil P level, to a depth of 15 cm across the site, of 3 mg kg⁻¹, very low by the standards of the MSU Cooperative Extension Service. Soil pH was approximately 8.0 throughout the site.

Four treatments consisted of annual P application rates of 0, 25, 50, or 100 kg P_2O_5 ha⁻¹ yr⁻¹. The fifth and sixth treatments were rates of application recommended by the MSU Soil Testing Laboratory, based on soil test results for samples taken from the upper 7.5 cm. One treatment, 'Bray', was based on the Bray P1 extraction procedure (9), and the other, 'Olsen', was based on the Olsen extraction procedure (9). The Bray P1 procedure is used most often in Michigan, but the Olsen procedure is recommended on soils with pH greater than 7.8, or in some cases when pH is greater than 7.2. The final two treatments were applied using water injection cultivation (WIC). This practice has been investigated in recent years as a means of delivering fertilizer to the rootzone while avoiding potential nutrient immobilization in the thatch layer. 'WIC1' received 50 kg P_2O_5 ha⁻¹ yr⁻¹, while 'WIC2' was fertilized at the rate recommended following a Bray P1 soil test.

Phosphoric acid (H_3PO_4) was used as the P source because it is highly soluble and avoids the complicating factor of having to match rates of other plant nutrients that are present in alternate sources. Applications of surface applied P were made with a CO_2 pressurized plot sprayer. Total application volume was 0.33 L per plot (25.4 L ha⁻¹). Injection treatments were made with a prototype WIC unit used for developmental research on WIC (Toro Co., Minn., MN). This unit produced pulses of water on 7.5 cm centers at a pressure of 34.5 MPa. Maximum injection depth was approximately 15 cm. Total volume of water delivered during injection was 10.9 L per plot (36.3 kL ha^{-1}). No attempt was made to equalize application volumes between surface applied and injected treatments. Treatment application was always followed immediately by 2.5 mm of irrigation. Application rates never exceeded 50 kg P$_2$O$_5$ ha^{-1} in a single application.

Soil test recommendations in 1992 prescribed annual application of 200 kg P$_2$O$_5$ ha^{-1} yr^{-1} for the Bray, Olsen, and WIC2 treatments. The first application to all eight treatments was made on 20 Apr 1992 (50 kg P$_2$O$_5$ ha^{-1} maximum, less for prescribed treatments). A second application was made on 6 Jun to the 100 kg ha^{-1}, Bray, Olsen, and WIC2 treatments. The Bray, Olsen, and WIC2 treatments received subsequent applications on 21 Jul and 24 Sep to yield the total annual rate. Soil samples collected in the spring of 1993 prescribed annual application rates to the Bray, Olsen, and WIC2 treatments of 50, 75, and 175 kg P$_2$O$_5$ ha^{-1}. Application dates in 1993 were 26 May (all treatments), 25 Jun (100 kg, Olsen, WIC2), 17 Aug (WIC2), and 24 Sep (WIC2). Nitrogen was applied at annual rates of 215 and 230 kg ha^{-1} in 1992 and 1993, respectively. Potassium was applied based on soil test recommendations at rates of 150 and 125 kg K$_2$O ha^{-1} in 1992 and 1993. Plots were irrigated to maintain optimum turf quality.

Soil samples were collected in Nov 1992 and 1993 and analyzed for pH and P content. From each plot, six cores 1.9 cm in diameter and approximately 25 cm in depth were collected. The verdure and thatch were removed (approximately 1.5 cm deep), and the cores were sectioned into depth increments of 0-7.5, 7.5-15, and 15-22.5 cm, measured from the bottom of the thatch layer. Because the primary interest of the experiment was vertical mobility through the soil, the verdure/thatch component was discarded in 1992. In 1993, this subsample was analyzed due to low P content of the underlying soil following the 1992 season. All samples were extracted using both the Bray P1 and Olsen extraction procedures, and P was quantified colorimetrically using a Lachat flow injection analyzer (Lachat Instruments, Mequon, WI). Data was analyzed as a split-plot design, with treatment as main plots and sample depth as the subplot. Statistical analyses were conducted using PROC GLM (SAS v. 6.12).

Results and Discussion.

Pesticide Leaching. Of the seven pesticides applied, only triadimefon was detected during the course of the study (Figure 1). Chlorothalonil, propiconazole, metalaxyl, isazofos, fenarimol and 2,4-D were not detected during the course of the study. The detections of triadimefon occurred multiple times in both lysimeters.

Triadimefon was applied on 21 Jul 1992 and detected later that summer in both lysimeters (Figure 1). The first detection occurred on 15 Sep 1992, 56 days after application. The next detection occurred in the 2nd lysimeter on 15 Oct 1992. While this difference in detection dates (31 days) seems large, only one other water sample was collected between 15 Sep and 15 Oct due to low percolation rates from the lysimeters at that time of year. A second pulse of triadimefon occurred later in that same year with detections in both lysimeters on 29 Nov and continued detections in lysimeter 1 on 14 Dec and 28 Dec of 1992.

The leaching of triadimefon is most likely the result of macropore flow (*10-11*). A review of rainfall during the summer of 1992 shows that July was a very wet month with total rainfall of 17.5 cm (Figure 2). Rainfall amounts following application were 0.28 cm on 22 Jul and 23 Jul, 0.9 cm on 25 Jul, and 4.5 cm on 30 Jul. In the 10 days prior to triadimefon application, 10.4 cm of rain fell. The pre-application rainfall ensured that antecedent soil moisture was high, permitting more rapid percolation of later rain events.

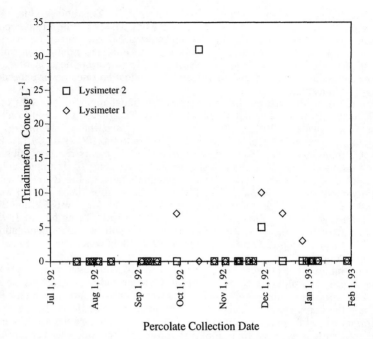

Figure 1. Detection of triadimefon in the percolate from lysimeters draining turfgrass soils.

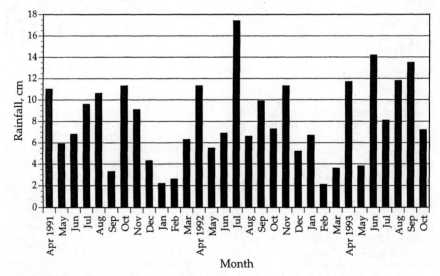

Figure 2. Monthly rainfall totals in East Lansing, MI from Apr 1991 through Oct 1993.

Other research has indicated that triadimefon is mobile when applied to turf (*12-13*). Petrovic and co-workers (*13*) found leaching of triadimefon, however, a greater quantity of triadimenol, a metabolite of triadimefon, was also found in the leachate samples they analyzed. We did not quantify triadimenol in this study. We did not confirm the identity of our detections by mass spectroscopy, but relied upon retention times to assign the peaks to triadimefon.

Triadimefon is not the most mobile of the pesticides applied during this experiment. Both 2,4-D and metalaxyl would be expected to be more mobile than triadimefon. Metalaxyl was applied on 5 Aug but no apparent macropore flow occurred with this application as it was not detected during the course of this study. Rainfall during Aug 1992 was low with only 6.6 cm collected during the month and 2.5 cm of that total was received on 27 Aug, too late to have any impact of macropore flow of metalaxyl.

The lack of observed leaching of the other pesticides is more understandable in light of the data developed concurrently on nitrate leaching. Labeled (^{15}N) nitrogen was applied to lysimeters on 26 Apr 1991 or 8 Nov 1991. While some incremental amounts of ^{15}N were detected in the leachate throughout the course of this experiment, the first significant increase in ^{15}N enrichment was detected on 31 Jul 1993, 826 days after application (Figure 3). Nearly 2.3 years were required for an unretained species, nitrate, to traverse the 1.2 m soil depth in the lysimeters. Pesticides, which are sorbed to soil, will take much longer to emerge from the lysimeter. Additionally, since irrigation was limited and applied only when the turf showed incipient drought stress, pesticide and nitrate leaching were less than may occur under more frequently irrigated turfgrass sites.

Because of the unique aspects of a well-managed turf, that is the high level of surface organic and the microbial activity associated with that organic matter, pesticides behave quite differently in soils under turf than in other agricultural cropping systems. Horst et al. (*3*) determined half-lives for pendimethalin, isazofos, chlorpyrifos (0,0-diethyl 0-(3,5,6-trichloro-2-pyridinyl) phosphorothioate), and metalaxyl of 6-15 , 5-11 , 6-12 , and 10-25 d, respectively, under turf conditions. A compilation of environmental properties of pesticides (*14*) lists half-lives of 90, 34, 30, and 70 d for pendimethalin, isazofos, chlorpyrifos, and metalaxyl, respectively. Pesticide half-lives in turf are reduced significantly compared to values obtained in field soils. This decrease in half-life reduces, but does not eliminate, the risk of groundwater contamination from pesticides used in turf. Because of the favorable data collected on the behavior of pesticides applied to turf, many turf practitioners, and perhaps researchers as well, tend to view pesticide leaching from turf as minimal or non-existent. Our data indicates that while leaching is minimal, it can occur. A better understanding, and real-world data, of pesticide contamination from turfgrass pest management practices is needed.

Lysimeters are an excellent tool for studying percolation of pesticides and fertilizers under natural conditions. However, the long transit times to be expected with soil columns of this depth mean that the additional studies will be expensive and time-consuming to conduct. Future studies of pesticide leaching using deep lysimeters should focus on long-term studies of single pesticides that are most likely to leach. The results of such studies would lead to a better understanding of the risk of groundwater contamination from pesticides routinely applied to turf.

Phosphorus Mobility. A significant main effect for depth with regard to pH was observed in 1992 (Table I). A significant treatment x depth interaction for pH was seen in 1993 (Table II). In both years, the treatment x depth interaction was significant for soil P content, regardless of the extraction procedure used. Data were analyzed for differences among treatments within each depth.

In 1992, the Olsen extraction procedure resulted in higher measured soil P content than the Bray P1 extraction (Table I). This is expected in soils buffered in the

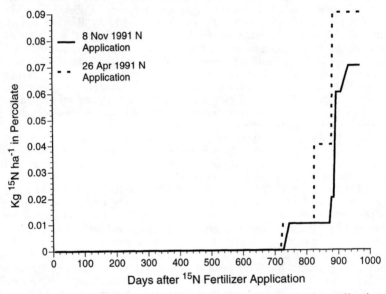

Figure 3. Cumulative ^{15}N recovered in percolate from a single application of ^{15}N-urea made on 26 April 1991 (Spring) or 8 November 1991 (Fall) at a rate of 39 Kg N ha^{-1}.

Table I. Soil phosphorus content in Nov 1992 following one year of P application.

Depth (cm)	pH	Treatment[a] (kg P_2O_5 ha^{-1})	Bray P1 extraction	Olsen extraction
			----------mg P kg^{-1} soil----------	
		0	1.8	3.5
		25	2.5	4.5
		50	2.6	4.6
0 - 7.5	7.5	100	4.8	6.6
		Bray	9.6	9.9
		Olsen	-[b]	16.6
		WIC1	2.4	4.9
		WIC2	5.8	6.8
		LSD (P=0.05)	1.6	3.2
		0	1.3	3.3
		25	1.3	3.5
		50	1.4	2.8
7.5 - 15	7.8	100	1.5	4.3
		Bray	1.6	3.6
		Olsen	-[b]	7.0
		WIC1	1.4	3.0
		WIC2	4.3	6.4
		LSD (P=0.05)	1.5	2.3
		0	1.5	5.6
		25	1.4	3.3
		50	1.5	3.3
15-22.5	7.8	100	1.8	4.1
		Bray	2.3	3.6
		Olsen	-[b]	8.7
		WIC1	1.5	4.4
		WIC2	3.1	4.6
	0.2	LSD (P=0.05)	NS	NS

[a]Phosphorous applications were made on 20 Apr, 6 Jun, 21 Jul, and 24 Sep 1992.
[b]Missing data.

Table II. Soil phosphorus content in Nov 1993 following two years of P application.

Depth	Treatment		Bray P1 extraction	Olsen extraction
(cm)	(kg P_2O_5 ha^{-1})	pH	----------mg P kg^{-1} soil----------	
	0	7.5	3.6	6.0
	25	7.3	12.9	15.6
	50	7.3	16.5	18.3
thatch	100	7.2	49.0	42.1
	Bray	7.2	43.0	35.1
	Olsen	7.1	59.4	45.5
	WIC1	7.3	14.2	16.2
	WIC2	7.3	32.3	31.2
	LSD (P=0.05)	0.2	10.8	9.8
	0	7.8	2.4	6.3
	25	7.8	4.1	5.9
	50	7.7	4.4	7.0
0 - 7.5	100	7.6	18.0	19.4
	Bray	7.6	25.3	25.0
	Olsen	7.6	32.9	25.6
	WIC1	7.8	6.9	9.4
	WIC2	7.6	47.6	46.6
	LSD (P=0.05)	0.1	12.4	12.5
	0	8.1	1.8	5.8
	25	8.2	2.8	3.3
	50	8.1	2.3	4.4
7.5 - 15	100	8.1	3.3	4.6
	Bray	7.9	10.6	11.9
	Olsen	8.0	14.9	16.0
	WIC1	8.1	6.6	9.9
	WIC2	7.9	19.5	22.6
	LSD (P=0.05)	0.2	5.0	5.8
	0	8.5	2.4	7.6
	25	8.5	2.4	3.3
	50	8.4	2.5	4.4
15 - 22.5	100	8.5	2.9	3.8
	Bray	8.5	4.1	6.1
	Olsen	8.5	6.0	8.6
	WIC1	8.3	2.8	4.3
	WIC2	8.5	7.3	10.4
	LSD (P=0.05)	NS	1.8	3.2

alkaline range (9). The trends were the same with both extraction procedures. As would be expected, soil P was higher in the 0 – 7.5 cm depth when higher P rates were applied. Bray, Olsen, and WIC2 treatments, which all received 200 kg P_2O_5 ha[-1], had the highest P contents. The same was true in the 7.5 – 15 cm depth when extracted with the Olsen procedure, but the differences were not detected as completely when the Bray P1 extractant was used. The ability of the Olsen procedure to extract more P at high pH makes it a more sensitive and more appropriate test under these conditions. Although the differences were not significant at the lowest depth, P content of the Olsen treatment was numerically higher. Injecting P did not result in significantly higher P levels at greater soil depths than for the other treatments. The WIC1 and 50 kg treatments received identical rates of P, and soil P levels were not significantly different. WIC2, Olsen, and Bray treatments also received identical annual application rates. Soil P for the WIC2 treatment was lower at all depths than for the Olsen treatment. Curiously, although the Bray and Olsen treatments received identical rates of P, the Olsen plots had higher P levels than the Bray plots, even when extracted with the same procedure. Mean P content for all treatments was 7.2, 4.2, and 4.7 mg kg[-1] (Olsen extraction) for the top, middle, and lower depths, respectively. Vertical movement of P did not appear to be severe during this first year, although the Olsen and WIC2 treatments did have higher concentrations at depths below 7.5 cm.

Because soil P remained low following the 1992 growing season, the thatch region removed from the top of the cores was analyzed in 1993. It was suspected that a large amount of P might be bound in this thatch layer. In general, soil P was much higher following the second year of P application. Means were 26.3, 18.2, 9.8, and 6.1 mg kg[-1] for thatch and the three underlying depth increments, respectively. Treatment differences were significant within each depth increment for P content, and in all but the lowest depth for pH (Table II). In 1993, application rates for the Bray, Olsen, and WIC2 treatments were 50, 75, and 175 kg P_2O_5 ha[-1], respectively. Soil P content appeared to be related to the cumulative P applications over two years. The Bray, Olsen, and 100 kg ha[-1] treatments, which received the highest surface application rates, had the highest levels of P in the thatch region. Below the thatch, the WIC2 treatment resulted in the highest soil P levels at all depths, although the Olsen treatment had similar soil P at 15 – 22.5 cm. Greater extraction of P with the Olsen procedure was apparent again in 1993, but primarily only when soil P levels were less than approximately 25 mg kg[-1]. Above this concentration, extraction efficiency was similar with the two procedures, even at high pH.

Soil P concentration, or P application rate, was inversely related to soil pH. Mean pH for all treatments at each of the four increasing depths was 7.3, 7.7, 8.1, and 8.5. Within depths, there was also a trend toward lower pH with increasing P application rate. This pH effect was due to the acidity present in the fertilizer solution. The lower pH in the thatch may also have been due to the buffering capacity of organic matter in the acidic range.

Soil P concentrations were much higher after two years of the higher rates of P application. They were proportionately higher than would be predicted by the application rates alone. Soil testing procedures are designed to estimate available plant nutrients. Soils have a finite limit to the amount of P that can be adsorbed (15). Once that adsorption capacity is reached, additional P remains in soil solution and is subjected to potential leaching. In this experiment, it appears that annual P rates averaging 100 kg P_2O_5 ha[-1] or higher may have exceeded the adsorption capacity of this sand, resulting in increased levels of available P. Lawson and Colclough (16) detected very little P in leachate from a sand-based rootzone material, but their application rate was low (22 kg P_2O_5 ha[-1] yr[-1]). Actual leachate P concentrations were not measured in our experiment, but it is reasonable to assume that the increased soil P levels resulting from higher fertilization rates creates increased pollution risk.

76

Injection of fertilizer P at high annual rates increases this risk further by decreasing the amount of P intercepted by thatch and placing more P deeper in the profile. Little downward P movement was seen during the first year, but after only two years soil P concentration increased dramatically. In a sampling survey conducted in Great Britain, over one half of the 1800 golf greens sampled had soil P concentrations exceeding 150 mg kg^{-1} (17), a value three times as high as any seen in this study. According to the MSU Cooperative Extension Service recommendations, if Bray P1 extractable P is over 30 mg kg^{-1} in a golf green (45 mg kg^{-1} for Olsen extractable P), application of fertilizer P is not recommended. This experiment shows that these levels can be attained over a relatively short period of time with regular addition of fertilizer P, even on a P deficient site. Turf managers must pay very close attention to soil test results and recommendations in planning phosphorus fertilization programs to limit the risk to the environment.

A well-managed turfgrass is an excellent system for retaining and degrading pesticides. Pesticide leaching by macropore flow represents the most likely pathway for most pesticides applied to turf to reach groundwater. For many pesticides, application to turf results in accelerated degradation and reduced leaching potential compared to a traditional pesticide application to bare soil. As an example, isozaphos has been assigned a GUS (Ground Ubiquity Score (18)) of 3.06 based upon a reported field half-life of 34 days and a Koc value of 100 mL gm^{-1} (14). Using the half-life values reported by Horst et al. (5), the computed GUS value for turf would range from 2.08 for a half-life of 11 days to 1.40 for a half-life of 5 days. According to the GUS screening model, a GUS of 2.8 corresponds to those pesticides that are expected to be detected in groundwater. A GUS of 1.8 to 2.8 is problematic and suggests that site conditions, rainfall, etc. will determine whether that pesticide reaches groundwater. A GUS of 1.8 or less indicates there is little risk the pesticide will reach groundwater. Thus application of isazofos to turf is either unlikely or only moderately likely to result in groundwater contamination, while applications of isazofos to bare soil will almost certainly result in groundwater contamination.

A well-managed turfgrass should not result in significant groundwater contamination from pesticides, nitrogen or phosphorus. However, the data discussed in this paper indicates that both phosphorus and pesticides can reach groundwater when applied to turf. It is essential that turf managers use IPM techniques, soil testing, and wisdom when applying agricultural products to turf because the key to reducing or eliminating movement of these products is in the hands of the turf manager.

Literature Cited.

1. Brown, K. W.; Duble, R. L.; Thomas, J. C. *Agron J.* **1977**, *69*, 667-671.
2. Starr, J. L.; DeRoo, H. C. *Crop Sci.* **1981**, *21*:531-536.
3. Sears, M. K.; Chapman, R. A. *J. Econ. Entomol.* **1979**, *72*, 272-274.
4. Branham, B. E.; Wehner, D. J. *Agron. J.* **1985**, *77*, 101-104.
5. Horst, G. L.; Shea, P. J.; Christians, N.; Miller, D. R.; Stuefer-Powell, C.; Starrett, S. K. *Crop Sci.* **1996**, *36*, 362-370.
6. Snyder, G. H.; Cisar, J. L. *Crop Sci.* **1996**, *36*, 1433-1438.
7. Snyder, G. H.; Cisar, J. L. *Int Turf Res. Soc. J.* **1993**, *7*, 978-983.
8. Miltner, E. D.; Branham, B. E.; Paul, E. A.; Rieke, P. E. *Crop Sci.* **1996**, *36*, 1427-1433.
9. Kuo, S. **1996**, *In* D. L. Sparks (ed.) Methods of soil analysis. Part 3. Chemical methods. SSSA Book Series:5. SSSA, ASA, Madison, WI.
10. Quisenberry, V. L.; Phillips, R. E. *Soil Sci. Soc. Am. J.* **1978**, *42*, 675-679.
11. Priebe, D. L.; Blackmer, A. M. *J. Environ. Qual.* **1989**, *18*, 66-72.
12. Petrovic, A. M.; Young, R. G.; Sanchirico, C. A.; Lisk, D. J. *Chemosphere* **1994**, *29*, 415-419.

13. Petrovic, A. M.; Young, R. G.; Ebel, J. G., Jr.; Lisk, D. J. *Chemosphere* **1993**, *26*, 1549-1557.
14. Wauchope, R. D.; Buttler, T. M.; Hornsby, A. F.; Augustijn-Beckers, P. W. M.; Burt, J. P. *Rev. Environ. Contam. Toxicol.* **1992**, *123*, 1-35.
15. Sposito, G. *The surface chemistry of soils.* Oxford University Press: New York, NY. **1984**.
16. Lawson, D. M.; T. W. Colclough. *J. Sports Turf Rsch. Inst.* **1991**, *67*, 145-52.
17. Stansfield, D. M. *J. Sports Turf Rsch. Inst.* **1985**, *61*, 136-40.
18. Gustafson, D. I. *Environ. Toxicol. Chem.* **1989**, *8*, 339-357.

Chapter 5

Potential Movement of Nutrients and Pesticides following Application to Golf Courses

L. M. Shuman, A. E. Smith, and D. C. Bridges

Department of Crop and Soil Sciences, University of Georgia, Griffin, GA 30223

Bermudagrass plots (3.7 m x 7.4 m) were established with a 5% slope to measure water runoff and associated nutrients and pesticides. Treatments included were dimethyl-amine (DMA) dicamba, DMA mecoprop, DMA and 2,4-D, chlorothalonil, dithiopyr, chlorpyrifos, benefin, and pendimethalin. Fraction of applied analytes transported from the plots ranged from 0.01% for benefin and pendimethalin to 14% for dicamba and mecoprop. The relationship between fraction transported and the negative log of the analyte solubility in water (pSw) was better fitted by a quadratic (R^2, 0.96) than a linear function (R^2, 0.86). For 1997 accompanying lysimeter leachate data for two working USGA golf greens showed that fertilizer NO_3-N required 20 to 30 days to appear in the leachate and PO_4-P required 30 to 50 days depending on fertilizer source and rainfall.

Watersheds delineate areas of interacting land and water resources and provide logical ecosystem boundaries. Relationships between landscape patterns and ecological processes are of particular relevance to the management of urban watersheds influenced by nonpoint source pollutants. Some of the most dramatic changes in landscape patterns in the U.S. are those associated with rapid urbanization. From 1980-1992 the U.S. population increased by 19.4%, to 255 million. Nearly 15% of that increase occurred in metropolitan areas. Expansion of metropolitan areas has spawned ever-increasing areas of turfgrass that is more intensively managed than ever before. There are over 14,000 golf courses in the U.S. and the number is increasing at the rate of one per day. Of the 700,000 ha of golf course area in the U.S. about 2% is in greens. These are typically 80% by volume coarse sand, to give a high percolation and water-removal rate.

The porous medium of golf course greens coupled with high inputs of fertilizer and irrigation water promotes leaching - not only of soluble nitrogen sources, but even of less soluble phosphate fertilizer. Sandy soils often have a deficiency of mineral elements (metal oxides and secondary clay minerals) that constitute the most common adsorbing surfaces for phosphate. Even the peat that is added to the sand of a typical golf green "mix"to increase water-holding capacity and give other desirable effects as a rooting medium, can supply soluble carbon species which can exacerbate phosphate losses through complexation and decreased adsorption. The fairway areas present different problems that lead to detrimental environmental effects through fertilizer losses. In the Piedmont region of the Southeast, the soils are often impervious causing high rates of runoff during heavy rainfalls, especially on sloped areas. As much as 70% of the rain can be lost as runoff. This can cause fertilizer losses through "floatoff" of recently applied particles and runoff of soluble species. Homeowner lawns are yet another large source of fertilizer elements being lost to surface waters. Residential and commercial lawn area is much greater in total than golf courses, and homeowners and commercial property managers are more prone to overwater and overfertilize.

Nitrogen and P added to turfgrasses if lost in runoff and subsurface flow can eventually find their way to potable water supplies. Added nutrients, especially P, can cause eutrophication of surface water, leading to problems with its use for fisheries, recreation, industry, or drinking water due to increases in the growth of undesirable algae and aquatic weeds (26). Phosphorus is usually the most limiting element for algae growth, since many blue-green algae are able to utilize atmospheric N_2 as an alternative source of N. Most P lost from grassed areas is also in the soluble, rather than particulate, form that is immediately available for algae growth (26). Thus, limiting both N and P losses from turfgrass areas is an important environmental issue.

Turfgrass is typically the most intensively managed biotic system in metropolitan landscapes. Increasing interest in the environmental effects of pesticides is, in general, a response to their increased use since the 1960s and due to advancements in technology allowing scientists to detect their presence at low concentrations. Many compounds, because of their constituents (many contain halogens or nitrogen), can be detected at sub-parts per billion levels. Concern for the impact of pesticides and nutrients on the environment is related to potential entrance into drinking water sources and their potential to adversely affect aquatic organisms and ecosystem functions. Although drinking water in rural areas primarily comes from groundwater sources, much of the drinking water in urban areas is derived from surface water sources such as reservoirs. It is estimated that as much as 95% of the drinking water for some major metropolitan areas comes from reservoirs. Contaminants are transported via runoff and on sediment. Erosion and runoff processes in relation to water quality and environmental impacts have been examined by Anderson et al. (1), Leonard (14), and Stewart et al. (32). Results of research conducted in the Piedmont Region of the southeastern United States has suggested that as much as 40-70% of the water from an average rainfall event on a turfgrass sod having a high water content (field capacity) and an average slope of 5% could leave the landscape as runoff water

(29). This suggests that there is a high probability for pesticides to be transported in surface waters.

Literature Review

Accurate estimates of pesticide usage in urban areas are difficult to obtain, because most pesticides produced for this market are available through a wide variety of outlets and may be applied by the homeowner, a pest control operator, or municipal governments. Reliable data on pesticide and fertilizer use in urban areas are limited, and estimation of pesticide loads in urban watersheds is complicated. Studies of pesticide use in urban areas suggest that large quantities of pesticides are frequently applied. For example, a study of homeowner use in three cities (Lansing, Dallas, and Philadelphia) showed that 341,000 kg of pesticides were used by suburban homeowners annually. Most pesticides were used in lawn care, with insects the main problem in Dallas, and weeds in Lansing and Philadelphia (*17*). In commenting on this use, the National Academy of Sciences emphasized that many pesticides used in lawn care are applied at rates as high as 5-10 kg ha^{-1}. If this survey in three cities is representative of use by American homeowners, the fact that a population of about 3 million used 341,000 kg suggests that outdoor use of pesticides around homes may involve as much as 25 million kg of pesticides annually.

As noted above, nitrates, in particular can leach through sandy golf greens, since they are soluble. In fact, the main concern for nitrogen affecting natural waters from greens is through leaching, and rather than through runoff. For example, Mancino and Troll (*15*) found nitrate concentrations as high as 45 mg L^{-1} in leachate from a sand-peat green. Petrovic et al. (*23*) found nitrate to be higher in leachate from fairways than the allowable drinking water standard of 10 mg L^{-1}. Owens et al. (*22*) also found nitrates above 10 mg L^{-1} in groundwater under a grassed pasture, mainly from downward flow and not in runoff. A factor in nitrate leaching can be the source of N applied. Brown et al. (*6*) indicated that nitrate losses were in the following order according to source: ammonium nitrate>12-12-12 fertilizer>milorganite> isobutylenediurea>ureaform-aldehyde. Some studies also show little of nitrate leaching losses, especially if moderate rates are used, excessive irrigation is avoided, and nitrates are applied to mature turf as opposed to land barely covered with vegetation (18, *21*).

Common-sense management practices exist which may limit leaching and runoff of nitrate fertilizer from golf greens and fairways. The fertilizer rates should be scheduled for light, frequent applications of a slow-release fertilizer and irrigation should be limited to turf needs (*22*). Brauen and Stahnke (*4*) reported nitrate leaching for turf plots that had immature turf. They suggested adding nitrogen often, at low rates, to minimize leaching. Likewise, Weed and Kanwar (*34*) found that nitrate leaching was minimized by applying N at rates just enough to meet crop N demand and at times close to the crop's peak uptake periods. Decreasing fertilizer rates during periods of slow growth and in the winter, especially when larger volumes of water are expected, may also help to limit nitrate losses from golf courses (5). Thus, management techniques have been shown to assist in lowering nitrate loses, but more information is needed, especially concerning the width of buffer zones and proper irrigation scheduling in relation to fertilizer application.

Phosphorus is transported to natural waters both in soluble form and adsorbed to soil particles, including both inorganic particles and organic matter. Since there can be large losses of soil from row-cropped agricultural lands, most P lost is in particulate form, but for grassed land and forests most P is transported in soluble form instead (26). Phosphorus must be added regularly at rates of at least 25 kg ha^{-1} annually for sand golf greens (5 kg ha^{-1} per month through the growing season of May to September) (9) and, although P runoff is less for grass than for cropped soils (24), P does build up and can become a problem due to leaching into groundwater and runoff into surface waters (25). In Australia, P was found in runoff as dissolved species and not associated with sediment (2). On undisturbed soil columns from golf fairways where P was leached during heavy irrigation, 35% of the added P was found below the 8-inch depth (31). The P that causes eutrophication of waters is that which is bioavailable. The bioavailable P is comprised of soluble P and a portion of the particulate P. Thus, when analyzing runoff and leachate for P, the P is commonly fractionated into soluble P, particulate P and, in some cases, bioavailable particulate P (27).

Soil properties can affect P movement. Phosphorus breakthrough data were modeled well by a four-parameter segmented exponential model, with P breakthrough found to be correlated with extractable Fe and Al, extractable P, and P sorption capacity (16). Kaiser and Zech (12) found that an increase in dissolved organic carbon (DOC) content of the leachate caused a decrease in the sorption of $H_2PO_4^-$, suggesting that $H_2PO_4^-$ had a higher sorption affinity than DOC. Onho and Erich (19) also showed that phosphate adsorption is inhibited by DOC derived from plant materials. Desorption of P can be rapid, occurring within one to four hours following a rainfall event. Thus, desorption and subsequent loss of P in runoff are a threat with every major rainfall event (>2.0 cm) for soils that have been recently fertilized. Phosphorus sorption and desorption, then, is an important factor affecting the amounts of P lost in runoff as well as by leaching, and determination of soil sorption/desorption characteristics likewise is important when attempting to predict P losses from soils.

Materials and Methods

Simulated fairways and greens. Twelve plots (3.7 x 7.4 m), separated by landscape timbers, were developed in a grid pattern with a 5% slope. The subsoil was a clay loam and the top soil was a sandy loam. A ditch was dug at the downslope end of each plot to install a trough for collecting runoff via a tipping-bucket sample collection apparatus. Plots were sprigged with 'Tifway 419' *Cynodon dactylon* (L.) Pers. x *C. transvaalensis* Burtt-Davy on May 17, 1993 and had become completely covered with sod by August 1, 1993. Wobbler off-center rotary-action sprinkler heads were mounted 7.4 m apart and 3.1 m above the sod surface. When operated at 138 kPa, the system produced a simulated rainfall at an intensity of 3.3 ± 0.4 cm/hr.

An outside lysimeter facility consisted of two greens (12 x 1.2 m) subtended with 10 round stainless steel lysimeters (55 cm diam x 52 cm deep) designed to direct of leachate and associated pesticides/nutrients to a central collection area. Prescribed rooting mixtures (sand and sphagnum peat moss) for the greens were based on the desired percolation rate for each grass species. Proportions were selected to give percolation rates of 39 or 33 cm/hr. Rooting media consisting of sand and sphagnum peat moss mixture of 85:15 and 80:20 (v/v) (87.7:2.3 and 86.8:3.2 by mass, respectively) resulted in the respective percolation rates. Lysimeters and areas around lysimeters were filled with sized gravel (10 cm), coarse sand (7.5 cm), and rooting mix (35 cm) in ascending sequence from the bottom, simulating USGA specifications for greens construction (*33*). Layers were packed into and around each lysimeter. The 85:15 mixture had a field capacity of 0.13 cm^3/cm^3, a wilting point of 0.03 cm^3/cm^3 and an effective saturated conductivity of 39.6 cm/hr. The 80:20 mixture had a field capacity of 0.15 cm^3/cm^3 and an effective saturated conductivity of 33.5 cm/hr. 'Penncross' *Agrostis stolonifera* L. was seeded into the 85:15 rooting media on October 15, 1991 and *C. dactylon* was sodded over the 80:20 rooting mix on March 10, 1992.

At the Cherokee Town and Country Club in Atlanta, two bentgrass putting greens also were each equipped with three lysimeters (stainless steel lysimeters 38 cm x 53 cm x 15 cm deep were placed 18 cm beneath each green surface). Stainless steel lines were run from the drain of each to the edge of the green where leachate was to be collected.

Treatments. There were 12 plots in the simulated fairway area on Cecil sandy loam grassed to 'Tifway 419' Bermudagrass. In most cases, nine plots were used, with all treated similarly as replications so that treatments differed in time rather than in space. The initial nutrient experiment involved rates of 10-10-10 fertilizer material at 0, 12.2, 24.4, and 48.8 kg ha^{-1} giving P rates of 0, 5.4, 10.7, and 21.5 kg ha^{-1}. Irrigation timing and amounts initially were 4 (5.0 cm), 24 (5.0 cm), 72 (2.5 cm), and 168 (2.5 cm) hr after treatment (HAT), with collections of runoff during and following each irrigation.

For pesticide experiments, plots were treated as follows: benefin (as Balan EC, 180 g a.i./L, DowElanco), chlorothalonil (as Daconil 2787, 502 g a.i./L, ISK Biosciences), chlorpyrifos (as Dursban Pro, 241 g a.i./L, DowElanco), dicamba (as Banvel, 480 g a.e. amine salt/L, Sandoz), dithiopyr (as Dimension, 119.8 g a.i./L, Rohm and Haas), 2,4-D (as Weedar 64, 456 g a.e. amine salt/L, Rhone-Poulenc), mecoprop (as Chipco Turf Herbicide MCPP, 240.6 g a.e. amine salt/L, Rhone-Poulenc), and pendimethalin (as Prowl, 480 g a.i./L, American Cyanamid). Treatments were applied between 1 June and 1 November 1993 and 1994 based on meteorological forecasts that projected for at least a 72-hr period with a low probability of rainfall. Rainfall was simulated at 24, 48, 96, and 192 HAT. Following simulated rainfall, normal rainfall events also were monitored until herbicides in runoff were no longer detected. Runoff was quantified, and subsamples were collected using the tipping-bucket apparatus.

Treatments for field lysimeters were placed on a 1.2 by 1.2 m area using a backpack sprayer and aqueous dilutions of chlorothalonil, chlorpyrifos, 2,4-D, dicamba, dithiopyr, dithiopyr-G (as Dimension G, 0.25% a.i., Rohm and Haas), and mecoprop. Dithiopyr-G was evenly distributed over a similar area. Each treatment was replicated 4 times and each experiment was repeated one or more times. An automatic track-irrigation system was developed for controlling irrigation rates and times (30). Watering nozzles traversed a horizontal track located above the sod, at a speed of 2.9 m/min. Flow rate of water was adjusted 1.82 mL/sec at 138 kPa. Daily irrigation with 0.625 cm of water plus a weekly irrigation of 2.5 cm were controlled using an automatic timer. These conditions simulated management practices and average rainfall events for golf course greens in the Piedmont region of the United States. An automatic moving rain shelter protected the greens during actual rainstorms.

There were no planned treatments at the Country Club putting greens, with the greens remaining managed as seen fit by the greenskeeper. Detailed records were kept on the amounts of nutrients and pesticides applied. Weekly leachate samples were continuously collected starting in January, 1995, and analyzed for N and P.

Measurements and analytical procedures. Nitrate and P were analyzed using a LACHAT flow analyzer for runoff and leachate samples. Special analysis for P included forms in the leachate of total, bioavailable, and particulate P. Total P (TP) was measured following perchloric acid digestion of unfiltered runoff or leachate. Soluble P (SP) was determined on a filtered (0.45 μm) sample. Particulate P was taken as the difference (TP-SP). Bioavailable P (BAP) was determined by extracting 20 mL of an unfiltered sample with 180 mL of 0.1 M NaOH for 17 hours shaking in an end-over-end shaker. The amount of P in the filtered extract was taken as the BAP. The bioavailable particulate P (BPP) was taken as the difference between BAP and SP (27). All P measurements were by a colorimetric method (molybdate blue-ascorbic acid) using a flow analyzer (LACHAT).

Dicamba, 2,4-D, and mecoprop were analyzed by procedures developed in our laboratory (13). Subsamples of 100 mL were transferred from the storage bottle into a 250 mL beaker. An internal standard (2,4,5-T) was added to the beaker and the mixed solution was acidified to a pH of 2 with 0.2 M HCl. Pesticides were extracted from the acidified solution by liquid-liquid partitioning into 200 mL diethyl ether. The diethyl ether was evaporated and analytes were esterified with triflouroethanol (TFE). Esters were quantified by gas chromatography (GC) using an electron capture detector (ECD). Detection limit for dicamba was 0.1 μg L^{-1} and for mecoprop was 0.5 μg L

Dithiopyr was extracted by solid phase extraction and analyzed by GC-ECD according to a method developed in our laboratory (*11*). The detection limit for dithiopyr was 0.4 μg L^{-1}. Benefin and pendimethalin were extracted from 50 mL aqueous subsamples by liquid-liquid partitioning into 150 mL dichloromethane. Dichloromethane was concentrated under vacuum to 1 mL and diluted with 1 mL toluene containing metribuzin as an internal standard. Analytes were quantified by GC-ECD. Chlorothalonil and chlorpyrifos were extracted by liquid-liquid partitioning into ethyl acetate. The ethyl acetate volume was reduced under vacuum to 1 mL and hydroxy metabolites of chlorpyrifos and chlorothalonil were methylated using 1 mL diazomethane and 0.034 g silica gel. Analytes and metabolites then were quantified with detection limits of 1 μg L^{-1}.

Results and Discussion

Simulated fairways. The first nutrient experiment involved applying rates of 10-10-10 fertilizer over time, with each treatment being replicated by the number of plots available. The first rate was 0.5 lb of N per 1000 sq ft., or 2.44 g N and 1.07 g P per m^2. Nitrate-N in runoff was low, because the form of nitrogen added was ammonium (Figure 1), and there was given too little time to revert to other forms. Future experiments will be carried out using different forms of N. The form of P reported is SP, in that the solutions were clear and TP and BAP were essentially the same as SP. Phosphorus concentration was as high as 2.7 mg L^{-1} during the first simulated rain event and decreased thereafter. Austin et al. (*2*) showed a similar result for flood irrigation, where the first runoff event resulted in high runoff compared to subsequent irrigations. Concentration at 168 hours did increase over that at 72 hours, but the volume of runoff was low at that time. This is the first experiment carried out for N and P at this site and as such the data should be considered to represent preliminary results.

As noted in the discussion above, nitrate concentrations in this runoff were low because the form of N applied was all ammonium (Table I). The 10-10-10 formulation included ammoniated phosphate, which is usually monoammonium phosphate, and ammonium sulfate. Thus, the recovery of N was only 0.72 % of that added, and much of this probably came from other sources including mineralized organic nitrogen. The phosphate concentration, as an average of all runoff events, was just less than 1 mg L^{-1}, with the percent recovered being approximately 11%. Although losses from agricultural land usually are <1 to 2 % of that added (*10, 27*), soluble P as high as 35% of that applied has been found during flood irrigation of grassed land (*2*).

Only samples collected over the first 192 HAT contained pesticide concentrations above the minimum detectible concentration (MDL) of 1 μg/L. The fraction of water leaving the plots at 24, 48, 96, and 192 HAT was 44.8, 72.1, 40.0, and 35.5%, respectively. The highest pesticide concentrations in the runoff water occurred during the first rainfall event, applied at 24 HAT (Table II). Concentrations of

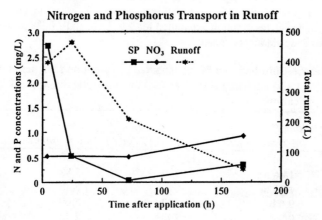

Figure 1. Phosphorus and nitrogen losses in runoff (24.4 kg/ha N and 10.7 kg/ha P were applied as 10-10-10 fertilizer).

Table I. Nitrate and P in runoff from 3.7 x 7.4 m bermudagrass field plots, including concentration, recovered mass, and percent of total applied.

Element	Formulation	Rate	Date Applied	Sample Duration	Irrigation & Precipitation	Runoff
		g ai m^2		days	mm	L
Nitrate	10-10-10	2.44	4/6/98	7	150	1250
Phosphate	10-10-10	1.07	4/6/98	7	150	1250

	Form	Date	Concentration	Element recovered Mass	Percent of Total
			$\mu g\ L^{-1}$	$\mu g\ m^{-2}$	(%)
Nitrate	MAP, DAS[a]	4/13/98	621	17,611	0.72
Phosphate	MAP	4/13/98	910	122,475	11.5

[a]MAP = monoammonium phosphate, DAS = diammonium sulfate.
2,4-D, dicamba, and mecoprop in the runoff water were 800, 360, and 810 µg/L, respectively, and analytes in runoff accounted for 9.6, 14.6, and 14.4% of the respective pesticide amounts applied.

Table II. Relationship between % transported in runoff and log octanol:water coefficient (K_{ow}) and log water solubility (S_w) for nine analytes applied as a broadcast application.

Analyte	Runoff (% Transported)	Concentration at 24 HAT (µg/L)	$\log S_w$ $(ppm)^a$	$\log K_{ow}$[b]
Nitrate-N	16.4 a[c]	12,500	6.00	-1.97
Dicamba	14.6 a	360	5.82	-1.82
Mecoprop	14.4 a	810	5.48	-1.77
2,4-D	9.6 b	800	5.09	-1.40
Chlorothalonil	0.8 c	290	2.78	1.53
Dithiopyr	1.9 c	39	-1.00	4.70
Chlorpyrifos	0.1 c	19	-1.48	4.97
Benefin	0.01d	3	-1.60	5.62
Pendimethalin	0.01d	9	-1.85	5.11

[a]log of the water solubility (ppm)
[b]log of the octanol:water partition coefficient
[c]Means in a column followed by the different letter are different ($P \leq 0.05$).

All analytes were applied in a water carrier which evaporated soon after application, allowing analytes to reside in close contact with the wax cuticle of the turfgrass. Simulated rainfall thus should result in a partitioning of analytes from the lipophilic wax into the transporting water. Therefore, the log of the octanol:water partitioning coefficient (K_{ow}) was determined according to the equation presented by Dao et al. (*8*).

$$logK_{ow} = \frac{logS_w - 4.184}{-0.922}$$

The log K_{ow} values and log values of the water solubility (S_w) were plotted against total % transported in runoff and fit to quadratic and linear equations (Table III). The resulting R^2 values were 0.97 and 0.96, respectively for quadratic equations and 0.86 and 0.84 for linear equations. The better fit of the data to the quadratic equations could have resulted from precipitation and crystallization of some analytes resulting in reduced dissolution of analyte molecules.

Table III. Equations and R^2 values for fraction (%) transported against log S_w and K_{ow} values for analytes included in the fairway runoff plots.

Y	X	Equation	R^2
% transported	log S_w	$y = a + bx + cx^2$	0.96
% transported	log K_{ow}	$y = a - bx - cx^2$	0.97
% transported	log S_{ow}	$y = a + bx$	0.84
% transported	log K_{ow}	$y = a - bx$	0.86

Concentration and fraction transported for these analytes with a relatively high water solubility was fairly high. However, it must be realized that runoff would probably be diluted many fold prior to reaching potable water systems. Additionally, simulated rainfall was instantly turned on at maximum constant intensity, in contrast to natural rainfall events which probably would not be at maximum intensity at the onset. However, these data still suggest that precautions must be exercised when applying more water soluble pesticides to golf course fairways, in the Piedmont region, having at least 5% slope.

Leachate from simulated greens. Only small quantities of pesticides were leached from lysimeters containing *A. stolonifera* (Ben) and *C. dactylon* (Ber) (Table IV). When present, analytes were leached during the 4- to 6-week period following treatment. No analytes were detected in leachate collected 7 weeks after treatment. Data were highly variable, since concentrations in leachate were close to the MDL of 1 µg/L. All quantities less than this concentration were expressed as 0, whereas concentrations ≥ 1 µg/L were expressed as actual numbers. Less than 0.5% of the applied analyte was leached from the lysimeters for all pesticides over the 70-day experiment. Chlorothalonil and its metabolite OH-chlorothalonil were present in leachate at the highest concentrations of the pesticides tested. OH-chlorothalonil was detected at a concentration of 43 µg/L. This is probably in response to the relatively high rates of chlorothalonil (9.53 kg/ha) applied twice monthly (4 applications for the duration of the experiment) and the higher water solubility of OH-chlorothalonil compared to chlorothalonil.

The Groundwater Loading Effects of Agricultural Management Systems (GLEAMS) model was used to predict the leaching potential of 2,4-D from lysimeter systems under *C. dactylon*. Input parameters are described in detail by Smith and Tillotson (*28*). The foliar and soil half life values were 8.3 and 10 days, respectively. The 2,4 D K_{oc} was 20 and the rooting mixture organic carbon was 5.8 % (mass basis). The GLEAMS model predicted a maximum concentration for 2,4-D transport of 48 μg/L at 4 weeks after treatment. The 2,4-D concentration in leachate from the lysimeters was not above the MDL of 1 μg/L throughout the 10 week collection period, however (Table 3). Inclusion of the thatch layer in the GLEAMS model did not alter predicted transport of 2,4-D in the aqueous effluent from the rooting media, compared to non-inclusion. The probable reason for no significant influence by the thatch layer is the thin thatch layer (1.40 cm) included in the model and the limited increase in organic matter used by the model for this layer compared to that for the rooting media (5.80 vs 2.26% w/w).

Table IV. Pesticide leached, over a 10-week period, through lysimeters subtending *A. stolonifera* (Ben) on 85:15 (sand:peat moss) rooting media and *C. dactylon* (Ber) on 80:20 rooting media.

Analyte	Application rate		Highest concentration[1]		Total transported	
	Ber	Ben	Ber	Ben	Ber	Ben
	(kg/ha)		(μg/L)		% applied	
2,4-D	1.12	0.28	0	0	0	0
Dicamba	0.28	0.07	2.6±22	1.2±1	0.2	0.5
Mecoprop	1.40	0.56	0	0	0	0
Dithiopyr	0.56	0.56	2.3±2	1.7±1	0.2	0.3
Dithiopyr-G	0.56	0.56	2.4±2	1.6±1	0.3	0.2
Chlorothalonil	[9.53 twice monthly]		15±6	10.5±4	0.03	0.02
OH-chlorothalonil[3]	-	-	43±7	39±9	0.1	0.2
Chlorpyrifos	[1.44 twice monthly]		0	0	0	0
OH-chlorpyrifos[4]	-	-	0	0	0	0

1. Highest concentration during weekly leachate collections.
2. Means ± SE.
3. Major metabolite of chlorothalonil.
4. Major metabolite of chlorpyrifos.

The USEPA is currently establishing drinking water standards of reference doses for surface and groundwater. Standards will be based on the same toxicological research used to establish reference doses (formerly called Acceptable Daily Intake, or

ADI) for food. These standards will become the maximum contaminant levels (MCLs) allowed for pesticide concentrations in potable water. The MCLs for only a few pesticides used on turfgrass have been recommended to date. The recommended MCL for 2,4-D is 70 µg/L. Even though the GLEAMS model greatly overestimated observed herbicide loadings in runoff, the maximum predicted concentration of 48 µg/L is still less than the recommended MCL standard (70 µg/L) for 2,4-D. It also must be realized that water from golf courses will be diluted many fold before reaching potable water systems.

Leachate from Country Club greens lysimeters. Leachates collected weekly from active putting greens were analyzed for nitrate N and phosphate P without filtering or other treatment. For 1997, nitrate-N generally increased throughout the year, being less than 2 mg L^{-1} early in the year and increasing to 4 to 5 mg L^{-1} by the fall before decreasing (Figure 2.). Trends for the two greens were similar, with several peaks occurring during the season. Peaks can be associated with N additions and were especially high after the nitrate form of N was applied. For example, nitrate N peaks were observed on days 30 and 148 after nitrate N was applied on days 9 and 128, respectively. Exact lag times between additions and appearance in leachate cannot be ascertained, because in many cases there was no leachate obtained between these two events. However, lag times were generally on the order of 20 to 30 days for this data. Brown et al. (5) found lag times of 10 to 25 days for N leaching from sand-peat greens, depending on rainfall and the N source applied.

Phosphate data showed a slight increase throughout the year and then a decreased once more in October, similar to the nitrate data (Figure 3). The two greens gave similar patterns, but the concentration of P was higher for green 2 than for green 1 for all dates after day 30. The peak on day 30 may have been due to an application in 1996. Two applications in 1997 gave discernable peaks from 30 to 50 days after application. Thus, it appears that P did leach, though somewhat slower than N. This result can be attributed to two effects. One is that P was added at a much lower rates than N, and the other is that P is more highly adsorbed by colloidal surfaces and is less soluble than nitrate. Highest P concentration was over 3 mg L^{-1}, which is high enough to cause algae growth (10). Phosphorus values found were higher than the 0.6 mg L^{-1} found by Culley et al. (7) for leachate from bluegrass sod. Thus, P in leachate from sand-peat greens may be higher than from turf grown on soil.

Lysimeters were installed and these greens were seeded to bentgrass in September, 1994, so the years recorded include the establishment period. Concentrations of nitrate N in leachates were fairly stable over the three years, though they were higher in 1997 (Table V). Maximum concentration of N found exceeded the drinking water standard of 10 mg L^{-1} set by the U.S. Public Health Service in four out of six cases, but averages for the period were all less than 3 mg L^{-1}. Total delivery (concentration x leachate volume), increased each year and was higher in 1997 than in other years. Note that the minima and maxima listed are for individual lysimeters, whereas data in Figs. 2 and 3 are averages for three lysimeters at each green.

Phosphorus concentrations were very high in 1995 compared to the other years (Table VI). This P may have come from that applied in 1994, since little was added in

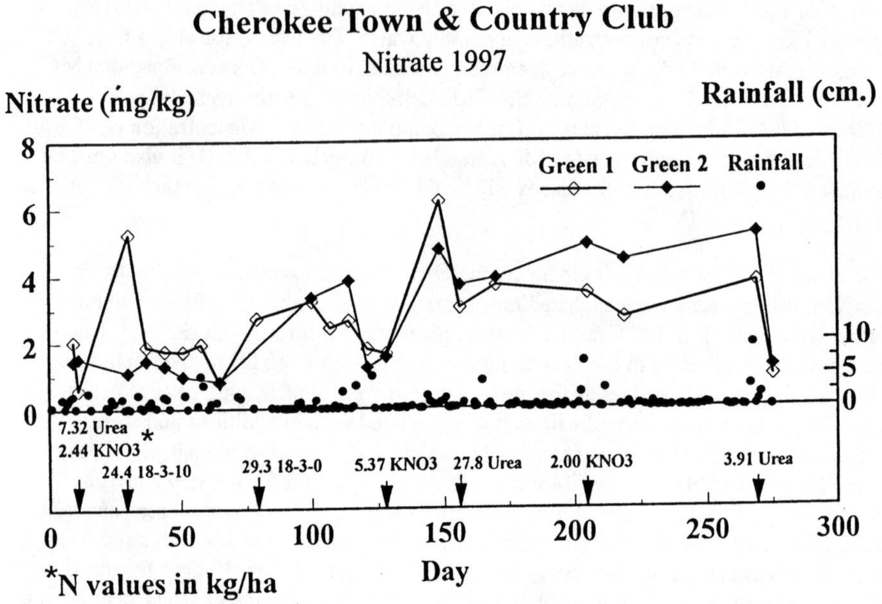

Cherokee Town & Country Club
Nitrate 1997

Nitrate (mg/kg)

Rainfall (cm.)

Figure 2. Average nitrate concentration in leachate for two golf greens in 1997. Lower axis indicates values for fertilizer amendments in kg/ha.

Cherokee Town & Country Club
Phosphorus 1997

Phosphate (mg/kg)

Rainfall (cm.)

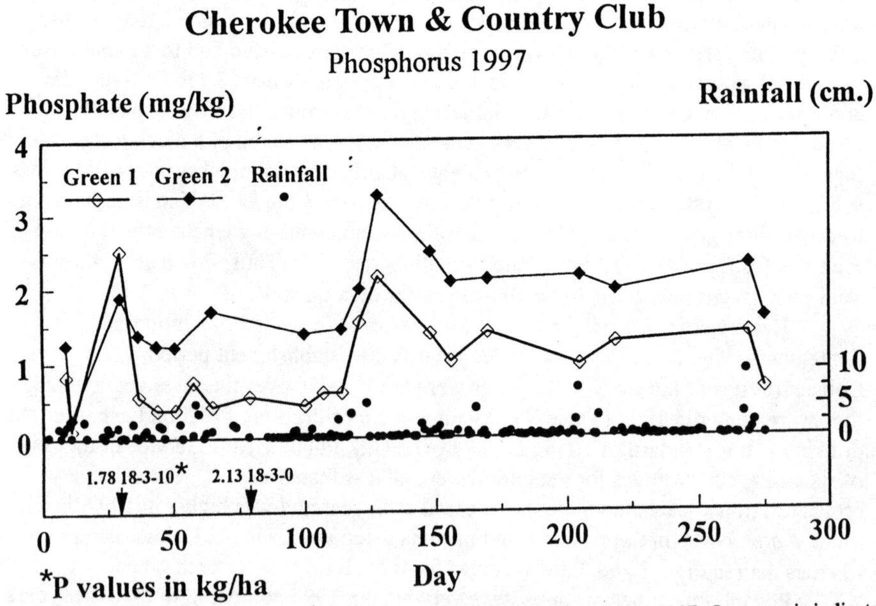

Figure 3. Average phosphate concentration in leachate for two golf greens in 1997. Lower axis indicates values for fertilizer amendments in kg/ha.

Table V. Nitrogen leachate concentration (mg/kg) and mass (mg) averages for collection dates, for three years for two USGA golf greens at the Cherokee Town and Country Club, Atlanta GA.

	Green 1			Green 2		
	Nitrate N concentration (mg/kg)					
	Mean	Min.	Max.	Mean	Min.	Max.
1995	2.23	0.01	13.61	1.14	0.01	8.80
1996	1.73	0.01	16.09	1.37	0.01	13.69
1997	2.63	0.01	12.99	2.60	0.42	8.67
	Nitrate N mass (mg)					
1995	3.73	0.01	33.90	1.63	0.01	30.09
1996	3.91	0.01	16.71	2.71	0.01	17.25
1997	7.88	0.02	36.96	6.55	0.07	33.65

Table VI. Phosphorus leachate concentration (mg/kg) and mass (mg) averages for collection dates, for three years for two USGA golf greens at the Cherokee Town and Country Club, Atlanta GA.

	Green 1			Green 2		
	Phosphate P concentration (mg/kg)					
	Mean	Min.	Max.	Mean	Min.	Max.
1995	3.21	0.65	6.07	8.53	5.55	13.27
1996	1.14	0.05	6.79	1.30	0.16	6.02
1997	0.93	0.05	5.34	1.72	0.15	4.11
	Phosphate P mass (mg)					
1995	7.06	0.04	23.03	22.34	0.06	77.04
1996	2.72	0.03	20.29	2.89	0.06	13.53
1997	2.90	0.16	13.14	4.41	0.10	15.75

1995 (26 kg ha⁻¹). This concentration would allow algae to grow even when diluted many times (*10*). The minimum for green 2 for 1995 was higher than the maximum in 1997. Thus, P levels were high during greens establishment and then decreased significantly thereafter. Nitrogen has been found to leach to a greater extent in a newly established green than afterward (*4*); thus, P may follow that trend, both because of lower uptake by the turf and higher relative application rates. The mass of P found was also high in 1995 and decreased over the other two years. Phosphorus added to the new green may have been more susceptible to leaching because of reduced uptake by the young turf and higher soluble organics in the new green mix, that could have exacerbated leaching. This latter process deserves further examination.

Acknowledgments

The authors thank Ray Pitts, Kathy Evans, Nehru Mantripragada and Hal Peeler for technical assistance, and Carole Layton for secretarial assistance. This research was supported by the United States Golf Association and by State and HATCH funds allocated to the Georgia Agricultural Experiment Stations.

Literature Cited

1. Anderson, J. L.; Balogh, J. C.; Waggoner, M. *Soil Conservation Service procedure manual: Development of standards and specifications for nutrient and pesticide management. Section I & II;* State Nutrient and Pest Management Standards and Specification Workshop, SUDA Soil Conservation Service July 1989: St. Paul, MN, 1989; pp 453.
2. Austin, N. R.; Prendergast, J. B.; Collins, M. D. *J. Environ. Qual.* 1996, *25*, 63-68.
3. Beard, J. B. *Turf management for golf courses;* Macmillan Publishing Co.: New York, NY.
4. Brauen, S. E.; Stahnke, G. K. *USGA Green Section Record*, 1995, *33*, 29-32.
5. Brown, K. W.; Duble, R. L.; Thomas, J. C. *Agron. J.* 1977, *69*, 667-671.
6. Brown, K. W.; Thomas, J. C.; Duble, R. L. *Agron. J.* 1982, *74*, 947-950.
7. Culley, J. L. B.; Bolton, E. F.; Bernyk, V. *J. Environ. Qual.* 1983, *12*, 493-498.
8. Dao, T.; Lavy, T.; Dragun, J. *Residue Reviews* 1983, *87*, 91-104.
9. Fry, J. D.; Harivandi, M. A.; Minner, D. D. *HortScience* 1989, *24*, 623-624.
10. Heckrath, G.; Brookes, P. C.; Poulton, P. R.; Goulding, K. T. *J. Environ. Qual.* 1995, *24*, 904-910.
11. Hong, S.; Lee, A.; Smith, A. *Journal of Chromatographic Science* 1992, *32*, 1-3.
12. Kaiser, K.; Zech, W. *J. Environ. Qual.* 1996, *25*, 1325-1331.
13. Lee, A.; Hong, M.; Smith, A. *Journal of the Association of Official Analytical Chemists International* 1995, *78*,1459-1464.
14. Leonard, R. A. In *Environmental Chemistry of Herbicides*; Grover, R. Ed.; CRC Press: Boca Raton, FL, 1988, Vol 1; pp. 45-87.
15. Mancino, C. F.; Troll, J. *HortScience* 1990, *25*, 194-196.
16. McGeehan, S. L.; Shafti, B.; Naylor, D. V.; Price, W. J. *Commun. Soil Sci. Plant Anal.* 1997, *28* 395-406.
17. National Academy of Sciences. *Contemporary pest control practices and prospects.* Vol. L. National Academy of Sciences: Washington, DC. 1975.
18. Neyland, J. *Golf & Sports Turf Australia* 1995, *Feb.*, 33-38.
19. Ohno, T.; Erich, M. S. *J. Environ. Qual.* 1997, *26*, 889-895.

20. Owens, L. B.; Edwards, W. M.; Van Keurch, R. W. *J. Environ. Qual.* **1994**, *23*, 752-758.
21. Petrovic, A. M. *Golf Course Management* **1989**, *Sept.*, 54-64.
22. Petrovic, A. M. *J. Environ. Qual.* **1990**, *19*, 1-14.
23. Petrovic, A. M. *USGA Green Section Record* **1995**, *33*, 38-41.
24. Sharpley, A. N. *J. Environ. Qual.* **1995**, *24*, 920-926.
25. Sharpley, A. N.; Chapra, S. C.; Wedephol ,R.; Sims, J. T.; Daniel, T. C.; Reddy, K. R. *J. Environ. Qual.* **1994**, *23*, 437-451.
26. Sharpley, A. N.; Menzel, R. G. *Advances in Agronomy* **1987**, *41*, 297-324.
27. Sharpley, A. N.; Smith, S. J.; Jones, O. R.; Berg, W. A.; Coleman, G. A. *J. Environ. Qual.* **1992**, *21*, 30-35.
28. Smith, A.; Tillotson, W. In *Pesticides in Urban Environments*; Racke, K., and Leslie, A., Eds. American Chemical Society: Washington, DC, 1993, pp 168-181.
29. Smith, A. E.; Bridges, D. C. In *Chemistry of Herbicide Metabolites in Surface and Ground Water*; Meyer, M. T., and Thurman, E. M., Eds.; American Chemical Society Press, 1996; pp. 165-177.
30. Smith, A.; Weldon, O.; Slaughter, W.; Peeler, H.; Mantri, N. *J. Environ. Qual.* **1993**, *22*, 864-867.
31. Starrett, S. K.; Christians, N. E.; Austin, T. A. *Journal of Irrigation and Drainage Engineering* **1995**, *121*, 390-395.
32. Stewart, B. A.; Woolhiser, D. A.; Wischmeier, W. H.; Caro, H.H.; Frere, M. *Control of water pollution from cropland: A manual for guideline development.* EPA 600/2-75-026a; SUEPA and USDA; Vol. L. U.S. Printing Office: Washington, D.C., 1975; 111 pp.
33. USGA Green Section Staff. In *Specifications for a method of putting green construction.* Bengeyfield, W. H., Ed.; U.S. Golf Assoc. 1989, 24 pp.
34. Weed, D. A.; Kanwar, R. S. *J. Environ. Qual.* **1996**, *25*, 709-719.

Chapter 6

Surface Runoff of Selected Pesticides Applied to Turfgrasses

T. L. Watschke[1], R. O. Mumma[2], D. T. Linde[3], J. A. Borger[1], and S. A. Harrison[4]

Departments of [1]Agronomy and [2]Entomology, The Pennsylvania State University, University Park, PA 16802
[3]Delaware Valley College, Doylestown, PA 18901
[4]Nittany Geoscience, State College, PA 16802

The potential for movement of pesticides in surface runoff from golf courses has brought about the need to determine the extent to which certain chemicals are transported by such means in addition to movement below the root zone. This study was conducted in 1992 and 1993 using a herbicide, an insecticide (applied only in 1993), and a fungicide applied at label rates to two turfgrasses maintained as golf course fairway turf. The turf species used were perennial ryegrass (Lolium perenne, L.) and creeping bentgrass (Agrostis stolonifera L. var. palustris (Huds.). Far.; syn. Agrostis palustris Huds.). The herbicide used was mcpp [(2-(2-methyl-4-chlorophenoxy)) proprionoic acid], the insecticide was isazofos (o-(5-chloro-1-(1-methyl-ethyl)-1H-1,2,4-triazol-3-yl)-o,o-diethyl phosphorothioate), and the fungicide was triadimefon 1-(4-chlorophenoxy)-3,3-dimethyl-1-(1H-1,2,4-triazol-1-yl)-2-butanone). The plots were sloped and irrigated at 152 min h-1 to force runoff within 24 hr of any pesticide application. Collection of runoff consisted of the following, first liter, second liter, and a composite that was collected at a rate of 16 ml min^{-1} throughout the runoff event. A leachate sample was collected at approximately 15 cm below the surface. In both years, residues of mcpp could be detected in the first two liters of runoff; but within two weeks, the concentration found was reduced by 99.5 percent. In 1993, after the turf was more mature, the concentrations found in first two liters were substantially lower and by the end of three weeks no residue could be detected.   Analyses for triadimefon also included the analyses for tridimenol, a metobolite. Residues of both the parent compound and the metabolite were found on all sampling dates in 1992. One day after treatment, residues ranged from 46 to 922 μg L^{-1}, but within four weeks, concentrations dropped by as much as 80 percent. By six weeks, the concentration of the metabolite exceeded that of the parent and by October the detection of both compounds was below 5 μg L^{-1}. In 1993 substantially lower concentrations were found (less than 17 μg L^{-1} for both compounds even in the first liter sampled). After 4 weeks, no residues were found in excess of 4 μg L^{-1}. Isazofos was applied only in 1993 and residues were found only for samples collected the first day after application. Concentrations of isazofos and tridimefon/ tridimenol in leachate were found to be very similar to the composite runoff samples (both were substantially lower than the concentrations found in the first two liters of runoff). The concentration of mcpp in the leachate was lower than for any other sampling point. These results suggest that certain pesticides applied to sloped plots of turfgrass can be transported in surface runoff when irrigation is applied more heavily than the norm within 24 hr of the pesticide application. Surprisingly, considering the irrigation rate, the level of detection for all compounds was very low even in the first two liters of runoff. Further, the application of pesticides to the turf was made within 15 cm of the weir collector for the runoff water.

The potential transport of pesticides applied to golf courses in surface runoff has been identified as a concern for the environment, particularly in areas where the land surrounding the golf course is ecologically fragile.   Limited research has been conducted with regard to determining pesticide transport from turfgrass areas specifically grassed and managed similar to a golf course fairway. The research that has been conducted has focused primarily on nutrient and sediment movement, the characterization of runoff from turf, and the differences in runoff between sodded versus seeded turf (Bennett, 1979; Gross et al., 1990, 1991; Harrison et al., 1993; Morton et al., 1988; Watson, 1985; Linde et al., 1994, 1995).   These studies consistently found that the volume of runoff from properly managed turfgrass sites was low (even when excessive irrigation was applied) and that sediment and nutrient transport was also very low (below public drinking water standards).

Although previous research has not found significant nutrient transport in runoff, it cannot be assumed that pesticides would follow a similar pattern of movement.   Previous research has also consistently found low concentrations of nutrients in leachate with the exception of the downward movement of nitrate from soluble sources applied to sandy soil conditions in conjunction with excessive irrigation. Limited research has been conducted to determine the leaching potential of pesticides applied to turf.   Bowman (1990) and Petrovic (1994) measured the downward migration of isazofos in different soil textural types and found that soils comprised of coarser particles allowed for greater migration.

The objective of this study was to determine the concentration in runoff and leachate water of three pesticides applied to slopes grassed and maintained like golf course fairways.

Materials and Methods

This experiment was conducted at the turfgrass runoff facility located at The Pennsylvania State University's Landscape Management Research Center on the University Park campus. These facilities have been characterized by Harrison et al. (1993) and more recently by Linde et al. (1995) which included the general maintenance practices that were used on these plots.

Water Collection Procedures. Water samples were collected from several runoff events, which were forced by irrigation at a rate of 152 mm hr^{-1}. The initial runoff event was typically within 24 hr of the pesticide application. The duration of irrigation varied due to environmental conditions and species being irrigated (runoff occurred more quickly and at greater volumes from perennial ryegrass than from creeping bentgrass). The difference in runoff between perennial ryegrass and creeping bentgrass has been reported and characterized by Linde et al. (1995). The sampling program used was as follows; runoff, first liter (#I), runoff, second liter (#II), runoff, composite 16 ml min^{-1} (#III), and a leachate composite that was collected from the sample collector at the conclusion of the irrigation event. The leachate sample was a composite from subsurface sample collection devices in four locations/plot. The timing of the various pesticide applications and the subsequent irrigation events can be found in Tables I and II.

Table I. Treatment schedule and watering/sampling events of turfgrass test plots in 1992.

Treatment			Watering		
Date	Chemical	Rate	Date	Type	Samples
		L ha^{-1}			
May 5	Fertilizer				
June 9	Mecoprop	0.78	June 10	Irrigation	28
June 23	Trichlorfon	1.44	June 24	Irrigation	28
July 14	Triadimefon	1.04	July 15	Irrigation	28
Aug 18	Triadimefon	1.04	Aug 19	Irrigation	28
Sept 8	Fertilizer		Sept 9	Irrigation	28
Oct 6	Fertilizer		Oct 14	Irrigation	28

Pesticide Application Procedures. All pesticides were applied at label recommended rates using a conventional boom-type sprayer calibrated to deliver a volume of 30.4 g ha^{-1} (Tables I and II). Each of the six slopes was entirely treated with the exception of a 15-cm band located directly adjacent to the weir. This 15-cm band was not treated to insure that drift from the treatment being applied would not come in direct contact with the weir, thus bypassing foliar interception by the turf.

Table II. Treatment schedule and watering/sampling events of
turfgrass test plots in 1993.

	Treatment			Watering	
Date	Chemical	Rate	Date	Type	Samples
		L ha^{-1}			
May 11	Dithiopyr				
May 25	Fertilizer				
June 8	Mecoprop	0.78	June 9	Irrigation	28
June 29	Isazofos	0.78	June 30	Irrigation	28
Aug 17	Triadimefon	1.04	Aug 18	Irrigation	28
Sept 7	Fertilizer		Sept 8	Irrigation	28
Sept 20	Trichlorfon	1.96	Sept 21	Irrigation	28
Oct 5	Fertilizer		Oct 6	Irrigation	28

Pesticide Analysis of Water Samples.

Mecoprop.

Solid Phase Extraction. Sample preparation for HPLC analysis was done using Baker High Capacity (HC) C18 solid phase extraction (SPE) cartridge (J.T. Baker; Phillipsburg, NJ). The cartridge was attached to a Visiprep 12 port SPE Vacuum Manifold (Model 5-7030; Supelco Inc., Bellefonte, PA) via the outlet end. A 75-mL polypropylene reservoir was inserted onto the inlet end of the SPE column. The SPE was conditioned by adding approximately 18 mL of methanol to the reservoir and letting the methanol percolate through the SPE by gravity until droplets formed at the manifold outlet. Then, a draw of 101 mm of mercury was applied to remove almost all the methanol before 10 mL of distilled-deionized water was added to the reservoir and drawn through with vacuum as above to almost dry. The sample was then added to the reservoir and a vacuum of 254 mm of mercury was applied and the flow rate through the individual sample SPE tube was adjusted to approximately 6-8 mL per minute.

The water was prepared for extraction by measuring out up to 200 mL of field sample that had been thoroughly mixed by shaking into an appropriate sized beaker and acidified to pH of 2.5-2.8 with trifluoroacetic acid (TFA). The sample water was then filtered through Whatman GF/F glass membrane filters (0.7 mm pore size) and added to the SPE column reservoirs. The vacuum was set at 178-254 mm of mercury and flow adjusted to 5-8 ml min^{-1}. After all the sample was eluted through the SPE, the SPE was rinsed with an additional 10 mL of distilled-deionized water. The SPE then was dried with a vacuum draw of 432 mm of mercury for 30 minutes. Additional clean up was performed by adding 4 X 500 μL of acetonitrile: water (50:50 v/v) with the pH adjusted to <2.8 with TFA and eluted under vacuum. MCPP was then eluted off the SPE in 4 X 500 μL of acetonitrile and the final volume adjusted to 2 mL with the pH <2.8 (50:50) acetonitrile: water.

Description of HPLC Apparatus and Chromatographic Conditions. The samples were analyzed on a Water's HPLC system, using a UV-Vis. Model 1481 variable wavelength detector at 230 nm, a Model 590 pump with a Model 712A auto sampler in line (Water's Co.; Milford, MA.). The analytical column used was a LC-8,

4.6 mm ID X 25 cm column (Supelco Inc; Bellefonte, PA). Data were collected and chromatograms generated using a Shimadzu Chromatopac C-R4AX data processor (Shimadzu Scientific Inst. Inc.; Columbia, MD). The mobile phase used was acetonitrile:water (50:50) with the pH adjusted to <2.8 using concentrated trifloroacetic acid. Column flow was 1.0 ml min⁻¹, the injection volume was 25 μL and the system back pressure was 750-800 psi. The detection limit was 1 μg L⁻¹.

Triadimefon and Triadimenol.

Extraction of Triadimefon and Triadimenol from runoff water. A liquid/liquid partitioning procedure was used to extract triadimefon and its metabolite, triadimenol, from the water matrix. To an 80-mL aliquot of aqueous sample was added 20 mL of saturated sodium sulfate solution, and this solution was extracted 3 X with 5 mL of methylene chloride. The combined methylene chloride extract was dried by elution through 10 g sodium sulfate contained in Whatman 2V filter paper. The methylene chloride was then evaporated to dryness under nitrogen and the residue re-dissolved in 3 X 2 mL of ethyl acetate with the ethyl acetate extracts combined. The ethyl acetate was evaporated to 1 mL with nitrogen and transferred to a septum vial for GC analysis.

GC Analysis. All analyses were carried out using capillary chromatography on a Varian Model 3400 CX GC equipped with a split/splitless injector, a Thermionic Specific Detector (NPD), and a Model 8200 Auto Sampler (Varian Instruments; Walnut Creek, CA). The capillary column used was a Rtx-35; 60 meter X 0.32 mm ID, 0.25 um film thickness (Restek; Bellefonte, PA). Helium was the carrier and make-up gas at 3 and 27 ml min⁻¹, respectively. Detector gases were hydrogen at 4.5±0.5 ml min⁻¹ and air at 175 mL min. The operating temperatures were: injector 250°C, detector 290°C, the oven was temperature programmed to hold at 90°C for 2 min after injection, ramp at 10°C min⁻¹ to 210°C, hold for 6 min, ramp at 15°C min⁻¹ to 285°C and hold for 4 min. The injector was set in the splitless mode for 1.5 min and then in the split mode for the rest of the GC run at a 50:1 split. The retention time under the above conditions for triadimefon was 6.75 and for triadimenol was 7.29 min. The sample injection volume was 2.0 μL. The detection limit was 1 μg L⁻¹.

Isazofos.

Extraction of Isazofos from Water. A 100-mL water sample was buffered (0.05M monobasic potassium phosphate and 0.002M sodium hydroxide) to a pH of 7 and then filtered through a Whatman GF/F glass membrane filter. Isazofos was extracted out of the water and concentrated on a methanol-activated 3-mL BakerBond C18 SPE cartridge in the same manner as previously described. After the sample passed through the SPE cartridge at a rate at 6-8 ml min⁻¹, the SPE tube were rinsed with 10 mL of distilled-deionized water and dried under vacuum for 15 min. The isazofos was eluted off the SPE tube in 3 mL of ethyl acetate and any excess water was removed with a disposable pipette. The eluant was further dried by adding 1.0 g of sodium sulfate to the adjusted 3 mL of ethyl acetate and vortexing. Approximately 1.0 mL of the ethyl acetate was decanted off into a septum sealed GC vial for analysis.

GC Analysis. All analysis were carried out using essentially the same GC conditions as with tridimefon except the oven was temperature programmed to hold at 70°C for 2 min after injection, ramp at 20°C/min to 210°C, hold for 4 min, ramp at 30°C/min to 280°C, and hold for 3 min. The retention time under the above conditions for isazofos was 15.93 min for a μg L^{-1} detection.

Results and Discussion

Residues of mcpp were found in samples collected within 24 hr of application in 1992 and 1993. In 1992, the first and second liters of runoff (I and II) had residues ranging from 4701 to 792 μg L^{-1}, however, two weeks later, these two samples had a 99.5 percent reduction in residue concentrations. Two months after application, no residues could be detected. In the more mature turf in 1993, residues were again detected in the samples taken within 24 hr of application, but were substantially lower than in 1992 (range from 102 to 167 μg L^{-1}) (Tables III and IV). Three weeks after the 1993 application, no residues could be detected in any sample. In 1992, the mcpp concentration in the first two liters of runoff from perennial ryegrass was considerably higher than from creeping bentgrass. Linde et al. (1995) reported that runoff from the perennial ryegrass plots occurred more quickly after the initiation of irrigation and at greater overall volume than from the creeping bentgrass. By 1993, the perennial ryegrass plots had reached maximum density (Linde et al. 1995) and hydrographs that were produced indicated that the magnitude of difference in the time lapse from the initiation of irrigation to runoff between the two species was not as great. The concentrations of mcpp in the first two liters of samples in 1993 were not found to be appreciably different between perennial ryegrass and creeping bentgrass. Leachate samples for 1992 showed higher residues in water collected under creeping bentgrass, but this phenomenon was not found for leachate samples analyzed in 1993. A rainfall event occurred on June 19, 1992 (nine days after application) which produced some runoff, but it did not contain detectable mcpp residues.

Triadimefon was first applied in July 1992. Samples were analyzed for triadimefon and for triadimenol, a metabolite. Resides of both parent and metabolite were found at all sampling times in 1992 (Table V). Samples collected within 24 hr of application had residues that ranged from 46 to 922 μg L^{-1}. After 4 wk, the residues detected decreased substantially (from 20 to 80 percent). The metabolite triadimenol, tended to be found in higher concentration than the parent in samples collected from a runoff event caused by natural precipitation on August 28, 1992. This trend continued in samples collected for a forced runoff event on September 9, 1992. By October, the concentration of residues for both the parent and the metabolite were less than 4 μg L^{-1}. The concentration in leachate samples tended to ˋ be very similar to those found in the composite runoff samples.

Table III. Residues (μg L^{-1}) of Mecoprop from runoff or leached water from sloped plots of creeping bentgrass and perennial ryegrass. Mecoprop (4S) applied at 0.78 k ha^{-1} of product on 6/9/92.

| Date | Sample | Turf | |
		Perennial Ryegrass	Creeping Bentgrass
6/10/92	Runoff #I	4101[a]	1128
	Runoff #II	3066	793
	Runoff #III	854	380
	Leachate	384	606
6/24/92	Runoff #I	26	11
	Runoff #II	17	4
	Runoff #III	7	14
	Leachate	11	20
8/19/92	Runoff #I	0	0
	Runoff #II	0	0
	Runoff #III	0	0
	Leachate	0	0

[a] Mean of three slopes in μg L^{-1}.

Table IV. Mean residues (μg L^{-1}) of Mecoprop from runoff or leached water from sloped plots of creeping bentgrass and perennial ryegrass. Mecoprop (4S) applied at 0.78 k ha^{-1} of product on 6/8/93.

| Date | Sample | Turf | |
		Perennial Ryegrass	Creeping Bentgrass
6/9/93	Runoff #I	142[a]	168
	Runoff #II	103	127
	Runoff #III	169	107
	Leachate	177	185
6/30/93	Runoff #I	0	0
	Runoff #II	0	0
	Runoff #III	0	0
	Leachate	0	0
8/18/93	Runoff #I	0	0
	Runoff #II	0	8
	Runoff #III	16	0
	Leachate	12	0
9/8/93	Runoff #I	0	0
	Runoff #II	0	0
	Runoff #III	0	0
	Leachate	0	0
9/21/93	Runoff #I	0	0
	Runoff #II	0	0
	Runoff #III	0	0
	Leachate	0	0
10/6/93	Runoff #I	0	0
	Runoff #II	0	0
	Runoff #III	0	0
	Leachate	0	0

[a] Mean of three slopes in μg L^{-1}.

Table V. Residues (μg L⁻¹) of Triadimefon/Triadimenol from runoff or leached
water from sloped plots of creeping bentgrass and perennial ryegrass.
Triadimefon (25WDG) applied at 1.04 L ha⁻¹ of product on 7/14/92 and 8/18/92.

Date	Sample	Turf			
		Perennial Ryegrass		Creeping Bentgrass	
		Triadimefon	Triadimenol	Triadimefon	Triadimenol
7/15/92	Runoff #1	922.1 [a]	146.2	395.5	46.4
	Runoff #2	643.6	63.9	380.1	59.3
	Runoff #3	359.3	115.2	299.2	68.5
	Leachate	326	53.1	213.2	62.1
8/19/92	Runoff #1	132	26.9	74.9	13.1
	Runoff #2	273.5	47.5	86.6	13.4
	Runoff #3	200.6	19.3	60	18.1
	Leachate	221	41.6	65.8	9.3
8/28/92 [b]	Runoff #1	12.8	56.6	3.8	19.4
	Runoff #2	3.2	58.1	--	--
	Runoff #3	21.7	78.5	22.8	24.7
	Leachate	17.2	52.9	9.1	28.1
9/9/92	Runoff #1	0.8	14.2	0	8.2
	Runoff #2	7	16.1	2.7	6.9
	Runoff #3	0.6	16.3	0	5.3
	Leachate	0.9	25.1	0	6.4
10/14/92	Runoff #1	0	3	0	1.2
	Runoff #2	0	1.9	2.4	0
	Runoff #3	0	2.1	3.8	0
	Leachate	1.3	4	0	0

[a] Mean of three slopes in μg L⁻¹.
[b] Rainfall produced runoff.

In 1993, less residues were found than in 1992 in all samples at all locations
(Table VI). For example, no runoff samples (even the first liter of runoff) had
residues in excess of 17 μg L⁻¹. In samples taken approximately 3 weeks after
application, no runoff samples contained residues greater than 3 μg L⁻¹. Two months
after application, no residues of either the parent or the metabolite could be detected.

Isazofos was applied only in 1993 and residues were found only in samples
collected within 24 hours of application (Table VII). Since isazofos residue samples
are not freezer stable, these samples were all analyzed within one month of sampling.
The concentration of isazofos detected did not appear to be influenced by the species
onto which it was applied. Clearly, for those pesticides applied in this study, the
residues of isazofos dissipate more rapidly than for the others. Although,

Table VI. Residues (μg L^{-1}) of Triadimefon/Triadimenol from runoff or leached water from sloped plots of creeping bentgrass and perennial ryegrass. Triadimefon (25WDG) applied at 1.04 L ha^{-1} of product on 8/17/93.

Date	Sample	Turf			
		Perennial Ryegrass		Creeping Bentgrass	
		Triadimefon	Triadimenol	Triadimefon	Triadimenol
8/18/93	Runoff #1	14.3 [a]	19	8.4	8.3
	Runoff #2	12.8	11.8	10	9.4
	Runoff #3	16.9	15.1	10.3	8.1
	Leachate	179.1	43.8	12.7	6.4
9/8/93	Runoff #1	1.1	2.7	1.2	.6
	Runoff #2	0.6	1.1	0	0
	Runoff #3	0	1.8	0	0.7
	Leachate	7.2	20.7	0	1.2
9/21/93	Runoff #1	0	1.1	3.4	0
	Runoff #2	0.5	1.5	0	0
	Runoff #3	0	2.1	0	0
	Leachate	0	0	0	0.5
10/6/93	Runoff #1	0	0	0	0
	Runoff #2	0	0	0	0
	Runoff #3	0	0	0	0
	Leachate	0	0	0	0

[a] Mean of three slopes in μg L^{-1}.

it cannot be assumed that isazofos would have dissipated as rapidly in 1992, as the other pesticides that were applied in 1992 could be detected longer than when they were applied in 1993. In another study (Petrovic et al., 1994) conducted to determine the leachability of isazofos for different soil textural types, they found leaching to be greater for sandy textural types than for textural types containing more fine particles. In a study conducted by Bowman (1990) to study the leaching and persistence of isazofos on different soils, it was found that there was a 50 percent disappearance time of 1.5 wk in a silt loam soil.

Conclusions

It appears that mcpp, triadimefon, and isazofos applied properly to established perennial ryegrass and creeping bentgrass managed as fairway-type turf can be found in very small amounts in runoff that has been forced with a very high irrigation rate (152 mm hr^{-1}). This irrigation is far in excess of any rate being used on any golf course. Over time, as turfgrass stands matured, it appeared that surface transport of the compounds used in this study decreased in their potential to migrate in surface

Table VII. Residues (μg L^{-1}) of isazofos from runoff or leached water from sloped plots of creeping bentgrass and perennial ryegrass. Isazofos (4EC) applied at 0.78 L ha^{-1} of product on 6/29/93.

| Date | Sample | Turf | |
		Perennial Ryegrass	Creeping Bentgrass
6/30/93	Runoff #I	678[a]	507
	Runoff #II	667	540
	Runoff #III	470	514
	Leachate	119	431
8/18/93	Runoff #I	0	0
	Runoff #II	0	0
	Runoff #III	0	0
	Leachate	0	0
9/8/93	Runoff #I	0	0
	Runoff #II	0	0
	Runoff #III	0	0
	Leachate	0	0
9/21/93	Runoff #I	0	0
	Runoff #II	0	0
	Runoff #III	0	0
	Leachate	0	0
10/6/93	Runoff #I	0	0
	Runoff #II	0	0
	Runoff #III	0	0
	Leachate	0	0

[a] Mean of three slopes in μg L^{-1}.

water. Clearly, the conditions used in this study (extremely high irrigation rate, collection and analysis of the first two liters of runoff typically within 24 hr of application, and application of the pesticide to within 15 cm of the weir) would not reflect actual practical circumstances. However, it is significant that turfgrass systems appear to have a very large capacity to retain and degrade compounds that are applied to them. Obviously, more research is needed on substantially more pesticides to continue to document their persistence and concentration in the water that leaves turfgrass sites.

References

1. Bennett, O. L. 1979. *Conservation*. In *Tall fescue*. Editors, R. C. Buckner and L. P. Bush. Agron. Monogr. 20. ASA, CSSA, and SSSA, Madison, WI. pp 319-340.

2.   Bowman, B. T. 1990. Environ. Toxicol. Chem. 9:453-461.
3.   Gross, C. M., J. S. Angle, R. L. Hill, and M. S. Welterlen. 1991. *Runoff and sediment losses from tall fescue under simulated rainfall.* *J. Environ. Qual. 20*, pp 604-607.
4.   Gross, C. M., J. S. Angle, and M. S. Welterlen. 1990. *Nutrient and sediment losses from turfgrass.* *J. Environ. Qual. 19*, pp 663-668.
5.   Harrison, S. A., T. L. Watschke, R. O. Mumma, A. R. Jarrett, and G. W. Hamilton. 1993. *Nutrient and pesticide concentrations in water from chemically treated turfgrass.* In *Pesticides in urban environments: Fate and significance.* Editors, K. D. Racke and A. R. Leslie. ACS Symp. Ser. 522. pp 191-207.
6.   Linde, D. T., T. L. Watschke, A. R. Jarrett, and J. A. Borger. 1994. *Nutrient transport in runoff from two turfgrass species.* In *Science and Golf II: Proc. 1994 World Sci. Congr. of Golf.* Editors, A. J. Cochran and M. R. Farrally. E. & F.N. Spon, London. pp 489-496.
7.   Linde, D. T., T. L. Watschke, A. R. Jarrett, and J. A. Borger. 1995. *Surface runoff assessment from creeping bentgrass and perennial ryegrass turf. Agron. J. 87*, pp 176-182.
8.   Morton, T. G., A. J. Gold, and W. M. Sullivan. 1988. *Influence of overwatering and fertilization on nitrogen losses from home lawns.* *J. Environ. Qual. 17*, pp124-130.
9.   Petrovic, A. M., R. A. Young, C. A. Sanchirico, and D. J. Lisk. 1994. *Migration of isazofos nematocide in irrigated turfgrass soils.* *Chemosphere 28*, pp 721-724.
10.   Watson, J. R., Jr. 1985. *Water resources in the United States.* In *Turfgrass water conservation.* Editors, V. A. Gibeault and S. T. Cockerham. Div. of Agric. and Nat. Resources, Univ. of California, Riverside, CA, pp 19-36.

Chapter 7

Mobility and Persistence of Pesticides Applied to a U.S. Golf Association Green

Pesticides in Percolate, Thatch, Soil, and Clippings and Approaches to Reduce Fenamiphos and Fenamiphos Metabolite Leaching

J. L. Cisar[1] and G. H. Snyder[2]

[1]Environmental Horticulture Department, University of Florida, Fort Lauderdale Research and Education Center, Fort Lauderdale, FL 33314
[2]Soil and Water Sciences Department, University of Florida, Everglades Research and Education Center, Belle Glade, FL 33340

The public is concerned about the impact of golf course agrichemical applications on the environment. From 1991-1994, the persistence and mobility of fenamiphos, fonofos, chlorpyrifos, izasophos, isofenphos, ethoprop, 2,4-D, and dicamba applied to a USGA green were determined. In most cases less than 1% of the organophosphates were found in clippings and with one exception, less than 0.1% of the organophosphates leached. Fenamiphos, and especially it's sulfoxide and sulfone metabolites, were sufficiently mobile to be observed in percolate waters. These compounds have also been detected in groundwater and surface waters in and around golf courses. Through 1997, three Best Management Practice approaches were investigated to reduce leaching and/or to improve efficacy of fenamiphos: 1) irrigation management, 2) surfactants, and 3) absorbents. A patent-pending cross-linked phenolic polyethylene adsorbent reduced fenamiphos and fenamiphos metabolite leaching.

Pesticides are applied to golf course greens to help provide a suitable aesthetic and playing performance surface for golf. There is a great deal known about pesticide efficacy in turfgrass systems (*1*). Yet, in comparison to other agricultural commodities, there is little information available in the literature that documents the environmental fate of pesticides applied to turfgrass (*2*).

There is increased public awareness of human health and environmental issues related to the potential impact of pesticide applications to turfgrass systems (*2-4*). Public and regulatory scrutiny has not solely focused upon potential misapplication of pesticides, rather, concerns have frequently been directed at the risks associated with routine or label use of pesticides in turfgrass (3-4). These concerns cannot be properly addressed because of a lack of research data about the fate of pesticides applied to turfgrass under routine maintenance conditions. Thus, from both a public and turfgrass manager's perspective, it is critical that research be conducted to assess various aspects of the environmental fate of pesticides in turfgrass, and particularly that occuring under routine or representative standard field conditions. One standard often accepted by golf courses is the United States Golf Association (USGA) putting green (5).

Pesticide mobility and persistence in putting greens is influenced by the composition of the soil used for the green construction. Since 1960, the USGA has provided recommended specifications for construction of largely sand-based putting greens in order to provide a predictable and acceptable putting surface (6). The green is designed to withstand heavy traffic under adverse climate conditions. The USGA green drains well, yet provides an acceptable level of water-holding capacity and a favorable root-zone depth. Initially, the USGA specifications recommended 20% (volume) organic matter, which should be helpful for retaining pesticides. However, in a humid climate considerable percolation can occur in this green, especially if irrigation is used excessively. Therefore, with regard to pesticide retention, the USGA green has both favorable and unfavorable characteristics. The potential effect of these characteristics on retention of standard pesticides applied to USGA needs to be determined.

Organophosphate (OP) pesticides are used for control of nematodes and insects and phenoxy and benzoic acid herbicides are used for weed control on putting greens throughout the United States. In Florida alone, annual expenditures for insecticides and herbicides constitute over 50% of the average golf course pesticide budget (7).

In 1991, the United States Golf Association initiated a large-scale environmental research program to ascertain the environmental impact of golf course agrichemical applications. Over a three year period, a series of research studies were conducted on a USGA green in sub-tropical south Florida to determine the mobility and persistence of pesticides applied to putting green. The primary objectives of these experiments were to determine the mobility and persistence of six OP pesticides and two herbicides (8-12). In those studies, pesticide in clippings, thatch, soil, and percolate water were obtained over several application cycles. Of all the pesticides applied, the OP pesticide fenamiphos and its' sulfoxone and sulfoxide metabolites had the greatest amount of leachate (9). In this symposium chapter, the results of the mobility and persistence studies from 1991 through 1994 are summarized and work conducted from 1994 through 1997 on best management practices to reduce fenamiphos leaching via irrigation management, and fenamiphos leaching via surfactants, and adsorbents, are presented.

Materials and Methods

Research site description. The mobility and persistence studies were conducted on a portion of a USGA Tifdwarf bermudagrass (*Cynodon dactylon* X *C. transvaalensis)* green built in 1990, at the University of Florida, Ft. Lauderdale Research and Education Center (FLREC) that was initially outfitted with six lysimeters for collection of percolate waters. The facility was expanded to twelve lysimeters in 1997 for absorbent–amended soil profile experiments (see "Pesticide Absorption with CPP" section). The green was mowed daily at a 4 mm mowing height with a walk-behind reel-type mower, with all clippings removed. At a 5 cm soil depth, the green root-zone mix had a dry bulk density of 1.78 g cm^{-3}, a predominant (59.3%) coarse (0.5 - 1.0 mm) sand size diameter, (18.2%) medium (0.50 - 0.25 mm) sand size diameter, fine sand (0.25 0.10 mm) size diameter, very low organic matter content (6.9 g kg^{-1}), a saturated hydraulic conductivity of 114 cm hr^{-1}, and a 45 cm water retention capacity of 5.4% (v/v). Complete physical analysis of the green soil profile has been previously reported (8).

Pesticide Applications for Mobility and Persistence Studies. Pesticides were applied at label rates. Fenamiphos, fonofos, and chlorpyrifos were applied to the green as a 10G, 5G, and 1G material, respectively at 1.13 g A. I. m^{-2}, 0.44 g A. I. m^{-2}, 0.117 g A. I. m^{-2}, respectively on 27 Jan. 1992. Chlorpyrifos was applied again on 21 April 1992 as a 2E liquid at 0.229 g A. I. m^{-2}, along with isazophos at 0.229 g A.I. m^{-2} and isofenphos at 0.229 g A.I. m^{-2} (Table II). All liquid formulations were diluted in water to provide a liquid application rate of 27.7 ml m^{-2}. The latter two materials, along with ethoprop at 2.245 g A.I. m^{-2}, were applied on 15 Sept. 1992. 2, 4-D and dicamba were applied on 3 and 10 Aug 1993 and again on 18 and 24 April 1994 as sprays at the per -application rates of 0.058 and 0.006 g A. I. m^{-2}, for 2,4-D and dicamba, respectively. Thatch, soil, percolate, and clipping samples were taken and processed as described in the following sections.

The turf foliage was dry at the time of each pesticide application. Following OP application on each date, the green was irrigated within 15 min to provide approximately–18 mm of water over a 22-minute period. Irrigation was not applied immediately after herbicide applications. Thereafter, routine irrigation and all other maintenance operations were controlled by an employee of the South Florida and Palm Beach Golf Superintendents Association, following instructions obtained from a committee of golf course superintendents. The employee irrigated frequently to replace evapotranspiration, which when combined with rainfall, produced appreciable percolate (Table II). Other than pesticide applications on treatment dates, pesticides were not applied to the experimental site.

Pesticide Recovery from Thatch and Soil. Approximately 3:00 P.M. on the afternoon of each application date, soil and thatch samples were taken in certain studies from the treated area using a 1.9 cm-diameter nickel-plated soil corer. Three sets of samples, each comprised of three thatch-soil cores composited over treated areas adjacent to two of the six lysimeters, were collected and divided into the thatch, 0-5, 5-10, and 10-15 cm soil depths. The samples (three core sections each) were placed in 475-ml wide-mouth glass jars closed with metal lids. Thatch and soil samples were similarly collected the day after pesticide application, and several more times during the first week. Thereafter for each application, samples were taken at least weekly through an application cycle (Table II). Soil samples were placed in a freezer (-20 °C) until they could be transported (generally within one week of sampling) on ice to a similar freezer in the Pesticide Laboratory at the University of

Florida, Everglades Research and Education Center (EREC) located in Belle Glade, FL, approximately 60 km distance from the FLREC. In the laboratory, OP pesticides in the thatch and soil samples were extracted with a sulfuric acid-methanol mixture, followed by a methylene chloride extraction. Dicamba and 2, 4-D samples were mixed with 53 g NaCl and 10 ml 2.25 M H_2SO_4 and then extracted with a 70:30 hexane:ether extraction solution.

Pesticide Recovery in Percolate. Percolate samples were collected from six previously-installed stainless-steel lysimeters (buried 10 cm below the surface of the green) following each pesticide application, which covered all six lysimeter locations and adjacent areas. The lysimeters were 36 cm inside diameter and 40 cm tall. A complete description of the lysimeters and off-site percolate collection building has been reported (8). Percolate samples generally were collected from the lysimeters twice weekly; more often during rainy periods and less often during dry periods when no percolate was expected. OP pesticides were extracted from percolate samples within 24 hours of sampling (generally on the same day) with methylene chloride, following saturation with NaCl. Herbicides were mixed with 53 g NaCl and 10 ml 2.25 M H_2SO_4 and then extracted with a 70:30 hexane:ether extraction solution. The extracts were stored in a freezer until they could be transported on ice to the Pesticide Labatory for analysis.

Pesticide Recovery fromClippings. Starting on the day after application of pesticides, a clipping sample was retained from each mowing of the treated area. A 20 g (fresh weight) portion of each sample was placed in a 475-ml wide-mouth jar and frozen until it could be transported on ice to the Pesticide Laboratory for analysis. The balance of the sample was weighed, oven dried (60 °C), and re-weighed to determine the moisture content. The sample area was noted (generally 23.47 m^2). In the laboratory, the clipping sample was macerated three times in a blender in the presence of a methanol-sulfuric acid solution and filtered. The combined OP filtrates were extracted three times with methylene chloride. Herbicide filtrates were extracted with a 70:30 hexane:ether solution.

    The extracts of thatch, soil, percolate, and clippings were concentrated in a vacuum-evaporator and analyzed by gas chromatography (Hewlett-Packard 5890A). All sampling and analytical procedures were carried out according to a quality assurance/quality control plan approved by a quality control officer appointed by the United States Golf Association, who submitted validation control samples to verify analytical procedures. Recovery of OP insecticides from sample matrix blanks fortified to provide 50% of full-scale signal routinely exceeded 95% during method development. Phenoxy acid recovery exceeded 90%. Pesticide amounts were calculated as weight of pesticide per unit of land surface area, and as weight of pesticide per weight of plant tissue in the case of clippings.

Fenamiphos Irrigation Management Study. An irrigation management study was undertaken on the prviously described USGA green to evaluate the hypothesis that reducing water percolation through prudent irrigation management following fenamiphos application could reduce pesticide leaching.

On 7 June, 1995 fenamiphos was applied as a 10G material at 1.13 g A. I. m^{-2}. Areas of one square meter centered over three lysimeters received irrigation by hand using a sprinker can twice a week at 12.5 mm per application (Low Irrigation), and three other areas recieved the same amount of irrigation on a daily basis (High Irrigation). The Low Irrigation plots were irrigated on 7, 10, and 13 June, and the High Irrigation plots were irrigated each day. For the first six days of the study, the plots were covered with plywood at night to prevent irrigation by the system used to irrigate the remainder of the green, and when rain appeared imminent. Thereafter, the differential irrigation treatments and rainfall protection were suspended and all plots receiving irrigation from the commercial irrigation system. Percolate was withdrawn from the lysimeters daily for the first week after fenamiphos application, and at 3 and 5 day intervals for the next two weeks.

The experiment was repeated beginning 16 Jan. 1996. Fenamiphos was applied in the above-described manner, except that lysimeters that were used for the Low Irrigation plots in 1995 were used for the High Irrigation plots in 1996 and lysimeters used for the High Irrigation plots in 1995 were used for the Low Irrigation plots in 1996. The low irrigation plots were irrigated at 12.5 mm per application on 19, 23, and 25 Jan., and the High Irrigation plots were irrigated at the same rate daily. Following the 25 Jan. application, all plots received routine irrigation from the commercial sprinkler system, and 52 mm of rainfall and irrigation on 29 Jan., 1996.

Surfactants to Improve Fenamiphos Efficacy and Distribution. Two studies were conducted to determine whether a commercially-available surfactant would enhance fenamiphos efficacy, and increase fenamiphos penetration through thatch. Both studies were conducted on a native sand soil (Hallandale fine sand, a Lithic Psammaquent) area planted in 'Tifgreen' bermudagrass (*Cynodon dactylon* X *C. transvaalensis*) adjacent to the USGA green, rather than on the green itself because the adjacent area had a high nematode population. Dr. Robin Giblin-Davis, nematologist at the FLREC, cooperated in the study.

Prior to the initiation of the studies, nematode precounts were conducted over the intended plot area on a 2 m^2 grid. Twelve 2.5 cm diameter by 10 cm deep cores were taken from each grid unit and mixed thoroughly. A 100 cm^3 subsample was extracted and read for nematode numbers. Grid units with less than 25 sting nematodes per 100 cm^3 were not used as plots in the study. In the first study, treatments were arranged in a randomized complete block design having six replications, with blocks being assigned on the basis of pretreatment counts of <u>Belonolaimus</u> <u>longicaudatus</u> (sting) nematodes. The second study was conducted in a similar manner, but used only 4 replications.

Where used, fenamiphos was applied to 2 x 2 m plots between 10:30 am and 12:30 pm on 3 April, 1995, as a liquid at 3.08 ml m^{-2} (1.1 g A.I. m^{-2}), applied in 1 L water m^{-2}. Each surfactant (SA) treatment consisted of applying "Primer" at 1.91 ml m^{-2} in 81.3 ml water m^2.

In the first study (SA Study 1), the following surfactant/fenamiphos treatment combinations were investigated:

1. Control: no treatment.
2. SA 1: SA applied 25 days prior to the date of fenamiphos application.
3. SA 2: SA applied just prior to fenamiphos application.
4. SA 1+2: Plots received both SA 1 and SA 2.
5. NEM: Fenamiphos application.
6. SA 1 + NEM: Combination of treatments 2 and 5, above.
7. SA 2 + NEM: Combination of treatments 3 and 5, above.
8. SA 1 + SA 2 + NEM: Combination of treatments 2, 3, and 5, above.

The plot area received 1.3 cm irrigation following the fenamiphos application. Plots were observed for phytotoxicity at one and two weeks post-treatment with fenamiphos. Turfgrass visual ratings and nematode samples were taken four weeks after the fenamiphos application. Samples were processed for nematode population as previously described.

For pesticide analysis, six 1.9 cm diameter cores were taken from each fenamiphos-treated plot and sectioned into thatch, and soil depths of 0-10 and 10-20 cm, with like portions being composited to make a single soil sample. Samples were taken one and three days after fenamiphos application. The thatch and soil were analyzed for fenamiphos and fenamiphos metabolites in the manner previously described.

The second surfactant (SA Study 2) experiment was conducted at the same time and in the general location as the first experiment. SA Study 2 compared application of surfactant just before fenamiphos application vs. application of fenamiphos mixed with surfactant to provide the same rates of both as were used in the first study. Other than there being only four replications, all other methods and materials were the same as in the first study.

All data was subjected to automated ANOVA procedures and significant means identified by Waller-Duncan's K-ratio t-test.

Fenamiphos Adsorption with CPP. In October 1997, silica sand was coated with a cross-linked phenolic polyether (CPP, University of Florida patent pending) at the rate of 10% by weight. The prepared material had a particle size range well within United States Golf Association (USGA) specifications (Table I). USGA specification sand must be at least 60% medium+coarse, and less than 20% fine+very fine sand. On November 7, 1997, the twelve lysimeters in a USGA green at the Ft. Lauderdale Research and Education Center were excavated to the gravel layer. Five cm of coarse sand, corresponding in size to that used in the original greens construction (8), was placed over the gravel layer. The CPP-sand was mixed at a rate of 20% by volume with freshly-obtained USGA rooting mix sand that corresponded in particle size to that used in the original greens construction (8) to provide a 10 cm deep layer over the coarse sand in six of the twelve lysimeters. Additional freshly-obtained rooting mix sand was used to completely refill the excavated hole, and this sand also was used over the coarse sand layer in the six lysimeters that did not receive the CPP-sand treatment. CPP-sand treated and untreated lysimeters were arranged in blocked pairs.

The cv. Tifdwarf bermudagrass sod cut from over the lysimeter was trimmed to a soil depth of approximately 4 cm and replaced over the lysimeter. The green was maintained using standard practices thereafter.

For SA study 1, the water collected in all lysimeters was evacuated and discarded on November 12, 1997. At approximately 1:00 PM, fenamiphos, as Nemacur 3E, was sprayed over the lysimeter area at the rate of 1.125 g A.I. m^{-2}, which is the label rate. Immediately after application, 1.7 cm of irrigation was applied to the area. Thereafter, irrigation was applied as needed to maintain the turfgrass. The first rainfall was recorded at 7:00 AM on November 17th (3.00 cm). Being a Monday, the rainfall may have occurred anytime between 7:00 AM on the 14th and 7:00 AM on the 17th. Rainfall recorded on the 21st and 24th was 0.08 and 1.65 cm, respectively.

Lysimeter water was evacuated on November 14, 17, and 20. Pesticide in the lysimeter water was extracted with methylene chloride and analyzed by gas chromatography (9). A determination of both fenamiphos and fenamiphos metabolite was performed.

For SA study 2 the area containing the lysimeters was maintained as a golf green since the previous study. Periodically, lysimeter water was evacuated and discarded. On the morning of July 28, 1998, the green was hollow-core aerified and topdressed, which are standard greens maintenance practices used to improve soil aeration and water penetration. In the afternoon, fenamiphos (Nemacur 3E) was mixed with 3 to 4 L of water and applied with a sprinkling can over 1 m^2 areas centered over the lysimeters to provide an application rate of 1.125 g A.I. m^{-2}. The plot area was irrigated to provide 0.8 cm water, and maintained as a golf green thereafter. Lysimeter water sampling began on July 31 and continued weekly or more often through September 14. The samples were analyzed for fenamiphos and fenamiphos metabolite. Data was subjected to automated ANOVA procedures to identify significant treatment effects.

Results and Discussion

Pesticide Recovery in Thatch and Soil. All OP pesticides were detected in the thatch/soil column for several months after application (8-10). Generally, OPs were found at very low concentrations within two months following pesticide applciation (9-10). OP concentrations in the thatch/soil layer generally decreased rapidly following application with the exception of fenamiphos metabolites which reached peak concentrations several days to one week following application (9-10). Most of the OP pesticides were found in the thatch rather than the soil again with the exception that fenamiphos metabolites were found in the soil portion approximately one week or more after application (9-10). The fenamiphos metabolites are more water soluble than fenamiphos and therefore more mobile than fenamiphos and other OP pesticides tested (9). The time required for peak concentrations of fenamiphos metabolites generally increased with soil-profile depth, whereas the magnitude of the peak concentrations generally decreased with increasing depth (9).

With the above-mentioned exceptions, our OP distribution results in thatch and soil are similar to that reported for OP pesticides by Sears and Chapman (13), and

Kuhr and Tashiro (14), yet somewhat at variance with values reported by Tashiro (15). Sears and Chapman (1978) reported that less than 10% of chlorpyrifos and isazophos, diazinon [0,0-Dimethyl 0-(2-isopropyl-6-methyl-4 pyrimidinyl], and chlordane (1,2,4,5,6,7,8,8-Octachloro-2,3,3a,4,7,7a-hexahydro-4,7-methano-1H-indene) applied to annual bluegrass (Poa annua L.) turf reached the root zone with primary distribution within the turf thatch layer. Kuhr and Tashiro (14) reported on the soil distribution of chlorpyrifos and diazinon of Kentucky bluegrass (Poa pratensis L.) grown in a sandy loam and found very low concentrations below 1.3 cm soil depth. However, Tashiro (15) reported approximately equal distribution of chlorpyrifos and diazinon in thatch and soil layers after application.

In our OP studies, pesticide residues in thatch and soil declined to extremely low values approximately two months after application. Sears and Chapman (13) reported similar low values for OP insecticides in soil 56 days after treatment.

In both phenoxy acid studies, peak concentrations of dicamba and especially for 2,4-D, in soil and thatch, were observable shortly after each of the two applications (11). However, concentrations rapidly declined during the two weeks following each application, but still persisted up to two months. Shortly after application, the percentages of 2,4-D and dicamba in the thatch were greatest, with somewhat lower thatch percentages, and therefore higher soil percentages, being observed several days after application. These data are in contrast to our results for the OP pesticides, and somewhat similar to what was seen for fenamiphos metabolites (9,12).

Pesticide persistence is a function of a number of factors including physical, chemical, and biological factors. Because repeat applications of pesticides often are needed to control pests in turf, the impact of enhanced biodegradation on the persistence of pesticides may be a critical environmental factor in turfgrass systems. With regard to enhanced biodegradation in turf, Ou et al. (16) reported results of a laboratory study that demonstrated complete propoxur [2-(Methylethoxy)phenol methylcarbamate] degradation within 24 hr after exposure to microorganisms isolated from turfgrass soils that had previous treatments of propoxur. Accelerated biodegradation of OP pesticides have also been noted in the literature. We speculated that enhanced biodegradation of fenamiphos and fenamiphos metabolite contributed to a lower amount of fenamiphos and fenamiphos metabolite residues found in percolate after a second application of the pesticide (9). Subsequently, Davis et al. (17) and Ou (18) reported upon accelerated biodegradation of fenamiphos in golf course turf after repeated application in field crops and golf course turf, respectively. In an earlier study, Niemcyzk and Chapman (19) suggested the turf thatch layer contained a degradative system capable of rapidly degrading the OP pesticide, isofenphos. Although determination of biodegradation was not a part of our experiments, less isofenphos and isazophos were found in turf thatch and soil after a second application, suggesting that enhanced biodegradation of these materials may have been a contributing factor (12). Greater quantities of chlorpyrifos were found in thatch and soil after a repeat application, which was consistent with the second application label rate of chlorpyrifos being nearly double the rate of the first granular application. As reported below, less pesticide in percolate (based on mean-weighted flow) was generally found after a second application.

Pesticide Recovery in Percolate. Total OP recovery in percolate was very low with percent losses generally below 0.1% of that applied for each application date regardless of the extensive quantities of percolate generated in a sub-tropical environment (Table III). The lone OP exception was fenamiphos metabolites in which over 17% and 1% of that applied on 13 Nov 1991 and 27 January 1992, respectively, were found in percolate after successive applications (Table II).

Although only 10% as much dicamba as 2, 4-D was applied (Table II), nearly 65% as much dicamba as 2,4-D was recovered in percolate water following the August 1993 applications, and over twice as much was recovered following the April 1994 applications (Table III). With approximately 10% of the applied dicamba found in percolate versus 0.5-1.5% of the applied 2,4-D found in percolate, clearly dicamba was more mobile than 2,4-D in the USGA green.

The leaching data are consistent with the data for thatch and soil. Since in most cases pesticide was found in the thatch, and little appeared to move into the underlying soil, it was reasonable that only small quantities were detected in the percolate waters. When pesticides were found to not remain in the thatch, more pesticide (as a percent of that applied), was recovered in percolate.

Pesticide Recovery in Clippings. The quantities of OP and phenoxy pesticides removed with the clippings were greatest for the first clipping after application, and decreased rapidly thereafter (9-11). These trends are in general agreement with previously reviewed research on OP pesticides residues on turf foliage, which demonstrated rapid decreases of OP pesticides on foliage within a very brief time period after application (2).

Although the amount of pesticide removed in clippings was generally very low, in one case, an appreciable amount (as a % of that applied) of OP pesticide was removed in the clippings over the sampling period following chlorpyrifos application (Table IV). The higher value (7.87%) for the first application (27 Jan. 1992) of chlorpyrifos may have been the result of granules having been removed with the clippings. The % chlorpyrifos recovery in clippings was 0.52% when applied as a liquid on the second application (21 Apr. 1992) even though the pesticide in liquid form was applied at approximately twice the rate of the granular material (Table IV). Clippings removed the day after application of chlorpyrifos as granules contained 1.82 mg chlorpyrifos kg^{-1} clippings (data not tabulated), expressed on a dry weight basis, whereas following application of chlorpyrifos, isazophos, and isofenphos as liquids the next-day clippings contained 0.12, 0.14, and 0.29 mg kg^{-1} of the pesticides, respectively. Total OP pesticide removed in clippings after granular applications of ethoprop, chlorpyrifos, fenamiphos, and fonofos, were 9.9 mg m^{-2}, 9.2 mg m^{-2}, 5.1 mg m^{-2}, and 4.3 mg m^{-2}, respectively, while the largest amount from a liquid application of pesticide removed in clippings was 2.0 mg m^{-2} of isofenphos. Percent recovery in clippings of phenoxy acids did not exceed 0.25% (Table IV). The quantity of phenoxy acid pesticides removed in clippings ranged from a low of 8.6 ug m^{-2} to a high of 221.6 ug m^{-2} (11).

Fenamiphos Irrigation Management. In both studies, the Low Irrigation treatment delayed but did not reduce cumulative fenamiphos and fenamiphos

Table I. Particle size range (mm) of CCP-coated silica sand.

>2.00	Very coarse 2.00 - 1.00	Coarse 1.00 - 0.05	Medium 0.05 - 0.25	Fine 0.25 - 0.10	Very fine 0.10 - 0.05	Silt+Clay <0.05
			%			
0.1	1.2	28.5	56.2	12.9	0.8	0.2

Table II. Pesticides used on the USGA green in persistence and mobility studies (1991-1994).

Pesticide Common Name	Form	Rate	Date Applied	Sample Duration	Irr.& Precip.	Percolation
		g A.I.m^{-2}		days	mm	mm
Fenamiphos	10G	1.125	11/13/91	75	149	340
Fenamiphos Metab.		1.125	11/13/91	75	149	340
Fonofos	5G	0.439	11/13/91	75	149	340
Fenamiphos	10G	1.12	1/27/92	85	191	269
Fenamiphos Metab.		1.12 5	1/27/91	85	191	269
Fonofos	5G	0.439	1/27/92	85	191	269
Chlorpyrifos	1G	0.117	1/27/92	85	191	269
Chlorpyrifos	2E	0.229	4/21/92	141	841	1370
Isazophos	4E	0.229	4/21/92	141	841	1370
Isofenphos	2E	0.229	4/21/92	141	841	1370
Isazophos	4E	0.229	9/15/92	125	376	777
Isofenphos	2E	0.229	9/15/92	125	376	777
Ethoprop	10G	2.245	9/15/92	125	376	777
2,4-D	4E	0.058	8/03/93	65	451	696
2,4-D	4E	0.058	8/10/93	65	451	696
Dicamba	4E	0.006	8/03/93	65	451	696
Dicamba	4E	0.006	8/03/93	65	451	696
2,4-D	4E	0.058	4/18/94	66|	439	508
2,4-D	4E	0.058	4/25/94	66|	439	508
Dicamba	4E	0.006	4/25/94	66	439	508
Dicamba	4E	0.006	4/25/94	66	439	508

Table III. Pesticide concentration (ug L^{-1}), recovered mass (ug m^{-2}), and percent of total applied.

Common Name	Application Date	Pesticide Recovered		
		Concentration	Mass	Percent of Total
		ug L^{-1}	g m^{-2}	(%)
Fenamiphos	11/13/91	2.06	701	0.06
Fenamiphos metab.	11/13/91	585.40	199,038	17.69
Fonofos	11/13/91	0.01	4	<0.01
Fenamiphos	1/27/92	1.56	419	0.04
Fenamiphos metab.	1/27/92	45.82	12,326	1.10
Fonofos	1/27/92	0.38	103	0.02
Chlorpyrifos	1/27/92	0.65	176	0.15
Chlorpyrifos	4/21/92	0.14	193	0.38
Isazophos	4/21/92	0.15	204	0.09
Isofenphos	4/21/92	0.04	53	0.02
Isazophos	9/17/92	0.05	41	0.02
Isofenphos	9/17/92	0.04	33	0.01
Ethoprop	9/17/92	1.46	1,138	0.07
2,4-D	8/3+10/93	2.59	1,808	1.60
Dicamba	8/3+10/93	1.68	1,173	9.70
2,4-D	4/18+25/94	1.20	611	0.50
Dicamba	4/18+25/94	2.53	1,290	10.80

Table IV. Cumulative amount of pesticide removed with clippings during two application cycles.

Study Period	Pesticide	Application Rate	Amount Removed
		g A.I. m^{-2}	% of applied
27 Jan. - 21 Apr.	Fenamiphos	1.125	0.38
	Fenamiphos metabolite	-	0.14[†]
	Fonofos	0.439	1.17
	Chlorpyrifos	0.117	7.87
21 Apr. - 5 June	Chlorpyrifos	0.229	0.52
	Isazophos	0.229	0.43
	Isofenphos	0.229	0.79
15 Sept. - 4 Jan.	Isazophos	0.229	0.38
	Isofenphos	0.229	0.89
	Ethoprop	2.245	0.44
3 Aug. - 5 Sept.	2,4-D	0.116	0.19
	Dicamba	0.012	0.25
19 Apr. - 24 May	2,4-D	0.116	<0.01
	Dicamba	0.012	0.07

[†]Calculated as a percent of fenamiphos applied.

metabolite leaching (*12*). Under the Low Irrrigation treatment, more fenamiphos and fenamiphos metabolites were observed in thatch and soil portions up to two weeks after application, a portion of which occurred during rainfall protected periods (*12*). However, in both studies appreciable rain events eventually led to statistically-equivalent total fenamiphos and fenamiphos metabolites leaching over both study periods (Table V). As in earlier studies on fenamiphos leaching (*9*) metabolite leaching accounted for the majority of the pesticide leached. Metabolite leaching amounted to 141 and 135 mg m^{-2} for the High and Low irrigation treatments, respectively, or 12.6 and 12.1% of the fenamiphos applied, respectively in 1995, and 17 and 38 mg m^{-2}, for the High and Low Irrigation treatments, respectively, or 3.3 and 1.5% of the fenamiphos applied in 1996 (Table V).

These results suggest that for the parent fenamiphos compound and its more mobile metabolites the use of a Best Management Practice approach such as irrigation control alone cannot provide assurance against leaching of fenamiphos metabolites in sub-tropical south Florida where appreciable rainfall can occur at any time of the year. However, irrigation management did reduce leaching of the parent compound rainfall restricted periods (Table V).

It is interesting to note that in the first study (1995), which was conducted on a portion of the green that had not been treated with fenamiphos for 3.5 years, metabolite leaching averaged 12.3% of the fenamiphos applied. In 1996, it averaged only 2.4% of the applied rate of fenamiphos. The observation of considerably less leaching of metabolite following a second fenamiphos application, is consistent with the observation that fenamiphos and presumably, fenamiphos-metabolite degrading microorganism populations increase following a fenamiphos application (*18*). The observation also is consistent our summary on pesticide mobility and persistence research on fenamiphos, in which metabolite leaching following an intial application of fenamiphos amounted to 17.7% of the parent compound applied, whereas following the second application, metabolite leaching amounted to only 1.1% of the applied rate of fenamiphos.

Effect of Surfactant on Fenamiphos Efficacy and Distribution. Surfactants are used to promote water penetration in soils, and may help especially to move water through thatch. In previous studies, it was observed that most OP pesticides were strongly adsorbed by thatch, and little moved through the soil with percolate waters. From a water-quality standpoint, such an observation appears beneficial. However, where control of soil-borne pests is desired, adsorption by thatch may reduce efficacy. If pesticide movement through the thatch can be promoted, perhaps lower rates or less frequent application of pesticides can be used, thereby reducing the potential for environmental contamination. Two studies were conducted to determine whether a commercially-available surfactant would enhance fenamiphos efficacy, and increase fenamiphos penetration through thatch.

No phytotoxicity was observed during the experiment for any of the treatments and the treatments had no significant (P <0.05) effect on visual quality four weeks after fenamiphos application (Table VI). Fenamiphos alone did not reduce B. longicaudatus nematode populations relative to the control (Table VI), i.e., applied at the current label rate the fenamiphos was ineffective. Furthermore, use of the

Table V. Percolation (Perc.) and leaching of fenamiphos and metabolite during a differential irrigation (Irr.)/rainfall protected period (1), during a period when rainfall and irrigation were the same on all plots (2), and for the Total of the two periods (Tot).

| Period | | | ----------1995------------ | | | ----------1996---------- | |
	Irr.	Perc.	Fenami-phos	Metab-olite	Perc.	Fenami-phos	Metab-olite
		mm	ug m^{-2}	mg m^{-2}	mm	ug m^{-2}	mg m^{-2}
1	L	49	3	2	7	1	0
	H	96	78	63	25	8	1
Signif.		*	*	*	+	NS	NS
2	L	138	74	133	62	48	38
	H	167	12	78	63	25	16
Signif.		*	NS	NS	NS	NS	NS
Tot.	L	187	77	135	69	49	38
	H	263	90	141	88	33	17
Signif.		*	NS	NS	NS	NS	NS

+, *, and NS represent P<0.10, 0.05, and P>0.10, respectively.

Table VI. Effect of a surfactant on efficacy of fenamiphos for reducing <u>Belonolaimus longicaudatus</u> nematode populations in bermudagrass turf four weeks after application.

Treatment	Visual quality[1]	Sting nematodes
		no. 100cc^{-1}
Control	5.0	116ab[2]
SA 1	5.8	164a
SA 2	5.0	153a
SA 1+2	5.4	90ab
NEM	5.3	109ab
SA 1 + NEM	6.0	99ab
SA 2 + NEM	5.8	108ab
SA 1 + SA 2 + NEM	5.5	66b
Significance	NS	*

[1] Rated on a 1 - 10 scale, with 10 being best possible and 6 being just acceptable.
[2] Values followed by the same letter are not different by the Waller-Duncan K-ratio t-test.

surfactant in any of the combinations investigated did not significantly improve fenamiphos efficacy relatvie to fenamiphos alone, (Table VI), although counts for the treatment involving two applications of surfactant were only 66% that of those for fenamiphos alone. However, visual turf quality ratings for the two treatments were nearly equal, indicating little difference in nematode control between the two treatments.

The major question of fenamiphos distribution in this study was whether use of the surfactant would reduce the amount of fenamiphos and metabolites in the thatch and increase the percent distribution in the soil layers. Such did not appear to be the case, based on samples made one day (Table VII) and three days (Table VIII) after fenamiphos application.

The second surfactant study (SA Study 2) compared application of surfactant just before fenamiphos application vs. application of fenamiphos mixed with surfactant to provide the same rates of both as were used in the first study. For SA Study 2, no significant differences in visual ratings or in nematode populations between the two treatments were observed four weeks after fenamiphos application (Table IX). Additionally, there was no difference in fenamiphos or fenamiphos metabolite distribution within the soil columns between the two surfactant application methods (Table X).

Effect of Adsorbents on Fenamiphos Leaching. Considerably more water was collected in the lysimeters than could be accounted for on the basis of rainfall, probable irrigation, and expected evapotranspiration. While the reason for this finding was not determined, it was possible that soil in the recently reconstructed lysimeter profile may have had a much greater hydraulic conductivity than that of the long-established surrounding green, and the surface of the soil over the lysimeters may have been somewhat lower that in the surrounding area. Both of these factors could have resulted in movement of surface water, possibly containing pesticide, from areas surrounding the lysimeters to the lysimeters, with subsequent percolation through the lysimeters.

For lysimeter water collected over the period from November 14 to 20, 1997 there was a 72% reduction in fenamiphos leached in the CPP-sand lysimeters as compared to the unamended lysimeters (Table XI). For the more water-soluble metabolite, the reduction in leaching was 54% (Table XI).

There was no significant (P>0.05) effect of CPP-sand on percolation, which averaged 57.7 cm during the seven weeks following fenamiphos application. Virtually no fenamiphos was leached in lysimeters containing CPP-sand (Table XI). Two weeks after application, total metabolite leaching was reduced 90% by CPP-sand (Table XI). The comparative reduction declined with time. Seven weeks after application total metabolite leaching was reduced 76% (Fig. 1). This apparently occurred because after the initial adsorption of metabolite by CPP sand, there was a slow desorption of the material, as evidenced by a gradual increase in accumulative metabolite leaching in the CPP-sand lysimeters over time (Fig. 1). Nevertheless, the initial high concentration of metabolite in percolate was prevented by the CPP-sand. The desorption characteristic of the CPP-sand might make it suitable as a "slow-release" carrier for pesticides.

Table VII. Amount and percentage distribution of fenamiphos and fenamiphos metabolite in thatch and soil one day after fenamiphos application.

Treatment	Thatch	Soil 0-10cm	10-20cm
		Fenamiphos	
SA 1	342.4	18.0	2.7
SA 1 + SA 2	250.0	14.3	2.1
No SA	369.9	15.2	2.6
SA 2	378.5	19.6	3.9
	NS	NS	NS
		----(%)----	
SA 1	94.4	4.8	0.8
SA 1 + SA 2	93.6	5.6	0.8
No SA	94.8	4.4	0.8
SA 2	93.8	5.3	1.0
	NS	NS	NS
		Metabolites	
		----(mg/m^2)----	
SA 1	132.5	23.6	2.1
SA 1 + SA 2	140.4	23.4	5.1
No SA	154.6	24.2	2.5
SA 2	145.6	28.2	2.9
	NS	NS	NS
		----(%)----	
SA 1	79.3	19.7	1.0
SA 1 + SA 2	79.9	16.6	3.5
No SA	81.4	17.3	2.5
SA 2	77.5	20.9	2.3
	NS	NS	NS

Table VIII. Amount and percentage distribution of fenamiphos and fenamiphos metabolite in thatch and soil three days after fenamiphos application.

Treatment	Thatch	Soil 0-10cm	10-20cm
		Fenamiphos	
	-------------(mg/m^2)-------------		
SA 1	171.6	8.6	1.0
SA 1 + SA 2	184.7	7.0	0.9
No SA	123.1	11.0	0.7
SA 2	173.8	6.3	1.9
	NS	NS	NS
	----------------(%)--------------		
SA 1	94.5	4.9	0.6
SA 1 + SA 2	95.7	3.8	0.5
No SA	81.5	17.5	1.0
SA 2	94.0	4.4	1.6
	NS	NS	NS
		Metabolites	
	----------------(mg/m^2)-------------		
SA 1	152.9	116.9	1.5
SA 1 + SA 2	241.9	147.0	3.6
No SA	180.8	128.4	7.7
SA 2	175.7	92.9	3.2
	NS	NS	NS
	----------------(%)----------------		
SA 1	51.7	47.6	0.8
SA 1 + SA 2	65.0	33.7	1.3
No SA	54.1	40.5	5.5
SA 2	60.6	37.6	1.8
	NS	NS	NS

Table IX. Effect of surfactant application method on efficacy of fenamiphos for reducing <u>Belonolaimus</u> <u>longicaudatus</u> nematode populations in bermudagrass turf four weeks after application.

Treatment	Visual Quality[1]	Sting Nematodes
		no. $100cc^{-1}$
SA just before fenamiphos	6.9	81
SA with fenamiphos	7.4	86
Statistical significance	NS	NS

[1] Rated on a 1 - 10 scale, with 10 being best possible and 6 being just acceptable.

Table X. Fenamiphos and metabolite in thatch and soil after surfactant application with fenamiphos or just before fenamiphos application.

		Soil	
Treatment	thatch	0-10cm	10-20cm
		fenamiphos (mg/m^2)	
SA before Nemacur	198.7	5.9	3.9
SA with Nemacur	333.6	7.9	1.3
	NS	NS	*
		fenamiphos distribution (%)	
SA before Nemacur	97.5	2.2	1.9
SA with Nemacur	97.4	2.3	0.4
	NS	NS	NS
		metabolite (mg/m^2)	
SA before Nemacur	106.0	16.1	4.5
SA with Nemacur	232.7	15.1	3.9
	NS	NS	NS
		metabolite distribution (%)	
SA before Nemacur	86.1	17.6	6.9
SA with Nemacur	90.4	7.9	1.7
	NS	NS	NS

Table XI. Effect of a cross-linked phenolic polyether (CPP)-coated sand in the lower 10 cm of a USGA green on accumulated leaching of fenamiphos and fenamiphos metabolite.

Study	Acumulation Period	CPP	Fenamiphos	Reduction due to CPP	Metabolite	Reduction due to CPP
	(weeks)		mg/m^2	%	mg/m^2	%
1	1	-	71.4		223.3	
		+	20.1	72	102.5	54
2	2	-	2.5		61.9	
		+	<0.1	100	6.1	90
	7	-	2.5		62.8	
		+	<0.1	100	15.3	76

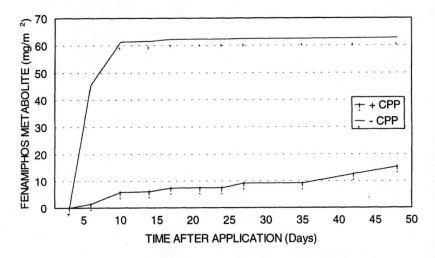

Figure 1 Effect of time on accumulated fenamiphos metabolite leaching in the August-September 1998 study (Study 2).

Conclusions

Balogh and Walker (2), concluded in their review on pesticide fate in turf that research on the subject was rather limited and also concluded that basic or applied research on turfgrass pesticide fate was a high priority. The USGA research initiative during the 1990s has helped to create a more comprehensive data base on pesticide fate in turfgrass systems. Research studies in sub-tropical south Florida summarized in this chapter were conducted over a six-year period focused on mobility and persistence of OP pesticides in a USGA green. Other pesticide fate pathways such as volatilization and dislodgeable residues and their potential risk to non-target organisms need further elucidation and are the focus of future research. Nevertheless, the research reported herein should help provide information necessary for the safe and effective use of OP pesticides applied to USGA greens. For each of the OP pesticides studied, with the exception of fenamiphos metabolites, less than 0.1% of the applied pesticide was recovered in the percolate water, in spite of large differences in rainfall and total percolation. Less than 1.0% was recovered in clippings, except for granular application of chlorpyrifos. With the exception of fenamiphos metabolites, little if any pesticide was observed in the soil underlying the thatch. By far, most of the applied pesticide remained in the thatch layer and disappeared with time, presumably by microbial degradation.

Three experimental approaches were evaluated in an attempt to reduce improve fenamiphos efficacy/distribution and reduce fenamiphos and fenamiphos metabolite leaching. A commercial surfactant did not improve fenamiphos efficacy or distribution in soil. Irrigation control delayed, but did not reduce cumulative fenamiphos and fenamiphos metabolite leaching in a rainy sub-tropical environment. However, the third approach, the addition of a patent-pending polymer-coated sand incorporated into the USGA green construction mix did significantly reduce fenamiphos and fenamiphos metabolite leaching from a USGA green.

Acknowledgments

Financial support by the Greens Section of the United States Golf Association is gratefully acknowledged. The authors also express their appreciation to the Palm Beach and Broward Chapters of the Florida Golf Course Superintendents Association for providing the USGA green and funding its maintenance, and to LESCO, Inc. for various agrochemicals, and to grants from Wedgworth Farms Inc. and the Florida Turfgrass Association for purchase of the gas chromatograph and miscellaneous equipment and chemicals. Technical assistance by Dr. Robin Giblin-Davis, Curtis Elliott, Esther Figueiras, Norman Harrison, Eva Greene, Marcus Prevatte, Theresa Sanford, and Karen Williams is very much appreciated.

References

1. Beard, J. B. 1973. Turfgrass Science and Culture. Prentice-Hall, Englewood Cliffs, NJ. 658 pp.

2. Balogh, J. C. and W. J. Walker. 1992. Golf Course Management and Construction. Environmental Issues. Lewis Publishers. Chelsea, MI.

3. Noah, T. 1994. Golf courses are denounced as health hazards. Wall Street Journal. Monday, May 2, 1994. pg. 1B.

4. Zaneski, C. T. 1994. Wildlife pays price for green fairways. The Miami Hearld. July 11, 1994. pg. 1A.

5. Hummel, N. W. 1993. Rationale for the revisions of the USGA green construction specifications. USGA green section record. 31(2):7-22.

6. USGA greens section staff. 1993. The 1993 revision: USGA recommendations for a methods of putting green construction. USGA Green Section Record. 31(2):1-3.

7. Hodges, A. W., J. J. Haydu, P. J. van Blokland, and A. P. Bell. 1994. Contribution of the turfgrass industry to Florida's economy, 1991/1992: A value added approach. Economics Report (ER 94-1). University of Florida, IFAS, Gainesville. 83 pp.

8. Cisar, J. L., and G. H. Snyder. 1993. Mobility and persistence of pesticides applied to a USGA-type green: I. Putting green fac. for monitoring pesticides. J. Int. Turfgrass Soc. 7:971-977.

9. Snyder, G. H. and J. L. Cisar. 1993. Mobility and persistence of pesticides applied to a USGA-type green: II. Fenamiphos and fonofos. J. Int. Turfgrass Soc. 7:978-983.

10. Cisar, J. L., and G. H. Snyder. 1996. Mobility and persistence of pesticides applied to a USGA Green. III: Organophosphate recovery in clippings, thatch, soil, and percolate. CropScience. 36(6):1433-1438.

11. Snyder, G. H. and J. L. Cisar. 1997. Mobility and persistence of turfgrass pesticides in a USGA-type green. IV. Dicamba and 2,4-D. J. Int. Turfgrass Soc. 8:205-211.

12. Cisar, J. L., and G. H. Snyder. 1997. Mobility and persistence of pesticides applied to a USGA-type green. V. Irrigation management and fenamiphos leaching. J. Int. Turfgrass Soc. 8:167-173.

13. Sears, M. K. and R. A. Chapman. 1979. Persistence and movement of four insecticides applied to turfgrass. J. Econ. Entomol. 72(2):272-274.

14. Kuhr, R. J., and H. Tashiro. 1978. Distribution and persistence of chlorpyrifos and diazinon applied to turf. Bull. Environ. Contam. Toxicol. 20:652-656.

15. Tashiro, H. 1982. Distribution and persistence of chlorpyrifos and diazinon in soil when applied to turf. In, Niemcyzk, H. D., and Joyner B. G. (eds.), Advances in Turfgrass Nematology. Hammer Graphics, Inc. Piqua, OH pp. 53-56.

16. Ou, L. T., P. Nkedi-Kizza, J. L. Cisar, and G. H. Snyder. 1992. Microbial degradation of propoxur in turf. J. of Environ. Sci. Health - Part B: Pesticides, Food Contaminants, and Agricultural Wastes. 27:545-564.

17. Davis, R. F., A. W. Johnson, and R. D. Wauchope. 1993. Accelerated degradation of fenamiphos and its metabolites in soil previously treated with fenamiphos. J. Nematol. 25(4):679-685.

18. Ou, L.T. 1991. Interactions of microorganisms and soil during fenamiphos degradation. Soil Science Society of America Journal. 55:716-722.

19. Niemczyk, H. D. and R. A. Chapman. 1987. Evidence of enhanced degradation of isofenphos in turfgrass thatch and soil. J. Econ. Entomol. 80(4):880-882.

Chapter 8

Environmental Fates of Fungicides in the Turfgrass Environment: A Minireview

W. V. Sigler, C. P. Taylor, C. S. Throssell, M. Bischoff, and R. F. Turco

Laboratory for Soil Microbiology and Turfgrass Science, Department of Agronomy, Purdue University, West Lafayette, IN 47907

Concerns surrounding the environmental impact of repeated fungicide use has prompted research aimed at determining the environmental fate of the applied fungicides. We will review the investigations focusing on fungicide degradation within the turfgrass canopy and soil. We also review the role of the physical characteristics of the turfgrass system such as adsorption, moisture content, soil organic matter and clay content, pH, and temperature, and how these parameters effect biotic fungicide degradation. Previous reports have been limited to the degradation of fungicides in soil, however little research has focused on leaf surface activity. Our data support previous observations concerning soil-fungicide interactions. Additionally, we can now define the role of the turfgrass canopy in terms of the sorption and degradation of fungicides. We suggest that this segment of the turfgrass profile exhibits a high level of sorptive behavior whereas little biotic fungicide degradation exists.

Fungicides are used to aid in the protection of agricultural crops and turf areas from infectious diseases. These compounds are used in the control of parasitic fungal pathogens that generally arise from the soil environment. Under the correct weather and environmental conditions, active fungi will invade and colonize the host plant tissue. Damage is caused as the fungus consumes vital contents of the host cells, disrupts metabolic processes, and/or blocks the flow of nutrients through the connective tissues of the plant. The first fungicides were used on grapevines in France in 1864 to protect the fruit from powdery mildew fungus. This discovery was quite by accident as the mixture was originally intended to discourage passing burglars by giving the grapes a visually less appealing look. The chemistry of modern day fungicides, however, is quite different from the original copper sulfate solution used on grapes in 1864. Fungicides used today are highly active, organic compounds that possess a wide range of toxicities, modes of action, and behaviors in the environment.

Why is fungicide degradation important?

Almost all plants grown for commercial purposes require the application of fungicides in order to prevent agronomic and/or economic loss. Concern has developed, though, over the widespread use of agricultural chemicals, including fungicides. In Iowa alone, approximately 54,000 pounds of fungicide active ingredient were applied to golf course turf surfaces in 1990 (Agnew 1993). Fungicides account for approximately 16% of the 500 million kg of pesticides used yearly in the United States (Pimentel and Levitan, 1986, Hatrik and Tekel, 1996). Much work has been conducted pertaining to increasing concerns over the environmental fate of these pesticides. Most of this research is part of an effort by the scientific community to develop reliable parameters capable of predicting a chemical's behavior in the environment.

Unlike the situation encountered in row-crop agriculture where single applications of chemicals are most often used, *repeated* fungicide applications to fruits, vegetables, and turfgrass areas are common. Specifically, golf course putting greens are subject to fungicide applications up to 20 times per year depending on geographic location and weather conditions. Questions have been raised surrounding the environmental impact of repeated and prolonged fungicide use. These questions include concern over ground water contamination, effects on soil and aquatic organisms, and the ultimate fate of the fungicide and its transformation products (Potter 1990, Tsuda 1992, Doneche 1993). Although fungicides are less often encountered than herbicides in groundwater, Cova and coworkers (1990) identified the fungicide vinclozilin as a compound often contaminating Italian groundwater samples.

Chemical-related damage to human lymphocyte chromosomes was shown in-vitro to be caused by the acylalanine fungicide, metalaxyl (Hrelia et al., 1996) illustrating the dangers of human exposure to fungicides. The combination of potential health effects and frequency of application to highly trafficked areas such as golf courses creates a demand for knowledge of the environmental fate of these chemicals.

Much research has been conducted surrounding the degradative fate of fungicides. Most investigations study fungicide interactions within the soil matrix and water systems whereas considerably less work has been initiated to look at reactions on the leaf surface. The purpose of this discussion is to review the various biological fates of several fungicides. An emphasis will be placed on the conditions and organisms that drive the degradative reactions.

General fungicide fate

While the following review will focus primarily on the biotic degradation processes that influence the disappearance of a fungicide from the environment, it must be noted that many other processes interact in effecting the ultimate fate of fungicides. A pesticide's fate is determined by mechanisms whose strength and importance depend on chemical properties and location in the environment. The action of a fungicide in the environment is founded on six physical/chemical processes. These include the chemical's:

1. Solubility-based movement in water;
2. Sorption/desorption to surfaces;
3. Abiotic degradation;
4. Biotic degradation;
5. Volatilization; and,
6. Plant uptake.

Loss of pesticide from the target site begins at application. The first route of loss is often due to the spray apparatus itself. As pesticides tend to be hydrophobic, they will adsorb to non-aqueous surfaces in an effort to gain greater thermodynamic stability. Spray equipment often has plastic or rubber components. Pesticides can adhere to these components reducing the amount of chemical actually applied. One goal of formulations is to bind the pesticides to carrier molecules with enough strength to prevent sorption to the spray apparatus.

Drift and volatilization are other routes of loss prior to pesticide deposition on the intended site. Chemicals are usually delivered in droplets to increase the uniformity of coverage. Wind and high temperatures can cause droplets to partially evaporate, decreasing their size and then allowing the resulting mist to move from the target area. Smaller droplet size also increases the likelihood of volatilization.

Fungicides, particularly when used in turfgrass, are applied directly to a plant canopy. In the canopy, contact fungicides act as protective coatings on the surface of the plant. While on the plant surface, the chemical may undergo photodecomposition, volatilize, breakdown microbially, or be washed off by rain or irrigation. Systemic fungicides are designed to enter the plant, following a foliar application, through the stomata or by absorption through the plant's cuticle. Once in a plant, the pesticide may be translocated to other areas or may undergo enzymatic breakdown, or both.

This review will focus only on the biotic degradation processes that influence the disappearance of a fungicide from the environment but it must be noted that all of the above-mentioned processes interact in effecting the ultimate fate of fungicides.

Fungicide degradation in the environment can be likened to a double-edged sword. First, from an ecological standpoint, there is a desire to promote the rapid disappearance of the fungicide or to transform it to a non-toxic metabolite. The conversion of a pesticide to a less toxic state is the goal of any ecologically friendly scheme whether it follows an industrial accident or a scheduled field application. Conversely, because a certain amount of activity and persistence is necessary for the adequate performance of a fungicide, unwanted or premature degradation of the chemical may not be desired (Jury 1983). Although highly recalcitrant fungicide active ingredients are no longer in use, repeated applications of commonly used fungicides to soils have resulted in alarmingly high rates of degradation. Increased degradation rates of vinclozilin and iprodione (Walker et al., 1987a and Slade et al., 1992), metalaxyl (Bailey et al., 1986), benomyl (Yarden, et al., 1985), and mancozeb (Doneche et al., 1983) have been reported to result from repeated applications.

Subsequently, a balance between prolonged activity and degradation in the environment is desired for safe, yet effective fungicide use.

Defining "degradation". The term "degradation" is quite ambiguous and for the purposes of this discussion, will require further definition. Biotic degradation processes refer to those degradation pathways that involve biological organisms that mediate the transformation of the xenobiotic material. Such reactions may include biotic hydrolysis, demethylation, dehalogenation, and hydroxylation. For example, the demethylation of the fungicide, chloroneb, results in the detoxification of the compound to a harmless metabolite (Hock and Sisler 1969). In general, both biotic and abiotic processes occur concurrently in the environment (Wolfe 1980). The dual nature of environmental breakdown lends complexity to the study and understanding of fungicide degradation.

In biological systems, degradation processes greatly change the physical properties of fungicides. From a microbial point of view, the ultimate goal of degradation is mainly to derive metabolic value from the fungicide carbon atoms. Carbon is utilized for both growth of the organism and for energy production. Several metabolic pathways are used in the conversion of organic carbon to biomass and energy. In biotic and degradation pathways, the complete metabolism of the fungicide to the mineral form of carbon, CO_2, is termed mineralization. Metabolism that is incomplete, that is, the fungicide is not converted completely to CO_2 but to an intermediate compound, is termed transformation. Transformation products of fungicides are often called metabolites, which can vary in toxicity and environmental properties (Clark 1978, Bailey 1986). "Dissipation" is a general term used to describe the loss of a pesticide from an environment. Dissipation results from a combination of all loss mechanisms and usually refers to loss of the parent fungicide.

Biotic degradation of fungicides. Bollag and Liu (1990) describe five processes that are involved in the transformation of pesticides in soils. They are;

1. Biodegradation;
2. Cometabolism;
3. Polymerization or conjugation;
4. Accumulation; and
5. Secondary effects of microbial activity such as pH and redox-driven changes.

Although all of the above are important in xenobiotic breakdown, biodegradative processes represent the most dynamic and well-studied impacts on fungicides in the environment. In order for biotic activity to take place in the transformation of a given fungicide, certain conditions must exist that optimize the potential for microbial metabolism. In the turfgrass environment, physical characteristics such as moisture content, organic matter and clay content, temperature, and pH play vital roles in determining the ability of microorganisms to degrade fungicides.

Microbial metabolism. Alexander (1994) defines biodegradation as the biologically catalyzed reduction in complexity of chemicals. As mentioned earlier, this reduction in complexity leads to a liberation of carbon-containing molecules, which are used by the degrading organism as an energy source or growth substrate. Microbial metabolism is important in the soil as it is the primary force in the transformation and degradation of chemicals (Munnecke 1982). Fungicide degradation through microbial (biotic) processes is generally proven through the comparison of a biologically sterilized matrix with a matrix that is non-sterilized and is still biologically active. Applications of the fungicide of interest are made to each matrix and the amount of degradation is noted. Most of the research that has been conducted in this area has revolved around soil or water as the microbial matrix. Sharom (1982) reported that metalaxyl degradation would occur in an unsterilized soil but not in a sterilized soil environment suggesting that microbial metabolic processes were mediating the breakdown. Biodegradation was shown to effect the fate of captan in soil as up to 25% of the parent compound was shown to be released as CO_2 after 60 days of laboratory incubation (Buyanovsky 1988). Martin et al. (1990) found that iprodione added to sterilized soil resisted degradation while additions to a similar unsterilized soil resulted in degradative activity and proof of biotic degradation. Vinclozilin degradation in soil was reported to be at least partially driven by *Pseudomonas spp.* and *Bacillus spp.* (Goleveva et al., 1991). Similarly, unidentified microbial communities were shown to be responsible for the degradation of oxadixyl in a turfgrass soil while no degradation was evident in sterilized soils (Anan'yeva et al., 1996).

Genetic diversity. The vast array of microorganisms living in specific environmental systems lends a high amount of genetic diversity capable of many metabolic processes. Genetic analyses have suggested that at least 1000 bacterial genome equivalents are present per gram of soil. It has been shown that levels of functional (pesticide degradation) diversity of an ecosystem increase with the number of different species (microorganisms) present (Tilman et al., 1997). This pool of talent affords a great probability that an organism exists in a given environment that is capable of transforming a specific fungicide. This point was illustrated by Bailey (1986) who found that single-species populations of soil bacteria degraded metalaxyl 30% *slower* than mixed populations in the same time period (52% vs. 75% degradation in 25 days). This example points to the powerful capabilities of the biological diversity in the soil with regard to fungicide degradation. More recently, Soudamini et al. (1997) reported that mixed fungal cultures isolated from active soils interacted synergistically to degrade metalaxyl faster than a single species.

Bacteria vs. fungi. In soils, fungi and bacteria are assumed to provide the bulk of the degradative potential. Both have been shown to be highly active in the breakdown and metabolism of fungicides. Helweg (1979) suggested an early indication of the biological fungicide degradative activity in soils in the degradation of a metabolite of benomyl. An increased rate of degradation was observed when an otherwise non-degrading soil is inoculated with a soil solution that has been treated with benomyl for six months. The resulting increase in degrading activity of the formerly inactive soil is believed to be microbial in nature After isolating them from degrading soils, Yarden (1985) reported the fungi, *Alternaria alternata* and *Bipolaris tetramera* were responsible for the degradation of carbendizam.

Triadimefon is also vulnerable to transformation by fungal organisms. In shake cultures, triadimefon was transformed to triadimenol by *Aspergillus niger* (Clark 1978). It is not disclosed as to from which environment the isolate was obtained but *A. niger* is known to be a widely distributed soil fungus (Radwan et al., 1991). In an investigation into metalaxyl behavior in avocado soils (sandy loam), Bailey (1985) reported that fungal and bacterial microflora were recovered that were capable of degrading the fungicide over a 45 day period. Sterilized soil in the same experiment displayed no metalaxyl disappearance.

Reports of bacterial-driven degradations of fungicides are quite common and it becomes obvious that bacterial sources provide most transformations of fungicides. Doneche (1983) isolated *Bacillus spp.* during in-vitro degradation studies that were capable of transforming the fungicide mancozeb and utilizing the chemical as a nutrient. *Bacillus spp.* are widely distributed in soils and, thus, may figure greatly in the widespread degradation of mancozeb in soil.

As indicated above, it should not be assumed that only one species of bacteria or fungi in a given environment is capable of fungicide degradation. Droby and Coffee (1991) isolated the fungi, *Trichoderma spp.*, *Fusarium spp.*, and *Penicillium spp.* along with the bacteria *Bacillus spp.* that were responsible for the biodegradation of metalaxyl in avocado soils. Goleveva (1991) successfully isolated 36 soil-inhabiting microbial strains capable of the transformation of vinclozolin. The most active of these were *Bacillus cereus, Bacillus brevis, and Pseudomonas fluorescens.* In addition to metabolizing vinclozolin, these bacteria were shown to utilize the fungicide as their sole source of carbon. Iprodione is a fungicide closely related to vinclozolin in chemistry and in breakdown products although different bacteria act in its metabolism. Arthrobacter spp. were found to degrade iprodione to 3,5-dichloroaniline and two other compounds in culture (Athiel 1995). This result was supported by evidence showing that the degradation of iprodione in soil could be positively impacted by the addition of *Arthrobacter* spp. (Mercadier et al., 1996). Rapid degradation of iprodione has also been attributed to the bacteria, *Pseudomonas putida* (Mitchell, et al., 1996). The isolation of soil fungi and bacteria capable of metabolism and subsequent degradation of fungicides is an important step in understanding the environmental fates of these chemicals. Future practices of fungicide application management, and bioremediation may well be served by the utilization of these organisms.

Moisture. Either air or water either occupies soil pores. The presence of water in soil greatly effects biological activity in the soil. Water is a catalyst for many biological processes that microorganisms perform as well as a solvent in metabolic activities. Fungicide degradation varies greatly with differing water regimes. Helweg (1979) noticed that in a sandy loam soil, there was an exponential increase in 2-aminobenzimidazole degradation when the moisture content of the soil was increased from 28% to 94% of field capacity. At water contents above 94%, the degree of degradation decreased. The decrease in degradation that accompanies high moisture contents may reflect the limits of aerobic microbial activity as oxygen becomes limiting.

Temperature. Temperature is a very important variable concerning biological processes. The pathways of biodegradation show a great dependence on temperature with regard to rate and intensity. Helweg (1979) reported on the variation in benomyl metabolite degradation based on the temperature of a sandy loam soil. Maximum degradation occurred at 22° C and was constant between 25° and 35° C. A decrease in degradation was noted at temperatures above 40° C. Heat sensitivity of microbial protein/enzyme systems are probably the main factor in the ultimate decrease in degradation at high temperatures.

pH. Soil pH affects fungicide dissipation rates. Many chemicals may undergo alterations depending on soil pH. A compound's charge determines the bonding reactions of which it is capable. Acidic soils may protonate a pesticide causing it to sorb to a greater or lesser extent to organic matter and mineral surfaces. Katayama, (1995) found that chlorothalonil underwent alkaline hydrolysis in aqueous solution at pH values greater than 8. Iprodione undergoes alkaline hydrolysis in aqueous solution at pH 8.7 in 14 hours (Cayley and Hide, 1980) to N-(3,5-dichloroanilinocarbonyl)-N-(isopropylaminocarbonyl) glycine (Belafdal et al., 1986). In soil, Walker et al., (1986a) observed that iprodione had a soil half-life of approximately 77 days when the soil pH was less than 5.5, and 23 days at pH 6.5. The degradation of vinclozolin has also been shown to be pH dependent (Walker, 1987a).

An extensive review of the environmental fate of chlorothalonil in relation to its use in Canada was recently published (Caux et al., 1996). This paragraph contains highlights from the review. Chlorothalonil is usually immobile in heavy textured soil, but can be moderately mobile in sandy soils. Soil half-lives typically range from 1 to 2 months. Chlorothalonil was detected in 4 out of 66 samples from field tile drains in the potato production region on Prince Edward Island and none of 752 rural domestic wells and 540 community wells tested in a national survey by the U.S. EPA. In solution chlorothalonil undergoes little photodegradation or hydrolysis at pH < 8.0. At pH values higher than eight, chlorothalonil undergoes a slow hydrolysis. Even though chlorothalonil undergoes extensive sorption to soil, it has been found that 90% of bacteria isolated from soil with a history of chlorothalonil use are capable of degrading the compound. However, when chlorothalonil is applied directly to a plant, 98% of the fungicide that reaches the plant remains unaltered.

Enhanced biodegradation. Often, single applications of a fungicide to an agronomic system will not provide prolonged protection against potential diseases. It has been reported, however, that repeated applications of fungicides will lead to a decrease in the level of disease control exhibited. This loss of control is often shown to be related to a degree of resistance that the target fungus develops to the fungicide. In other cases, the biodegradation of the fungicide is enhanced, leading to an accelerated disappearance of the chemical (Alexander 1994). The increased rate of biodegradation of a xenobiotic from a given environment resulting from its repeated application is termed enhanced biodegradation. The cause of this phenomenon may be the result of the build-up of a microbial population that is capable of utilizing the fungicide as a carbon source. Once a degrading population is established, additions of fungicide add carbon to the system and promote increases in the degrading population. Subsequent fungicide applications will ultimately be met by larger and larger populations of degrading organisms leading to the more rapid breakdown of the chemical. *Bacillus spp.* responsible for the degradation of mancozeb were found to increase 100-fold per gram of soil after only the second fungicide application (Doneche 1983).

Many studies review the concept of enhanced degradation and its effect on fungicide persistence. Yarden (1985) reported the enhanced degradation of the benomyl metabolite, carbendizam. Soil that had experienced a history of benomyl treatments exhibited shorter carbendizam persistence than soils with no benomyl history. Treatments of metalaxyl to previously treated soils proved to be more rapidly metabolized than treatments to soils unexposed to metalaxyl (Bailey 1986; Droby and Coffee, 1991). The production of the common vinclozilin and iprodione biodegradation product, 3,5-dichloroaniline, was shown to be accelerated in lettuce soils with a history of iprodione or vinclozilin applications (Martin 1991).

Enhanced degradation can cause dramatic changes in disappearance times of fungicides from the environment. Doneche (1983) reported 20-30% of applied mancozeb remained in a sterile loamy sand soil after 90 days. Application to unsterilized soil resulted in the complete disappearance of the parent compound after the same time period. The quarter-life of mancozeb was only 90 hours after a second application was made to the unsterilized soil, adding validity to the power of enhanced degradation.

In a laboratory study to investigate the degradation of iprodione and vinclozilin in a silt loam soil, Walker (1986b) suggested that enhanced degradation caused a shift in half-life from 23 days to 5 days after the first and second fungicide treatments, respectively. The same experiment conducted under field conditions showed 3% of the parent material remaining after 77 days following the first application. The second application resulted in less than 1% remaining after 18 days. In a similar study, iprodione in silty clay loam soils was found to exhibit a half-life of 30,12, and 4 days after the first, second, and third treatments, respectively (Walker 1987). A simultaneous study using vinclozilin resulted in half-lives of 30, 22, and 7 days after first, second, and third treatments, respectively.

Related studies on the enhanced degradation of iprodione and vinclozilin have further related the increased disappearance of the compounds from the soil to repeated applications. Slade (1992) reported similar results to those of Walker with half-lives of vinclozilin and iprodione in clay loam soil after first and second treatments to be 22 and 3.5 days, and 32 and 2 days, respectively.

In a 10 year experiment, Takagi et al., (1991) found that the rate of chlorothalonil degradation, applied at a rate of 57 mg kg^{-1} yr.$^{-1}$, in the plow layer initially increased as a result of repeated applications over years, suggesting an enhancement in degradation rate. This rate decreased when chlorothalonil was applied in conjunction with γ-BCH lindane and increased when applied in combination with organic compost. However, late in the experiment, the rate of chlorothalonil degradation decreased in soils receiving chlorothalonil alone, but the rate remained high in soils receiving chlorothalonil and compost. They hypothesized that in the non-composted soils, chlorothalonil was not completely degrading in a period of a year. As a result, chlorothalonil was building up and becoming toxic to the microbes capable of degrading it. This effect is suppressed with the addition of compost which lowers chemical concentrations in soil solution due to sorption, an increase of available nutrients, and supplies a fresh influx of microbes. In the later years of the experiment, Takagi et al. found that subsurface concentrations of chlorothalonil were high under the plots with decreased decomposition, but low under the compost amended plots. Recent studies have shown repeated applications of chlorothalonil result in a *decreased* degradation rate, which is different from most other pesticides (Motonaga et al., 1996). The lack of easily degradable organic matter in the soil was assumed to be the reason for the rate change. Interestingly, further applications resulted in a complete recovery of the degradation rate.

Fungicide degradation on turfgrass leaves. The decay rate of a fungicide on turfgrass leaves provides a measure of persistence as well as a means of chemical comparison. The decay rates of chloroneb and vinclozilin on turfgrass leaves were determined to be 0.01 and 0.09 day-1, respectively (Frederick et al., 1994). This relates to a canopy half-life of 69.3 d and 7.7 d, respectively. The dissipation rate of a fungicide on turfgrass leaves is generally lower than in the soil. As a comparison, chloroneb and vinclozilin half-lives in soil were reported to be, 2.66 and 0.81 d, respectively. Average iprodione, metalaxyl, and triadimefon half-lives in the turfgrass canopy were found to be 3.3, 3.7, and 3.6 d, respectively (Sigler, original data). These values are similar to those reported previously. Frederick et al. (1994) reported a canopy half-life of 1.98 d for triadimefon. Taylor (unpublished) reported values of 4.65 and 1.65 d for the half-lives of metalaxyl and triadimefon, respectively. Although the suggested half-lives appear to be indicative of high microbial activity, unfavorable conditions for microbial activity define the turfgrass canopy and create an inhospitable environment for microbial adaptation. It is our suggestion that the relatively small half-lives are likely the result of vegetation removal (organic surface partitioning) combined with some microbial activity.

Sorption to surfaces

Sorption, which includes adsorption and absorption, is an important force in determining the distribution of fungicides in the environment (see Table 1 for typical sorption values). Adsorption refers to the attachment of a material (fungicide) to soil particles and/or organic matter. Of the three types of soil particles; sand, silt, and clay, clay provides the surfaces with the most adsorptive potential. Soil organic matter, along with clay, provides numerous sites for the attachment and possible sequestering of xenobiotic compounds. As stated by Weber and Miller (1989),

Table I. Fungicide dissipation and retention characteristics for different environmental systems.

Fungicide	Medium	Loss %	Time[a]	Mechanism / Notes	Sorption[b]	Literature Cited
anilazine	soil	50	0.5 - 1			Nelson 1996
"	orthic luvisol	<2	110	from K_{ow}, K_{ow} by HPLC mineralization	K_{oc} 126	Dell et al. 1994 / Heitmann-Weber et al. 1994
"	soil, bound material	69		retained after 0.01 M CaCl$_2$ rinse		Mittelstaedt et al. 1987
"	soil, bound material	90	100	retained after 0.01 M CaCl$_2$ rinse		"
benomyl	mixture of peat moss, manure, clay & dolomite sand	99.97		from K_{ow}, K_{ow} by HPLC material used as filter	K_{oc} 3130	Dell et al. 1994 / Flaim & Toller 1989
"	soil	detected	<3 mo. >2yr.			Li & Nelson 1985
"	loam, 0-15 cm	detected	21 wks.			"
"	sandy loam	not detected	13 wks.			"
"	soil	50	4 wks.			Nelson 1996
"	soil	50	90 - 360			Willis & McDowell 1987
"	apple foliage	50	5.2-7.2			
benzimidazole	turfgrass thatch			proportional to concentration applied	% bound 19.9 - 93.2	Liu & Hsiang 1996
"	turfgrass leaves				% bound 46.2 - 56.9	
"	turfgrass thatch	50	2.5 wks.	bioassay		"

Table I (cont). Fungicide dissipation and retention characteristics for different environmental systems.

Fungicide	Medium	Loss %	Time[a]	Mechanism / Notes	Sorption[b]	Literature Cited
benzimidazole	turfgrass thatch			proportional to concentration applied	% bound 19.9 - 93.2	Liu & Hsiang 1996
"	turfgrass leaves				% bound 46.2 - 56.9	"
"						"
bromacil	turfgrass thatch	50	2.5 wk.	bioassay		Jury et al. 1987
"	soil	50	350		K_{oc} 0.072	Buyanovsky et al. 1988
captan	high clay, low pH soil	25	60	mineralization		Dell et al. 1994
"	high clay, low pH soil	36	60	all processes	K_{oc} 68.7	Giles & Blewett 1991
"	strawberry foliage	50	6.29	from K_{ow}, K_{ow} by HPLC		"
"	strawberry foliage	50	4.61	dislodgable residue study		"
"	strawberry foliage	50	5.6	dislodgable residue study		Jury et al. 1987
"	soil	50	3	dislodgable residue study	K_{oc} 0.033	Li & Nelson 1985
"	soil	not det.	1 wk.			Willis & McDowell 1987
"	strawberry foliage	50	9			
carbendazim	cellulose				K_d 4	Barak et al. 1983
"	ethyl cellulose				K_d 36	"
"	pepper lignin				K_d 700	"
"	pine lignin				K_d 590	"
"	protein (BSA)				K_d 7	"
"	soil	50	20			Yarden et al. 1985
"	soil, disinfected	50	40-55			"
"	soil, pre-treated	50	11-4.0	enhanced degradation.		"

Continued on next page.

Table I (cont). Fungicide dissipation and retention characteristics for different environmental systems.

Fungicide	Medium	Loss %	Time[a]	Mechanism / Notes	Sorption[b]	Literature Cited
chloroneb	turfgrass thatch				K_{oc} 526	Dell et al. 1994
"	soil				K_{oc} 714	"
"				from K_{ow}, K_{ow} by HPLC	K_{oc} 213	Frederick et al. 1994
"	soil under turf	50	2.7			"
"	thatch, KBG[c]	50	8.7			"
"	leaves, KBG	50	69.3			"
"	soil	50	10 - 180			Nelson 1996
chlorothalonil	aerobic sandy loam	50	1-2 mo.			Caux et al. 1996
"	aerobic water/sediment	50	<2 hr			"
"	grape leaves	50	10 - 15 d			Caux et al. 1996
"	non-sterile soil	50	10.3-36.5			"
"	potato foliage	50	3.6-21			"
"	sandy loam				K_d 1000	"
"	sandy soil				K_d 794	"
"	silt				K_d 6310	"
"	silt loam & clay loam	≥97	168 d,	photodegradation		"
"	silty clay loam				K_d 1259	"
"	sterile soil	50	18-213			"
"	water	1.8% d^{-1}		hydrolysis pH>8		"
"	soil	50	68	from K_{ow}, K_{ow} by HPLC	K_{oc} 1380	Gustafson 1989
"	soil				K_{oc} 201	Dell et al. 1994
"	soil		7		K_{oc} 1.38	Jury et al. 1987
"	soil, no fertilizer	92.3	14	abiotic + biotic dissipation		Katayama et al. 1995
"	soil, chemical fertilizer	93.6	14	abiotic + biotic dissipation		"
"	soil, farmyard manure	99.95	14	abiotic + biotic dissipation		"

Table I (cont). Fungicide dissipation and retention characteristics for different environmental systems.

Fungicide	Medium	Loss %	Time[a]	Mechanism / Notes	Sorption[b]	Literature Cited
chlorothalonil	soil	50	5 - 90 d			Nelson 1996
"	sand from cranberry bog				K_d / K_{om} 25 / 568	Reduker et al. 1988
"	apple foliage	50	4.1±1.5			Willis & McDowell 1987
cymoxanil	soil		<2 wk.			Cohen & Coffey 1986
cyproconazole	soil	50	80 - 100			Nelson 1996
dicloran	rinsate	53	24 hr.	photolysis, anatase (TiO_2) high fung. concentrations		Chiarenzelli et al. 1995
"	rinsate + soil	57	24 hr.	photolysis, anatase (TiO_2) high fung. concentrations		"
dihydroxy-anilazine		<8	110	mineralization		Heitmann-Weber et al. 1994
diniconazole	Ca-Montmorillonite, pH 4				K_{oc} 750	Weber & Swain 1993
"	Ca-Montmorillonite, pH 7				K_{oc} 134	"
"	Ca-OM, pH 3				K_{oc} 3965	"
"	Ca-OM, pH 4				K_{oc} 2140	"
etridiazole	soil	50	20		K_{oc} 555-1568	Nelson 1996
ETU	soil column	50	2.5	ETU = ethylenethiourea		Calumpang et al. 1993
EU	soil column	50	4.8	EU = ethyleneurea		Calumpang et al. 1993
fenamiphos	basil mineral media & putting green soil	70	7			Ou & Thomas 1994
"	soil	50	38 - 313			"
"	soil extract from putting green soil, culture	70	3			"

Continued on next page.

Table I (cont). Fungicide dissipation and retention characteristics for different environmental systems.

Fungicide	Medium	Loss %	Time[a]	Mechanism / Notes	Sorption[b]	Literature Cited
fenarimol	cellulose				K_d 20	Barak et al. 1983
"	ethyl cellulose				K_d 102	"
"	pepper lignin				K_d 2300	"
"	pine lignin				K_d 660	"
"	protein (BSA)				K_d 211	"
"	Al-montmorillonite	70	38	100°C with CO_2 & NH_3		Fusi et al. 1983
"	no clay	no change	several days	100°C with CO_2 & NH_4		"
"	KBG, leaves				K_d / K_{oc} 155 / 144	Lickfeldt & Branham 1995
"	KGB, thatch				K_d / K_{oc} 86.5 / 218	"
flutolanil	soil	50	20 - 365			Nelson 1996
"	soil	50	40 - 60			"
fosetyl-Al	Osaka soil				K_d 7.94	Uchida et al. 1982
"	soil	50	< 1.5 hr.			Nelson 1996
iprodione	Soil 1.7-4.7% OM	50	< 1.8 hr			Cohen & Coffey 1986
"	sandy loam				K_d / K_{om} 7.9 / 790	Cayley & Hide 1980
"	loam				K_d / K_{om} 26 / 369	"
"	Fen peat				K_d / K_{om} 101 / 253	"
"				from K_{ow}, K_{ow} by HPLC	K_{oc} 261	Dell et al. 1994
"	lettuce soil	81		field effectiveness		Martin et al. 1991
"	lettuce soil	58		field effectiveness		

Table I (cont). Fungicide dissipation and retention characteristics for different environmental systems.

Fungicide	Medium	Loss %	Time[a]	Mechanism / Notes	Sorption[b]	Literature Cited
iprodione	soil	50	7 - 160			Nelson 1996
"	clay loam	50	>35			Slade et al. 1992
"	clay loam, 3 treatments	50	2	enhanced degradation		"
"	soil, pH 6.5 1st appl.	50	30	enhanced degradation		Walker 1987
"	soil, pH 6.5 2nd appl.	50	12	enhanced degradation		"
"	soil, pH 6.5 3rd appl.	50	4	enhanced degradation		"
"	soil	90	3 - 24			Walker 1987 (2)
"	grassed soil, no history of use	50	10, 30			Walker & Welch 1990
"	soil, pH 5.7	50	60			"
"	soil, pH 6.5	50	30-35			"
"	soil, no history of use	3	77			Walker et al. 1986
mancozeb	soil column	50	2.9			Calumpang et al. 1993
"	soil	50	6 - 139			Nelson 1996
maneb	microagroecosystem	50	36			Calumpang et al. 1993
"	soil	50	4-8 wk.			"
"	snap pea foliage	50	2.8			Willis & McDowell 1987
"	tomato foliage	50	3.2			
metalaxyl	soil	50	28 - 14	enhanced degradation	K_d 0.2 - 0.6	Bailey & Coffey 1985
"	soil, previous treat.	50	14			Bailey & Coffey 1986
"	soil	14 - 16	84	^{14}C-CO_2 formation only		"
"	soil, 0-15 cm	50	30			"
"	soil, 15-30 cm	50	15			Cohen & Coffey 1986
"	soil, 30-45 cm	50	18			"

[a] Units are days unless otherwise indicated
[b] Retention coefficient (K_{oc}, Koc, Kom, % bound)
[c] KBG = Kentucky bluegrass

sorption can affect the relative fraction of the organic chemical that is in each phase (solid, aqueous, and vapor). The batch isotherm, used to determine partition coefficients, or K_d values, is usually the method of choice for sorption isotherms based on its ease, speed, and acceptance by the EPA (Green et al., 1980). This procedure involves equilibrating a known amount of pesticide with a sorbent, such as soil, plant material, or organic solvent, in a liquid solution which is usually aqueous. Once equilibrium has been reached, the amount of pesticide present in the aqueous phase is determined. The amount of pesticide sorbed is then calculated as the difference between the initial and final amount of pesticide in solution. Commonly, data is fitted to the Fruenlich equation from which the partition coefficient of the fungicide can be determined:

$$x_m = K_f C_e^{\ n}$$

where x_m = amount of pesticide sorbed (μmol kg^{-1}), C_e = pesticide concentration at equilibrium (μ mol L^{-1}), n = describes the type of sorption reaction, and K_f = partition coefficient (μmol $^{(1-n)}$ L^n kg^{-1}). If n = 1, the system is linear and K_f is often replaced by K_d (L Kg^{-1}). Insights into the nature of the n value are available by Weber and Miller (1989). Despite general acceptance in the methodology of sorption, there is still little agreement on the best standard procedure. Chemicals are thought to sorb to soil from the soil solution or water films which coat soil particles. In the field, pesticide sorption to plants involves wetting and drying cycles which may cause significant differences in chemical behavior. Another consideration under scrutiny is the reliability of sorption data calculated from loss of pesticide from solution instead of actual chemical mass on the sorbent. Spieszalski et al. (1994) determined sorption coefficients for the insecticides, chlropyrifos and fonofos, on thatch and soil. They performed the experiments using the traditional batch isotherm method modified to account for the amount of pesticide on the solid phase. They were able to tabulate a chemical mass balance. The modified method tended to give lower K_d values than the traditional method. For example, the K_d of chlropyrifos determined with the mass balance method averaged 40 - 85% of the K_d values determined by the traditional method. They suggested that volatilization, degradation and non-uniform centrifugation were contributing to chemical loss from the soil solution. The losses from the solution phase are unaccounted for with the traditional method.

Partitioning of non-ionic compounds has been successfully correlated with octanol-water partition coefficients in many plant and soil systems. Uchida et al. (1982) compared sorption coefficients determined in the lab for buprofezin (an insecticide) and flutolanil to those predicted from previously formulated equations based on octanol-water partition coefficients and observations with a set of neutral fungicides in soil and rice plants. The log K_d values determined experimentally, 1.48 and 0.9 in soil and 2.02 and 1.81 in rice plants for buprofezin and flutolanil, respectively, correlated well with those predicted from the equations. They concluded that the equations derived from neutral fungicides had extended use to compounds with unrelated structure. Predicting the behavior of ionic compounds is more difficult, as the chemicals react as a function of the sorbent and the pH in the system. Weber and Swain (1993) found that sorption coefficient (K_{oc}) of diniconazole for Ca-organic matter and Ca-montmorillonite almost doubled when solution pH was lowered from 4 to 3.

By controlling sorption, two vital constituents of soil, organic matter and clay effect many processes in the biodegradation of fungicides in the soil. The primary impact of the presence of clays and organic matter results from a physical/chemical reaction, not a biological one, in which the solution levels of the chemical are lowered. In general, a fungicide located on the soil surface has a lesser chance of being transformed biotically than one located in the aqueous environment of the soil solution. Adsorptive surfaces present non-aqueous-type environments will tend to attach fungicides and prevent them from being biotically degraded although a bound fungicide may be persistent in the environment and be subject to a more long-term degradative fate. Retention on surfaces will not necessarily alter the eventual biodegradative fate of the fungicide but may effect the rate of degradation.

Adsorption coefficients have been reported for several fungicides on plant surfaces, thatch, and in soil (Dell et al., 1994; Taylor, unpublished data; and Horst et al, 1996). The overall amount of fungicide adsorption in soils is complicated by variation in differing soil organic matter and clay contents. Adsorption usually increases as a function of increasing clay and organic matter percentage. The benomyl derivative carbendizam showed enhanced adsorption in soils as organic matter content increased (Helweg 1977). Cayley (1980) reported that iprodione and vinclozolin adsorption was strong for peat-type soils and loams, but weak in sandy soils possessing lower organic matter content. It was also reported that both fungicides increased in adsorptive behavior as organic matter content of the soils was increased. The half-life of benomyl in soils ranges from 10.5 months in a silt loam soil to 6 months in fine sand containing lower organic matter and clay (Baude 1974). It has also been suggested recently that benomyl degradation is inversely related to the organic matter and clay content of soil (Liu 1994). Doneche (1983) described the effect of clay content on the biodegradation of mancozeb. In between 30 and 90 days of incubation of mancozeb in loamy sand, 17.2% of the fungicide was degraded as opposed to 19.2% degradation when there was no clay present. The somewhat small increase in degradation can be partially attributable to the small amount of clay present initially (7.4%) when compared to the sand percentage (85.6%).

If organic material is not present or present in only small amounts, hydrophobic chemicals will sorb to other materials if the reaction provides a more energetically favorable microenvironment. When studying chemical mobility in soil columns, soil type was found to be a factor in the disappearance of dichloran, iprodione and vinclozolin (Elmer and Stipes, 1985). Pesticide dissipation, measured by soil extraction and bioassay, occurred more rapidly in a loam than a sandy loam soil for all three compounds. It was hypothesized that vinclozolin, which was detected at 25 cm in the loam soil and 35 cm in the sandy loam soil, may sorb to the clay in the loam soil because of the pesticide's high cationic charge. Sukop and Cogger (1992) found that sorption of metalaxyl to finer textured subsurface soils was greater than or equal to that of surface soils. They pointed out that if the movement and leaching potential of metalaxyl was established from K_{oc} values derived solely from surface soils, potential chemical movement would be overestimated.

Not all fungicides have been reported to behave similarly in differing clay/organic matter environments. Iprodione degradation was shown to actually correlate quite poorly (r = -0.181) with the amount of clay present in the soil (Martin 1991). Similarly, metalaxyl sorption is not well correlated to K_{oc} in low organic carbon soils and it was suggested that sorption to mineral phases was significant (Sukop and Cogger, 1992). This observation is consistent with the high variability reported for movement of metalaxyl in soil columns (Horst et al., 1996). Valverde-Garcia et al. (1988) determined that although sorption of thiram was correlated to soil organic matter, adsorption of dimethoate (an insecticide) was more closely related to total soil surface area and cation exchange capacity. These observations suggest caution in making predictions using K_{oc} values for estimating movement and availability of fungicides in the environment. These effects can be extended to the areas of fungicide environmental toxicity. In a study to determine the effects of benomyl and carbendazim on earthworms, Lofs-Holmin (1981) correlated earthworm toxicity with soil texture and mineral phases as well as organic matter.

The attachment of fungicides to soil organic matter and clay presents a complex scenario when determining a fungicide's degradative fate. Because K_{oc} values are indicative of the mobility of chemicals associated with organic matter, they are useful in predicting the possibility of microbial degradation of fungicides. Cova et al. (1990) reported that pesticide K_{oc} values of less than 500 are indicators of potential groundwater contamination, thus, the fungicides mentioned thus far represent varying environmental risk. K_{oc} values higher than 500 are inherent in chemicals that are relatively non-mobile from the site of deposition and exhibit limited off-target movement. Through understanding the adsorption characteristics of specific fungicides, a realistic grasp of biotic fungicide degradation can be achieved.

The role of turfgrass leaves and thatch in fungicide fates

Higher rates of fungicide dissipation in soils are thought to be attributable to an increased level of microbial activity. However, unlike the soil environment, the conditions in the plant canopy are not conducive to great microbial activity. Exposure to ultra-violet light and low water availability combine to create an inhospitable environment for microbial activity and population. Exposure to UV light has been shown to damage nucleic acids (Miller and Kokjohn, 1990) and will effect microorganisms exposed to radiation (Gascon et al., 1995). Additionally, large changes in free water at the leaf surface may effect microbial populations. Electron microscopy has shown that microcolonies of bacteria are commonly found in crevices of the leaf surface (deCleene et al., 1989). It is speculated that the microbial need for water and associated nutrients drives this behavior. These conditions have an effect on the numbers of organisms that inhabit the leaf surface. The number of bacteria per gram of turfgrass leaves is comparable to that of soil (10^7-10^8), however, the greater density of soil equates into much larger bacterial populations inhabiting the lithosphere (Sigler, unpublished). This is not to discount the presence of biotic activity on the leaf surface.

Although the soil is a significant site of processes that determine pesticide fate, the sink-like nature of the turfgrass canopy would presumably predispose itself to the types of biodegradation and enhanced biodegradation commonly found in the lithosphere. Due to its extreme density, the turfgrass canopy and thatch represents a considerable barrier to the penetration of fungicides into the soil as well as a possible site of fungicide accumulation. Turfgrass is unique in that it has three perennial sorption matrices: leaves, thatch, and soil. Adsorption to leaf surfaces and to turfgrass thatch has a large impact on the environmental fate of organic chemicals, including fungicides (Lickfeldt and Branham, 1995). Turfgrass thatch is a layer of partially decomposed plant material found between the living plants and soil, and is high in lignin and cellulose (Hurto et al., 1980). Thatch is of particular importance as it has been shown to be very effective in retaining pesticides (Dell et al., 1994, Niemczyk et al., 1987, Spieszalski et al., 1994). Using bioassay, Liu and Hsiang (1996) found that 19.5% to 93.2% of benomyl was bound to thatch and 46.2% to 56.9% to creeping bentgrass clippings. Starrett et al. (1996) recovered 19% isazofos, 79% chlorpyrifos, 24% of metalaxyl and 71% pendimethalin applied to thatch. Despite high K_d values, partition coefficients normalized to organic matter are often lower than those for soils. This may reflect fungicide sorption to non-organic surfaces discussed previously. When studying sorption of pesticides to Kentucky bluegrass, Lickfeldt and Branham (1995) showed lower K_{oc} values on thatch than for leaves, for ethoprop, 1,2,4-TBC, fenarimol, and phenanthrene, but not acetanilide which was more water soluble. Thatch K_{oc} coefficients were smaller than literature soil K_{oc} values, a phenomenon also reported by Dell et al. (1994) and Spieszalski et al. (1994). Moreover, Lickfeldt and Branham (1995) warned that equations derived from soil systems did not work well for thatch or leaves. Thatch can inhibit pesticide movement to such an extent that only small amounts of pesticides may reach the soil surface. In a field experiment, Frederick et al. (1996) measured concentrations of vinclozolin initially after spraying at around 700 μg g^{-1} on leaves, 17 μg g^{-1} on thatch, and less than 1 μg g^{-1} in the soil. Rueegg and Siegfried (1996) found similar trends of distribution when measuring difenoconazole and penzonasole plus captan residues on grass and soil in a Switzerland apple orchard. Nightingale (1987) conducted a three-year study designed to detect the presence of organic pollutants in urban recharge basins. Three of the five basins studied were completely covered with turf. They detected no loading of organophosphorus pesticides and concluded that outside of the insecticide chlorodane, the pollution potential of the basins was insignificant.

The adsorptive nature of turfgrass leaves and thatch was investigated by Liu and Hsiang (1996) in separate studies. Benomyl was found to bind heavily to both substrates. Between 46.2% and 59.9% of applied benomyl was bound by turf leaves while 20% - 93% was bound to thatch. The levels of bound residues were determined by bioassay and are not necessarily indicative of field conditions. Additionally, metalaxyl was determined to be seven times more prevalent in thatch than in the verdure of a Kentucky bluegrass turf system throughout a 113 day post-application monitoring period (Horst et al.,1996). Iprodione (K_{oc}~392), metalaxyl (K_{oc}~30), and triadimefon (K_{oc}~171) were shown to exhibit sorptive behavior on turfgrass clippings (Taylor, unpublished). Although leaf K_{oc} values were generally on the lower end of published values of soil K_{oc} (iprodione~253-1300, metalaxyl~29-

287, and triadimefon~73-345), adsorptive capacity of fungicides to the leaf surface presents an indication of at least temporary fungicide binding. Further research indicated that only 35, 32, and 43% of applied iprodione, metalaxyl, and triadimefon was solvent-extractable from turfgrass leaves after a 16-day incubation period (Sigler, unpublished), suggesting that a strong binding/uptake mechanism guides the adsorption of fungicides to the leaf surface.

While limited information exists concerning the populations of degradative microorganisms on plant material one could probably assume that some microbial activity would take place in this environment given the proper conditions. Turfgrass leaves have been reported to intercept a large proportion of applied pesticides. Stahnke et al. (1991) showed that a dense turf canopy retained 95% of an applied pesticide and, thus, is a major sink for fungicide deposition. However, the turfgrass canopy has been sparingly investigated and little information exists concerning this site and the processes that govern the biodegradation of applied chemicals.

Future direction

Information concerning the microbial population of specific environments is useful in understanding life processes such as the metabolism of substrates, and degradative pathways. By combining traditional techniques of biodegradation analysis with new molecular methods of population fingerprinting, the ecology and activity of a given environmental system can be assessed. Chlorothalonil and captafol were found to decrease bacterial and fungal populations for one week in a sandy loam (Tu, 1993). Dormanns-Simon (1996) found that soil treated with dithane actually had increased levels of bacteria and actinomycetes, probably due to decreasing competition from fungal communities. However, mancozeb decreased nitrification and reduced levels of bacteria, fungi, and actinomycetes for three months in a fraction of vineyard soils tested (Doneche et al., 1983). This trend for minor and temporary shifts in populations has been noted for systems under a variety of management practices. For example, captan had no substantial effect on fungal communities in conventionally tilled or no-till soil or residue (Beare et al., 1993). Mancozeb when applied at recommended rates and at 10x, had no lasting or notable effects on total numbers of ciliates, testaceans, rotifers or nematodes in a spruce forest soil although ciliate community structure remained slightly shifted 90 days after application (Petz and Foissner, 1989). Captan, but not benlate, inhibited NH_4^+ oxidation in soil used for pot grown tomatoes (Somda et al., 1991). Fungicide application has even been shown to maintain microorganism communities on the surface of a vegetable crop. Results from a two year field study using maneb resulted in no effect on bacterial populations of the lettuce phylloplane (Mercier and Reeleder, 1987).

Characterization of bacteria capable of degrading fungicides is important to the efficacy and persistence of the chemical. An avenue of future research should include the use of molecular techniques for investigating the microbial ecology of soil systems in which degradative capabilities have been detected. The methods currently available allow for the detection of active organisms in their respective environments, especially the niche-type environments that are always present in plant/soil systems.

Acknowledgments. This work was supported by grants from the United States Golf Association, the Midwest Regional Turf Foundation, and the Purdue Agricultural Experiment Station.

Literature Cited.

Agnew, M.L; Lewis, D. A survey of pesticides used on Iowa golf courses in 1990. Iowa State University: Ames, IA. 1993. FG-460.

Alexander, M. *Biodegradation and Bioremediation*; Academic Press, Inc.: San Diego, CA, 1994. pp. 8-16.

Anan'yeva, N.D.; Demkina, T.S.; Blagodatskya, Ye. V.; Abelentzev, V.I. *Eurasian Soil Sci.* **1996**, *28*, pp. 157-167.

Athiel, P.; Alfizar, Mercadier, C.; Vega, D.; Bastide, J.; Davet, P.; Brunel, B.; Clayet-Marel, J.-C. *J. Appl. Environ. Micro.* **1995**, *61*, pp. 3216-3220.

Bailey, A. M.; Coffey, M. D. *Pytopath.* **1985**, *75*, pp. 135-137.

Bailey, A. M.; Coffee, M. D. *Can. J. Microbiol.* **1986**, *32*, pp. 562-569.

Baude, F. J.; Pease, H. L.; Holt, R. F. *J. Agric. Food Chem.* **1974**, *22*, pp. 413-418.

Beare, M. H.; Pohlad, B. R.; Wright, D. H.; Coleman, D. C. *Soil. Sci. Soc. Am. J.* **1993**, *57*, pp. 392-399.

Belafdal, O.; Bergon, M.; Calmon, J. *Pestic. Sci.* **1986**, *17*, pp. 335-342.

Bollag, J. M.; Liu, S. Y. *Pesticides in the soil environment*; SSSA Book Series; Soil Sci. Soc. Am.: Madison, WI, 1990; No. 2, pp. 169-211.

Buyanovsky, G. A.; Pieczonka, G. J.; Wagner, G. H.; Fairchild, M. L. *Bull. Environ. Contam. Tox.* **1988**, *40*, pp. 689-695.

Caux, P. Y.; Kent, R. A.; Fan, G. T.; Stephenson, G. L. *Crit. Rev. Environ. Sci. Tech.* **1996**, *26*, pp. 45-93.

Cayley, G. R.; Hide, G. A. *Pest. Sci.* **1980**, *11*, pp. 15-19.

Clark, T.; Clifford, D. R.; Deas, A. H. B.; Gendle, P.; Watkins, D. A. M. *Pestic. Sci.* **1978**, *9*, pp. 497-506.

Cova, D.; Molinari, G. P.; Rossini, L. *Ecotox. Environ. Safe.* **1990**, *20*, pp. 234-240.

DeCleene, M. *EPPO Bull.* **1989**, *19*, pp. 81-89.

Dell, C. J.; Throssell, C. S.; Bischoff, M.; Turco, R. F. *J. Environ. Qual.* **1994**, *23*, pp. 92-96.

Doneche, B.; Seguin, G.; Ribereau-Gayon, P. *Soil Sci.* **1983**, *135*, pp. 361-366.

Dormanns-Simon, E. *J. Environ. Sci. Health. B.* **1996**, *31*, pp. 573-576.

Droby, S.; Coffee, M. D. *Gen. Appl. Biol.* **1991**, *118, pp.* 543-553.

Elmer, W. H.; Stipes, R. J. *Plant Disease.* **1985**, *69,* pp. 292-294.

Frederick, E. K.; Bischoff, M.; Throssell, C. S.; Turco, R. F. *Bull. Environ. Cont. Toxicol.* **1994**, *53,* pp. 536-542.

Frederick, E. K.; Bischoff, M.; Throssell, C. S.; Turco, R. F. *Bull. Environ. Cont. Toxicol.* **1996**, *57,* pp. 391-398.

Gascon, J.; Oubina, A.; Perez-Lezaun, A.; Urmeneta, J. *Cur. Micro.* **1995**, *30*, pp. 177-182.

Golovleva, L. A.; Finkelstein, Z.I.; Polyakova, A. V.; Baskunov, B. P.; Nefedova M. Y. *J. Environ. Sci. Health. B.* **1991**, *26*, pp. 293-307.

148

Green, R. E.; Davidson J. M.; Biggar J. W. In *Agrochemicals in Soil*; Banin, A.; Kafkafi, V., Ed.; Pergamon Press: New York, NY, 1980, pp. 241-251.

Hatrik, S.; Tekel, J. *J. Chrom. A.* **1996**, *733*, pp. 217-233.

Helweg, A. *Pest. Sci.* **1977**, *8*, pp. 71-78.

Helweg, A. *Water, Air Soil Poll.* **1979**, *12*, pp. 275-281.

Hock, W. K.; Sisler, H. D. *J. Agric. Food Chem.* **1969**, *17*, pp. 123-128.

Horst, G. L.; Shea, P. J.; Christians, N.; Miller, D. R.; Stuefer-Powell, C.; Starrett, S. K. *Crop Sci.* **1996**, *36, pp.* 362-370.

Hrelia, P.; Maffei, F.; Fimognari, C.; Vigangni, F.; Cantelli-Forti, G. *Mutat. Res.* **1996**, *369*, pp. 81-86.

Hurto, K. A.; Turgeon, A. J.; Spomer, L. A. *Agron. J.* **1980**, *72*, pp. 165-167.

Jury, W. A.; Spencer, W. F.; Farmer, W. J. *J. Environ. Qual* **1983**, *12*, pp. 558-564.

Katayama, A.; Mori, T.; Kuwatsuka, S. *Soil Biol. Biochem.* **1995**, *27*, pp. 147-151.

Lickfeldt, D. W.; Branham, B. E. *J. Environ. Qual* **1995**, *24*, pp. 980-985.

Liu, L. X.; Hsiang, T. *Soil Biol. Biochem.* **1994**, *26*, pp. 317-324.

Liu, L. X.; Hsiang, T. *Pest. Sci.* **1996**, *46*, pp. 139-143.

Lofs-Holmin, A. *Swedish J. Agric. Res.* **1981**, *11*, pp. 141-147.

Martin, C.; Davet, P.; Vega, D.; Coste, C. *Pest. Sci.* **1991**, *32*, pp. 427-438.

Martin, C.; Vega, D.; Bastide, J.; Davet, P. *Plant Soil* **1990**, *127*, pp. 140-142.

Mercadier, C.; Garcia, D.; Vega, D.; Bastide, J.; Coste, C. *Soil Biol. Biochem.* **1996**, *12*, pp. 1791-1796.

Mercier, J.; R. K. Reeleder. *Can. J. Microbiol.* **1987**, *33*, pp. 212-216.

Miller, R. V.; Kokjohn, T. A. *Ann. Rev. Micro.* **1990**, *44*, pp. 365-394.

Mitchell, J. A.; Cain, R. B. *Pest. Sci.* **1996**, *48*, pp. 1-11.

Motonaga, K.; Takagi, K.; Matumoto, S. *Biol. Fert. Soils.* **1996**, *23*, pp. 340-345.

Munnecke, D. M.; Johnson, L. M.; Talbot, H. W.; Barik, S. In *Biodegradation and detoxification of Environmental pollutants*; Chakrabarty, A. M., Ed.; CRC Press: Boca Raton, FL, 1982; pp. 1-33.

Niemczyk, H. D.; R. A. Chapman. *J. Econ. Entomol.* **1987**, *80*, pp. 880-882.

Nightingale, H. I. *Soil Sci.* **1987**, *144*, pp. 373-382.

Petz, W.; Foissner, W. *Biol. Fertil. Soils.* **1989**, *7*, pp. 225-231.

Pimentel, D.; Levitan, L. *BioSci.* **1986**, *36*, pp. 86-91.

Potter, D. A.; Buxton, M. C.; Redmond, C. T.; Patterson, C. G.; Powell, A. J. *J. Econ. Entom.* **1990**, *83*, pp. 2362-2369.

Radwan, M. A.; Salama, A. K.; Ahmed, A. H.; Alexandria J. *Agri. Res.* **1991**, *36*, pp. 315-325.

Rueegg, J.; Siegfried, W. *Crop Prot.* **1996**, *15*, pp. 27-31.

Sharom, M. S.; Edgington, L. V. *Can. J. Plant Path.* **1982**, *4*, pp. 334-340.

Slade, E. A.; Fullerton, R. A. *Pest. Sci.* **1992**, *35*, pp. 95-100.

Somda, Z. C.; Phatak, S. C.; Mills, H. A. *J. Plant Nut.* **1991**, *14*, pp. 1187-1199.

Soudamini, M.; Awasthi, M. D.; Mohapatra, S. *Pest. Res. J.* **1997**, *9*, pp. 62-66.

Spieszlski, W. W.; Niemczyk, H. D.; Shetlar, D. J. *J. Environ. Sci. Health. B.* **1994**, *29*, pp. 1117-1136.

Stahnke, G. K.; Shea, P. J.; Tupy, D. R.; Stougaard, R. N.; Shearman, R. C.. *Weed Sci.* **1991**, *39*, pp. 97-103.

Starrett, S. K.; Christians, N. E.; Austin, T. A. *J. Environ. Qual.* **1996**, *25*, pp. 566-571.

Sukop, M.; C. G. Cogger. *J. Environ. Sci. Health. B* **1992**, 27, pp. 565-590.

Takagi, K.; Wada, H.; Yamazaki, S. *Soil Sci. Plant Nutrition* **1991**, *37*, pp. 583-590.

Tilman, D.; Knops, J.; Wedin, D.; Reich, P.; Ritchie, M.; Siemann, E. *Sci.* **1997**, *277*, pp. 1300-1302.

Tsuda, T.; Aoki, S.; Kojima, M.; Fujita, T. *Comp. Biochem. Physiol.* **1992**, *1*, pp. 63-66.

Tu, C. M. *J. Environ. Sci. Health. B* **1993**, *28*, pp. 67-80.

Uchida, M.; Nishizawa, H.; Suzuki, T. *J. Pesticide Sci.* **1982**, *7*, pp. 397-400.

Valverde-Garcia, A.; Gonzalez-Pradas, E.; Villafranca-Sanchez, M.; Del Rey-Bueno, F.; Garcia-Rodriguez, A. *Soil Sci. Soc. Am. J.* **1988**, *52*, pp. 1571-1574.

Walker, A. *Pestic. Sci.* **1987**, *21*, pp. 219-231.

Walker, A. *Soil. Pestic. Sci.* **1987**, *21*, pp. 233-240.

Walker, A.; Welch S. J. In *Enhanced biodegradation of pesticides in the environment.* Racke, K. D.; Coats, J. R., Ed.; ACS Books; Am. Chemical Society: Washington D.C., 1986, pp. 53-67.

Walker, A.; Brown, P. A.; Entwistle, A. R. *Pestic. Sci.* **1986**, *17*, pp. 183-193.

Weber, J. B.; Miller, C. T. In *Reactions and Movement of Organic Chemicals in Soils.* Soil Sci. Soc. Am. Madison, WI, 1989; Spec. Pub. 22, pp. 305-334.

Weber, J. B.; Swain, L. R. *Soil Sci.* **1993**, *156*, pp. 171-177.

Wolfe, N. L.; Mingelgrin, U.; Miller, G. C. In *Pesticides in the soil environment*; Soil Sci. Soc. Am. Madison, WI, 1989; Spec. Pub. 22, pp. 103-168.

Yarden, O.; Katan, J.; Aharonson, N.; Ben-Yephet, Y. *Phytopath.* **1985**, *75*, pp. 763-767.

Chapter 9

Application of DMI Fungicides to Turfgrass with Three Delivery Systems

G. L. Schumann[1], J. Marshall Clark[2], J. J. Doherty[2], and B. B. Clarke[3]

Departments of [1]Microbiology and [2]Entomology, University of Massachusetts, Amherst, MA 01003
[3]Department of Plant Pathology, Rutgers University, New Brunswick, NJ 08903

Three DMI fungicides (fenarimol, propiconazole, and triadimefon) were applied to Kentucky bluegrass using a standard boom sprayer, a high-pressure injector, and a modified slicer/seeder with a gravity-feed applicator. Applications were followed immediately with 1.3 cm of irrigation. The fungicides and one breakdown product were monitored in the leaves, roots, thatch, and soil profile for 28 days post-application. No fungicide was detected in the roots in any appreciable amount. No fungicide was detected below the top 5.1 cm of the soil, however, triadimefon was converted to the more mobile triadimenol fairly rapidly in all four matrices.

Ectotrophic, root-infecting (ERI) fungi cause severe injury to turfgrasses under modern management conditions and are some of the most difficult pathogens to control. These fungal pathogens grow along the outside of turfgrass roots and crowns and infect these tissues when plants are stressed. Aboveground symptoms include patches or rings of dead and dying plants that do not appear until root deterioration is advanced. Diseases caused by these fungi, often called "patch diseases," include necrotic ring spot (*Leptosphaeria korrae* J.C. Walker & A.M. Sm.), spring dead spot (*Ophiosphaerella herpotricha* (Fr.:Fr.) L. Holm and other ectotrophic species), summer patch (*Magnaporthe poae* Landschoot & Jackson), take-all patch (*Gaeumannomyces graminis* (Sacc.) Arx & D. Olivier var. *avenae* (E.M. Turner) Dennis) and others. Necrotic ring spot and summer patch are diseases of bluegrasses (*Poa* spp.) and some fine fescues (*Festuca* spp.); spring dead spot is a disease of bermudagrass (*Cynodon* spp.); take-all patch is a disease of bentgrasses (*Agrostis* spp.).

Diseases caused by ERI fungi are increasingly important in intensively managed turf. These turfgrass swards are frequently of similar genotype. They are also subjected to numerous stresses such as frequent low mowing height, soil compaction, temperature and moisture extremes, and the non-target effects of pesticides (*1*). According to Dernoeden

(2), there are at least four reasons why ERI fungi cause such destructive diseases: (a) limited availability of disease resistant germplasm; (b) incomplete understanding of pathogen biology and disease etiology; (c) few field studies have been conducted to investigate the effects of cultural practices on patch disease severity; and (d) inconsistent levels of disease control with fungicides.

General cultural practices employed to reduce the effects of diseases caused by ERI fungi include avoiding low mowing heights, especially during stress conditions, improved drainage, reduced soil compaction through various cultivation methods, watering deeply and infrequently, use of slow release nitrogen sources, and maintenance of soil pH near 6.0 (2,3).

Chemical control of diseases caused by ERI fungi is both erratic and expensive because multiple applications are usually required (2). Fungicides are generally more effective when applied in large (800-1000 L H_2O ha^{-1}) volumes of water (4). This may indicate a problem in delivering the fungicide to the target area, the top 8 cm of the soil, where these fungi survive. The timing and rate of fungicide application also influence efficacy (5-11).

There are several physical barriers that impair the effective delivery of fungicides to the roots in a turfgrass system. With traditional foliar (boom sprayer) application, the first barrier is the leaf canopy. Penetrant fungicides used to control ERI fungi are absorbed by the epicuticular wax and cuticle of the plant (12). Since these compounds are translocated primarily acropetally (i.e., root to shoot), they do not enter root tissue in sufficient concentrations following foliar applications to elicit fungicidal action in the roots (13). The next physical barrier is the thatch layer which is high in organic matter content (14). The organic matter content is the most important factor affecting the binding of pesticides in environmental matrices, thereby retarding the movement of pesticides to groundwater (15-17). Because the thatch layer associated with turfgrass has a very high organic matter content (14), it has been shown to adsorb many hydrophobic pesticides (18,19), including the fungicide triadimefon (20).

Fungicides commonly used to manage summer patch are the benzimidazoles (i.e., benomyl and thiophanate-methyl), and the sterol biosynthesis inhibiting (SBI) fungicides (2). Many SBI fungicides inhibit demethylation by binding to the cytochrome P-450 component of the C-14 demethylase and are known as demethylation inhibitors or DMI fungicides (13). When applied in 800-1000 L H_2O ha^{-1} prior to symptom expression, the DMI fungicides generally provided better summer patch control than the benzimidazoles (4,5,6,8,10). Three DMI fungicides, fenarimol (Rubigan), propiconazole (Banner), and triadimefon (Bayleton), were chosen for comparison in this study. Although similar in their modes of action, these fungicides vary in their water solubility, vapor pressure, environmental persistence, and mammalian toxicity (Table I).

By analyzing the various compartments of the turf system (i.e., foliage, thatch, roots, and soil) this study attempted to elucidate the main barriers to effective fungicide delivery. The problems associated with the efficient delivery of fungicides to the target root zone of turfgrasses have stimulated interest in the use of alternative delivery technologies to foliar (boom sprayer) application. Two novel application technologies were used in the current study: high-pressure injector and a slicer/seeder modified for

Table I. Physical properties, environmental characteristics, application rates, and toxicities of selected DMI fungicides.

Fungicide	Molecular Weight[a] (g mol^{-1})	Vapor Pressure[a] (mm Hg)	Water Solubility (mg L^{-1})	K_{oc}[c]	K_{ow}[f]	Application Rate (kg a.i. ha^{-1})	$T_{1/2}$ Soil[h]	$T_{1/2}$ Water[h]	Acute Oral LD$_{50}$ (Rats)[a] (mg/kg)
Fenarimol	331.2	4.88×10^{-7}	13.7[a]	600[d]	4900[a]	0.76	> 365 d[a]	> 365 d[a]	2500
Propiconazole	342.2	4.20×10^{-7}	100[a]	1000[d]	5248[a]	1.67	40-70 d[a]	25-85 d[a]	1517
Triadimefon	293.8	1.50×10^{-7}	64[a]	300[a]	1288[a]	3.05	6-18 d[a]	>1 y[a]	1000
Triadimenol	295.8	$<7.50 \times 10^{-6}$	120[b]	252[e]	182[g]	NA	110-270 d[a]	>1 y[a]	700

[a] The Pesticide Manual, 1994. C. Tomlin, ed. British Crop Protection Council Surrey, UK.

[b] The Merck Index, Eleventh Editon. 1989. S. Budavari, ed. Merck and Co., Inc. Rahway, NJ.

[c] K_{oc}- organic carbon/water partition coefficient. Ratio of absorbed chemical per unit weight of organic carbon to the aqueous solute concentration.

[d] SCS/ARS/CES Pesticide Selected Properties Database USDA Nematodes, Weeds and Crops Research Unit, Tifton, GA.

[e] HPLC Screening Method for the Determination of the Adsorption-Coefficient on Soil-Comparison of Different Stationary Phases. W. Kordel, J. Stutte, and G Kutthoff. Chemosphere, 1993, 27(12):2341-2352.

[f] K_{ow} - octanol/water partition coefficient.

[g] Estimated using CLOGP (version 3.4), Medicinal Chemistry Project, Pomona College, Claremont, CA.

[h] $T_{1/2}$ – half-life of fungicides in soil and water, respectively.

pesticide application. The two delivery systems were compared to a standard boom sprayer method to determine if any were superior for delivering the DMI fungicides to the root zone without increasing the possibility of groundwater contamination.

Methods and Materials

Experimental Site. Three 12.2 m x 12.2 m (0.0149 ha) plots of Kentucky bluegrass (*Poa pratensis* L.) were selected at the University of Massachusetts Turfgrass Research Center in South Deerfield, MA. The turf was established in 1985 and overlay a Hadley silt loam soil (coarse, silty, mixed, nonacid, mesic, Typic Udifluvent, pH 6.8). Thatch was approximately 3 cm thick. The turf was maintained at a height of 5.1 cm, and irrigated as needed to prevent drought stress.

Fungicide Application. The DMI fungicides were applied to the plots one time on 22 Aug 1992, and the treatments were not replicated. The fungicides were fenarimol {α-(2-chlorophenyl)-α-(4-chlorophenyl)-5-pyrimidinemethanol}, propiconazole {1-[[2-(2,4-dichlorophenyl)-4-propyl-1,3-dioxolan-2-yl]methyl]-1H-1,2,4-triazole}, and triadimefon {[1-(4-chlorophenoxy)-3,3-dimethyl-1-(1H-1,2,4-triazol-1-yl)-2-butanone}.

The fungicides were applied using three types of applicators: 1) a standard boom sprayer (F.E. Myers and Bro. Co., Ashland, OH); 2) a high-pressure injector (Cross Equipment, Albany, GA); and 3) a modified slicer/seeder (Cushman/Ryan, Lincoln, NE). Each fungicide was applied at the recommended rate for summer patch control. Fenarimol (Rubigan 1AS, Dow AgroSciences), Indianapolis, IN) was applied at a rate of 0.76 kg active ingredient (ai) ha^{-1}, propiconazole (Banner 1.1 EC, Novartis, Greensboro, NC) at 1.67 kg ai ha^{-1} and triadimefon (Bayleton 25 DF, Bayer, Kansas City, MO) at 3.05 kg ai ha^{-1}.

The boom sprayer was used to apply all three fungicides as a mixture to the first plot. The boom sprayer consisted of nine T-jet "9010" flat fan nozzles applying 621 L ha^{-1} at an operating pressure of 1925 kPa and generated a swath width of 4.6 m. The sprayer was calibrated, but the tank mix was not analyzed.

The high-pressure injector consisted of four 60 L tanks connected to a skid-mounted boom with 18 nozzles at a 7.2 cm spacing. The three fungicides were applied individually in three separate passes to the second turf area. They were injected directly into the thatch as an aqueous stream at a pressure of 11,100 kPa in 880 L ha^{-1}.

The slicer/seeder was modified for pesticide application by placing a small plastic tank on top of the machine with 10 coulters spaced 5.0 cm apart leading to a point directly behind each slicer. The slicer created a furrow approximately 1.3 cm deep into which the tank mixture flowed via the gravity-fed coulters at a rate of 408 L ha^{-1}. Only propiconazole was applied with the modified slicer/seeder to the third turf area because it was an experimental unit and there was concern about applying a mixture of fungicides with the gravity-feed mechanism. The boom sprayer application was put on last to minimize absorption of the fungicides by the foliar canopy and was immediately followed by 1.3 cm of irrigation water applied to all three experimental plots.

Sampling. Each of the three application mechanism plots was divided into sixteen 3.1 m x 3.1 m subplots that were further subdivided into nine 1 m^2 sections, one of which was randomly selected for each sampling period. It is estimated that each square meter of the plot contained approximately 455 g of leaf tissue (±15.7%), 11,989 g of thatch (±8.4%), 56,700 g of soil (±3.9%) (1 m x 1 m x 5.1 cm deep), and 85 g of root tissue (±11.3%) (1 m x 1 m x 5.1 deep) (all on a wet weight basis). These calculated quantities were used to determine the percentages of applied fungicides recovered. The amount of fungicide applied to each turfgrass matrix was calculated by assuming 100% of the applied fungicide was deposited in each matrix. The percent of applied fungicide recovered was determined by dividing the actual amount recovered by this number. All matrices were sampled immediately before and after the applications and at 1,3,7,10,14,21, and 28 days post-application. At each collection period, leaf tissue was collected from each of the sixteen subplots by cutting the plants at the soil line. The tissue was placed in 0.95 liter plastic Zip-Loc freezer bags for storage. Sixteen soil cores, one from each subplot, were then taken with a 2.5 cm x 30.5 cm JMC Zero Contamination Tube Corer (Clements Assoc., Inc., Newton, IA). The thatch layer of each core was delineated as the deepest point at which rhizomes were visible and was marked. Samples were then placed in ice chests and transported to the Massachusetts Pesticide Analysis Laboratory (MPAL) where they were stored at -20°C until analysis.

In the laboratory, the thatch layer was removed from the top of each soil core and the remainder of each core was divided into five 2.5 cm sections and bulked. The bulked samples were screened through a #10 sieve and the roots or thatch were separated out individually by hand. The soil, roots, and thatch components were placed into individual plastic freezer bags for storage in a -20° C freezer for analysis.

Storage Samples. On each sampling date, nontreated (i.e., no fungicide applied) leaf tissue and soil cores were collected from an adjacent untreated area to be used as fortified storage samples (amended with known amounts of each fungicide at the time of storage), concurrent fortified samples (amended with known amounts of each fungicide at the time of analysis), and concurrent nontreated samples.

Analytical Procedures

Leaf. The determination of fungicide residues was based on a method developed by Newsome and Collins (*21*). Five g of leaf tissue were cut up using scissors and placed in a 250 ml red actinic Erlenmeyer flask. Fifty ml of acetone were added to the flask and it was shaken for 30 min on a rotary table top shaker (Eberbach) at low speed (180 rpm). The liquid was then decanted through 13 g anhydrous sodium sulfate into a 500 ml flat bottom boiling flask. The leaf tissue was re-extracted with an additional 50 ml acetone that was likewise quantitatively transferred to the boiling flask. The combined extract was reduced to approximately 3 ml under vacuum, quantitatively transferred in acetone to a 50 ml centrifuge tube, and the sample volume adjusted to 10 ml under a stream of nitrogen. The sample volume was then readjusted to 50 ml with distilled water. Prior to loading the sample extracts, C_{18} solid phase extraction cartridges (500 mg, J&W Scientific, Folsom, CA) were sequentially washed with 10 ml aliquots of methanol, methylene chloride, and methanol, to remove any contaminants within the cartridge. The cartridge was activated with 25 ml distilled water. The sample extract was passed through the activated cartridge, followed by a wash step that consisted of 10 ml of a 40% solution of methanol in distilled water. The retained fungicide was eluted from the cartridge with 5 ml of methanol. Following elution, a 4 ml aliquot of the eluent was transferred to a 15 ml centrifuge tube and the volume adjusted to 10 ml with distilled water. A liquid/liquid extraction was performed in the centrifuge tube by adding 5 ml toluene and gently shaking the tube. The fungicide partitioned into the toluene (i.e. top layer), which was then removed for analysis. Each sample analysis was replicated three times using three subsamples.

Thatch. Fifty ml of acetone were added to 1 g of thatch in a 120 ml amber glass jar and the sample was homogenized twice for 20 seconds at 20,000 rpm with a Brinkmann Polytron PT3000 homogenizer (Littau, Switzerland). The homogenized sample was shaken for 30 min on a rotary table top shaker, and the acetone decanted into a 500 ml flat-bottomed boiling flask. The extract was reduced to approximately 3 ml under vacuum, quantitatively transferred in acetone to a 50 ml centrifuge tube and the sample volume adjusted to 10 ml under a stream of nitrogen. The sample volume was then readjusted to 50 ml with distilled water. The C_{18} cartridges were washed and activated as outlined for leaf tissue. The sample was passed through the cartridge followed by a 10 ml wash solution of 40% methanol in distilled water. The sample was eluted from the cartridge with 10 ml acetone into a 13 ml centrifuge tube. The sample eluent volume was adjusted to 2 ml under a stream of nitrogen for analysis. Each sample analysis was replicated three times using three subsamples.

Soil. Fifty ml of acetone were added to 25 g of soil in a 250 ml red actinic erlenmeyer flask and shaken for 30 min on a rotary table top shaker at low speed (180 rpm). The liquid was decanted through 13 g anhydrous sodium sulfate into a 500 ml boiling flask. The sample was re-extracted with 50 ml acetone and the extract quantitatively transferred to the boiling flask containing the first extract. The sample was reduced to approximately 3 ml under vacuum. The remainder of the procedure followed that outlined for thatch above.

Roots. Twenty ml of acetone were added to 0.5 g of root tissue in a 60 ml amber glass jar. The sample was homogenized and shaken as previously described for thatch. The acetone was decanted into a boiling flask. The sample was re-extracted with acetone and the extract quantitatively transferred to the boiling flask containing the first extract and reduced to approximately 3 ml. The remainder of the procedure followed that outlined above for thatch.

Instrumental Analysis. Fungicide residues were analyzed using a Varian Model 3400 gas chromatograph equipped with a nitrogen-phosphorus detector (NPD) and a DB-5 glass megabore column (53 mm i.d. X 15 m, J & W Scientific, Folsom, CA). Operating conditions were: injection volume, 1.0 µl; injector temperature, 250°C; detector temperature, 300°C; column oven temperature 160°C for 2 min, ramped at 15°C min^{-1} to 235°C and held for 3 min. Gas flows were: helium, 10 ml min^{-1}; hydrogen, 3.5 ml min^{-1}; nitrogen, 20 ml min^{-1}; and air 115 ml min^{-1}.

Quality Control/ Quality Assurance (QC/QA). Calibration procedures for the chromatographic system that were used for analysis involved developing standard curves consisting of at least three concentrations for each fungicide of interest using high purity analytical standards (>99%). Triadimefon standards were provided by Mobay Corp. (now Bayer Corp.) (Kansas City, MO). Fenarimol and propiconazole standards were provided by Riedel de Haen (Seelze, Germany). Standard curves consisting of at least three concentrations of each analyte were included at the beginning of each day of analysis. Area counts for each standard concentration (n=3) were averaged to generate a calibration curve (r^2> 0.99).

For each day of analysis, corresponding fortified storage samples and nontreated leaf, thatch, roots, and soil samples amended with the fungicides of interest and not amended (i.e., concurrent fortified and nontreated samples, respectively) were extracted and analyzed to determine recovery efficiency and daily method performance, respectively. Reagent and matrix blanks also were included in each analysis.

Results and Discussion

None of the application methods allowed the three fungicides to reach the roots in any appreciable amount, principally due to their high affinity for thatch (Figures 1-4). Due to space limitations, only summary graphs are presented. Results of analyses of each fungicide in each matrix from each sampling with statistical analysis have been published (*22*). Regardless of application type, the majority of the fungicides (40-80%) were associated with the thatch layer. Additionally, none of the fungicides was ever found below the top 5.1 cm of the soil.

Triadimefon (Figures 3A, 4A) was rapidly converted to triadimenol (Figures 3B, 4B) in all four matrices, and was not detectable at the end of the experiment (28 days). Triadimenol was the only compound detected in the roots in any appreciable amount (0.04 - 0.19% of the applied triadimefon). Triadimenol would appear to pose a greater threat to groundwater than triadimefon because it is more water soluble, has a greater half-life in soil, a lower organic carbon/water partition coefficient, and greater mammalian toxicity than triadimefon (Table I). However, triadimenol was never detected below the top 5.1 cm of the soil profile over the course of the experiment (28 days).

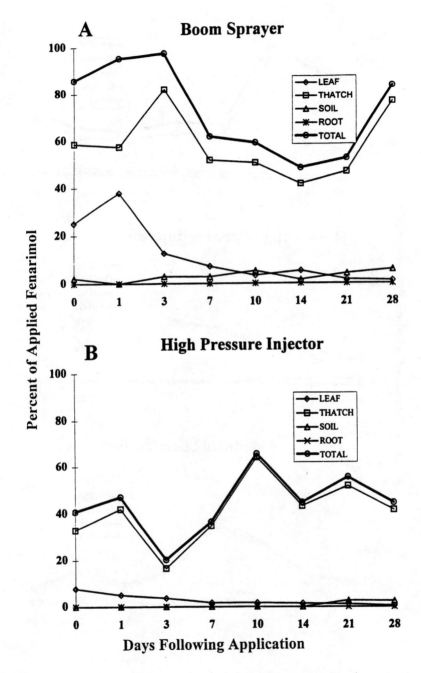

Figure 1. Percent of total applied fenarimol recovered from all matrices (total) and from leaf, thatch, roots and soil for each day from Day 0 to Day 28 following application with the boom sprayer (A) and high-pressure injector (B).

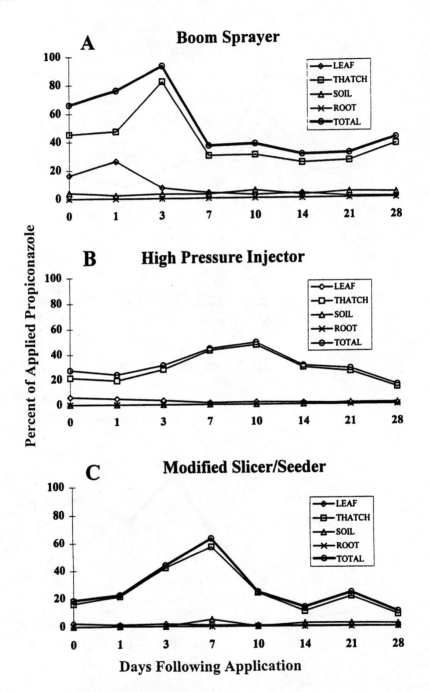

Figure 2. Percent of total applied propiconazole recovered from all matrices (total) and from leaf, thatch, roots and soil for each day from Day 0 to Day 28 following application with the boom sprayer (A), high-pressure injector (B), and modified slicer/seeder (C).

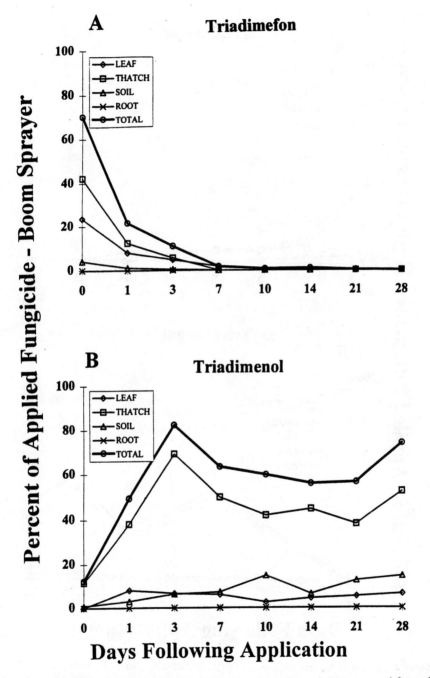

Figure 3. Percent of total applied triadimefon (A) and triadimenol (B) recovered from all matrices (total) and from leaf, thatch, roots, and soil for each day from Day 0 to Day 28 following application with the boom sprayer.

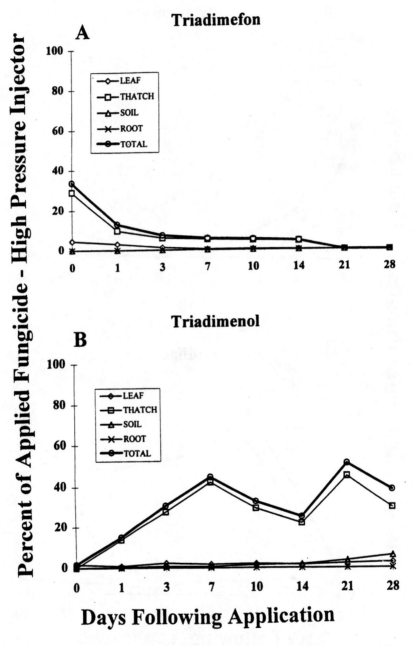

Figure 4. Percent of total applied triadimefon (A) and triadimenol (B) recovered from all matrices (total) and from leaf, thatch, roots, and soil for each day from Day 0 to Day 28 following application with the high-pressure injector.

Recoveries for fungicides applied by boom sprayer ranged between 81% and 109% over the first three days following application, between 7% and 47% for the high-pressure injector, and 18% to 44% for the modified slicer/seeder. Immediately following application with the boom sprayer, 40-60% of the total applied fungicides was recovered from the thatch, another 18-25% was recovered from the leaf tissue, and 1-6% was covered from the soil. Immediately following application with the high-pressure injector, 20-30% of the total applied fungicides (but 78-86% of the amount recovered) was associated with the thatch. Only 5-8% of the fungicides applied by high pressure was recovered from the leaf tissue, and none was recovered from the roots or soil. Immediately following application using the modified slicer/seeder, 16% of the total applied propiconazole (but 87% of the propiconazole recovered) was associated with the thatch, 2% was found associated with the leaf, and none was recovered from the roots or soil. Obviously, thatch is a major reservoir for these fungicides and functions as a barrier to their efficacious delivery to the turfgrass root system. As a barrier, thatch probably also restricts the movement of these fungicides into the groundwater.

Considerably more of the fungicides were recovered from the samples in the boom sprayer plot compared to the two alternative application methods. Both the high-pressure injector and the slicer/seeder deposited the fungicides in discreet bands primarily in the thatch. The high-pressure injector nozzles deposited streams of the fungicides 7.2 cm apart. The coulters of the slicer/seeder were 5 cm apart. The internal diameter of the soil corer used to collect turfgrass samples was 2.5 cm. Due to this disparity in size, it is possible that the some turfgrass cores were removed from areas between the fungicide-treated bands. Because of this sampling method, an approximation of mass balance was achieved only for the residues recovered following boom sprayer application. The low and varied recoveries may indicate that the sampling methodology was inadequate for the high-pressure injector and slicer/seeder applications. These findings also suggest a lack of lateral movement of these SBI fungicides, which is supported by anecdotal evidence concerning other lipophilic pesticides (P. Vittum, personal communication).

This study was initiated to identify barriers to delivery of fungicides applied to control patch diseases caused by ectotrophic root-infecting fungi and to determine the potential for two new mechanisms to improve delivery of the fungicides to the roots. The apparent lack of lateral movement of the fungicides makes the new mechanisms unlikely as improved alternatives for delivery to turfgrass roots in their current configuration. The results also explain the many reported failures of chemical control for these diseases, especially in lawn-type turfgrass where a significant thatch layer is present.
The results also suggest that a more formal fate study might elucidate some valuable information for improved delivery of fungicide to turfgrass roots without increasing the risk of groundwater contamination.

Because a large proportion of the applied fenarimol, propiconazole, and the triadimefon metabolite triadimenol remained associated with the thatch at the end of the study, a longer experimental time frame should be examined including spring and fall recharge

162

events in future research. Fate studies should also include turfgrass maintained as golf putting greens since mowing heights are much lower than in the current study, and thatch is generally less developed. Comparisons could be made between soil and sand-based growing media and should include additional DMI fungicides, such as the currently registered compounds, cyproconazole and myclobutanil. In addition, a new fungicide group, the strobilurins, [e.g., azoxystrobin (Heritage) and related compounds] is now commonly used for management of diseases caused by ERI fungi, however there is a paucity of information regarding their environmental fate after application.

Acknowledgments

Research support is gratefully acknowledged to the following organizations: Bayer Corporation, Dow Agrisciences, Golf Course Superintendents of New England, Inc., Massachusetts Turf and Lawn Grass Council, Inc., and Novartis (then Ciba-Geigy Corporation), and USDA Cooperative Regional Project NE-169. We also acknowledge the Pesticide Bureau, Massachusetts Department of Food and Agriculture, Boston MA for hosting this project. Use of trade names does not imply endorsement of the products named nor criticism of similar ones not mentioned.

Literature Cited

1) Endo, R.M. *The Biology and Utilization of Grasses*; Younger, V.B., McKell, C.M., Eds.; Academic Press: New York, 1972; pp. 171-202.

2) Dernoeden, P.H. *Turfgrass Patch Diseases Caused by Ectotrophic Root- Infecting Fungi*; Clarke B.B, Gould, A.B., Eds.; APS Press: St. Paul, MN, 1993; pp. 123-161.

3) Thompson, D.C.; Clarke, B.B.; Heckman, J.R. *Plant Dis.* 1995, *79*, pp. 51- 56.

4) Landschoot. P.J.; Clarke, B.B. *Fungic. Nematic. Tests* 1989, *44*, p. 245.

5) Dernoeden, P.H. *Fungic. Nematic. Tests* 1989a, *44*, p.255.

6) Dernoeden, P.H.; Minner, D. *Fungic. Nematic. Tests* 1981, *36*, p.145.

7) Dernoeden, P.H.; Nash, A.S. *Fungic. Nematic. Tests* 1982, *37*, p.151.

8) Hartman, J.R.; Clinton, W.; Powell, A.J. *Fungic. Nematic. Tests* 1989, *44*, p. 248.

9) Dernoeden, P.H. *Integrated Pest Management for Turfgrass and Ornamentals*; Leslie, A.R.; Metcalfe, R.L., Eds.; U.S. Environmental Protection Agency: Washington, DC, 1989b; pp. 273-296.

10) Plumley, K.A.; Clarke, B.B.; Landschoot, P.J. *Phytopathology* 1991, *81*, p. 124.

11) Smiley, R.W. *Advances in Turfgrass Pathology*; Larsen, P.O.; Joyner, B.G., Eds.; Harcourt Brace Jovanovich: Duluth, MN, 1980, pp. 155-175.

12) Baker, E.A.; Hayes, A.L.; Butler, R.C. *Pestic. Sci.* 1992, *34*, pp. 167-182.

13) Kuck, K.H.; Scheinpflug, H.; Pontzen, R. *Modern Selective Fungicides*; Lyr, H., Ed.; Fisher Publishing: New York, NY, 1995, pp. 205-258.

14) Hurto, K.A.; Turgeon, A.J.; Spomer, L.A. *Agron. J.* 1980, *72*, pp. 165-167.

15) Bollag, J.; Myers, C.J.; Minard, R.D. *Sci. Total Environ.* **1992**, *123/124*, pp. 205-217.

16) Senesi, N. *Sci. Total Environ.* **1992**, *123/124*, pp.63-76.

17) Singh, R.; Gerritse, R.G.; Aylmore, L.A.G. *Aust. J. Soil Res.* **1989**, *28*, pp. 227-243.

18) Lickfeldt, D.W.; Branham, B.E. *J. Environ. Qual.* **1995**, *24*, pp. 980-985.

19) Niemczyk, H.D.; Krueger, H.R. *Advances in Turfgrass Entomology*; Niemczyk, H.D.; Joyner, B.G., Eds.; Hammer Graphics, Inc.: Piqua, OH, 1982, pp. 61-63.

20) Dell, C.J.; Throssell, C.S.; Bischoff, M; Turco, R.F. *J. Environ. Qual.* **1994**, *23*, pp.92-96.

21) Newsome, W.H.; Collins, P. *J. Chrom.* **1989**, *472*, pp. 416-421.

22) Doherty, J.J. M.S. Thesis, University of Massachusetts, 1998.

Chapter 10

The Effect of Salinity on Nitrate Leaching from Tall Fescue Turfgrass

D. C. Bowman[1], D. A. Devitt[2], and W. W. Miller[2]

[1]Department of Crop Science, North Carolina State University, Raleigh, NC 27695
[2]Department of Environmental and Resource Science,
University of Nevada, Reno, NV 89512

A column study was conducted to determine the impact of saline irrigation water on NO_3 leaching from turfgrass. Tall fescue turf was grown in columns filled with sand and outfitted with a vacuum drainage system. Treatments consisted of three N levels (25, 50 and 75 kg NH_4NO_3-N ha^{-1} $month^{-1}$) and three irrigation salinity levels (0, 1.5 and 3 dS m^{-1}) in a 3 X 3 factorial arrangement. Irrigation was scheduled to provide a 30% leaching fraction. Leachate was collected quantitatively after every irrigation and analyzed for salts and NO_3-N. Clippings were collected and analyzed for total N. Nitrate concentrations in the leachate were very low, averaging approximately 1.0 mg N L^{-1}. Clipping yield and N content were unaffected by salinity, while root mass was increased. These data indicate that moderate levels of rootzone salinity do not increase NO_3 leaching, nor do they impair growth or N absorption. This suggests that moderately saline irrigation water may be used to irrigate tall fescue turf without increasing NO_3 contamination of groundwaters, as long as leaching is adequate to control rootzone salinity

Soil salinity is a significant problem in the western United States. This is primarily due to the natural occurrence of soluble salts in many desert soils and the use of moderately saline water for irrigation. In southern Nevada's Las Vegas Valley, irrigation in excess of plant water requirement has leached native salts below the rootzone, creating a perched saline aquifer with an electroconductivity (EC) of approximately 9 dS m^{-1} and a volume estimated at approximately 100,000 acre feet. If properly managed, this water supply could be used as an alternative or supplemental irrigation source, decreasing the demand on high-quality water while reducing the potential for contaminating the primary aquifer.

One concern regarding the use of saline water for turf irrigation is the possible negative impact on turfgrass N nutrition. Nitrogen is the most heavily used nutrient in turfgrass management, with typical applications ranging from 50 to 600 kg N ha^{-1} yr^{-1}. When application rate exceeds turfgrass demand, excess N may be lost from the soil, becoming free to interact with other segments of the biosphere. In the arid southwest, and in southern Nevada in particular, it is essential that irrigation exceed evapotranspiration to facilitate leaching and maintain a favorable salt balance within the soil profile. However, this same downward movement of water becomes a prime pathway for nitrogen loss, as NO_3 is highly mobile and subject to leaching. Work by Devitt et al. (1) and Letey et al. (2) on agricultural crops indicates that mass emission of nitrogen is more closely related to the amount of percolating water than to the amount of fertilizer applied. Since salinity reportedly inhibits N uptake in a number of species (3-5), leaching losses from turfgrasses could be increased considerably where saline irrigation waters are used in excess of evapotranspirational demand.

Although no data are available on the effects of salinity on NO_3 leaching, numerous studies have examined other factors and management practices as they affect NO_3 leaching from turfgrasses. For example, Brown et al. (6) measured concentrations of NO_3 as high as 74 mg N L^{-1} in the leachate below bermudagrass following application of 163 kg N ha^{-1} as NH_4NO_3, with total leaching loss of 23% of the applied N. Snyder et al. (7) found peak NO_3-N concentrations between 20 and 40 mg N L^{-1} in the soil solution below the turf rootzone 5 to 10 days after applying 50 kg N ha^{-1} as NH_4NO_3. Up to 56% of the applied N was lost during a 3 week period. However, minimizing the downward movement of water by carefully controlling irrigation with tensiometers reduced losses from 56% to 2% (8). Similar results have been reported by Morton et al. (9). De Nobili et al. (10) applied urea at a rate of 160 kg N ha^{-1} preplant to a turfgrass mixture and measured NO_3 leaching over a four month period. Total leaching losses amounted to 9.5% of the applied N, with an average leachate concentration of 13 mg N L^{-1}. Cohen et al. (11) monitored NO_3-N in wells at four golf courses on Cape Cod over a two year period. At three of the four courses, NO_3 concentrations in the shallow groundwater was only slightly higher than background.

In contrast to the above studies, Rieke and Ellis (12) found little effect of fertilization on NO_3 leaching following application of 290 kg N ha^{-1} yr^{-1} to a mixed turf. Starr and Deroo (13) reported similar low leaching losses. Mancino and Troll (14) investigated NO_3 leaching from a creeping bentgrass turf under conditions favoring heavy leaching losses (sand rootzone, soluble nitrate-based fertilizers, and 46% leaching fraction). When the fertilizers were applied at a low rate of 9.76 kg N ha^{-1} weekly, NO_3 leaching averaged less than 0.5% of the applied nitrogen. With a heavier application of 49 kg N ha^{-1}, cumulative losses averaged 3.5% for the NO_3 sources. Gold et al. (15) reported a maximum flow-rated NO_3-N concentration of 1.62 mg N L^{-1} in the leachate from a home lawn fertilized with 244 kg N ha^{-1} yr^{-1}. Approximately half of the leachate samples had concentrations at or below 0.1 mg N L^{-1}. Bowman et al. (16) suggested that rapid biological immobilization, both by the turf and soil microorganisms, may reduce leaching losses from turf by limiting the period of time that the fertilizer N is resident in the soil.

The objectives of this research were to determine the effects of salinity and N application rate on NO_3 leaching and N mass emission from, and N uptake by tall fescue turf under greenhouse conditions.

MATERIALS AND METHODS

Plant Culture. Research was conducted in a greenhouse maintained at 28/18°C day/night. Experimental conditions were chosen to maximize the potential for leaching, i.e. a porous sandy soil, fertilization with NH_4NO_3 at relatively high rates, and irrigation considerably in excess of plant demand.

Lysimeters were constructed from polyvinyl chloride (PVC) columns 15 cm in diameter and 60 cm deep. Each column was equipped with a porous ceramic cup (2.2 cm diameter by 7 cm long) which was embedded in 3 cm of diatomaceous earth layered at the bottom of the columns. The ceramic cup was connected via tubing to a 2 L collection bottle, which in turn was connected to a manifold vacuum line and pump equipped with a pressure gauge. Each column was packed with a sand (Table I) to a bulk density of 1.52 g cm^{-3}.

Columns were seeded with 'Monarch' tall fescue (*Festuca arundinacea* Schreb.) at a rate of 400 kg ha^{-1}. A fertilizer solution was applied at seeding to supply the equivalent of 50, 80 and 100 kg ha^{-1} N, P and K, respectively. The turf was established for six months, during which time it was mowed every 7-12 days and irrigated twice each week. Ammonium nitrate was applied monthly at 50 kg N ha^{-1} during establishment.

Table I. Particle size distribution of the sand used in the columns.

Size Class	Percent of total
Sand	
Very Coarse (1-2 mm)	1.2
Coarse (0.5-1 mm)	6.4
Medium (0.25-0.5 mm)	42.4
Fine (0.1-0.25 mm)	28.5
Very Fine (0.05-0.1 mm)	14.4
Silt/Clay	7.1

Treatments. Salt treatments were initiated January 13 and continued until final harvest of the columns on December 18. Treatment solutions were formulated to salt levels of 0, 15 and 30 meq L^{-1} using tap water (EC 0.1 dS m^{-1}) and NaCl:CaCl$_2$ in an 8:1 molar ratio. Each column was irrigated once with 1000 ml of the appropriate salt solution to rapidly equilibrate the columns. Thereafter the columns were irrigated approximately twice per week with a volume of solution calculated to maintain a

30% leaching fraction. The amount of irrigation was determined for each column by gravimetric mass balance. After each irrigation a vacuum of 0.010 MPa was applied to the ceramic cups for 16 hr to promote drainage. The collected leachate was weighed and subsampled for analysis.

Nitrogen treatments were initiated February 12. Solutions of NH_4NO_3 were prepared using the saline irrigation waters described above, and were applied to the columns monthly to supply 25, 50 or 75 kg N ha^{-1}. Application volume was 300 ml column^{-1}, equivalent to a depth of 1.6 cm. To determine N allocation to new leaf growth, $^{15}NH_4^{15}NO_3$ (9.98% enrichment) was applied on September 11. Clippings collected during the subsequent month were pooled for ^{15}N analysis. Additional P and K were added at rates of 80 and 100 kg ha^{-1}, respectively on June 12. The experimental design was a two factor randomized complete block, with three salt levels, three nitrogen levels and four replicates. Significant differences were determined according to LSD ($P<0.05$).

Analyses. Leachate samples were collected after each irrigation and analyzed for NO_3-N and salt content. Nitrate was determined by the rapid diffusion method (*17*) and salt content with an EC meter. Mass emission of NO_3-N was calculated as the product of leachate volume and concentration. Clippings were collected at every mowing (every 7-14 days), dried, weighed and ground. Total N in the clippings was determined by a modified Kjeldahl procedure (*18*). The ^{15}N enrichment of the Sept./Oct. clippings was analyzed commercially (Isotope Services Inc., Los Alamos, NM). Tissue from the June 23 harvest (approximate midpoint of the experimental period) was extracted under vacuum with 0.5 N HNO_3. Sodium and K were determined by flame emission spectrophotometry, Ca and Mg by atomic absorption spectrometry, and Cl with a chloride specific electrode (Orion Research Inc., Boston, MA). All columns were harvested on December 18. Above-ground tissue was separated into new leaves and verdure. Roots were sampled using a single 2.5 cm dia. core taken from the center of each soil column and sectioned by depth at 0-15, 15-30, 30-45 and 45-60 cm. Roots were washed from the soil, dried and weighed to determine distribution. Tissues were analyzed for total N and ion concentrations as above. Tissue ion concentrations (using bracket notation, eg. [Na]) are expressed as percent of dry weight.

RESULTS AND DISCUSSION

Electroconductivity of leachate samples was measured to determine when the columns had reached a steady-state salt profile. Salt content of the leachate rose steadily during the first four months of the experiment and then leveled off, indicating that a constant salt profile had been attained (Figure 1). Leachate EC's for the 0.1, 1.5, and 3.0 dS m^{-1} salt treatments stabilized at approximately 0.4, 4.5, and 9 dS m^{-1}, respectively. These values are close to those predicted based on a leaching fraction of 30%.

Soil extracts from the final harvest indicated a uniform salt profile within the soil column (Figure 2), suggesting either that water uptake was relatively constant

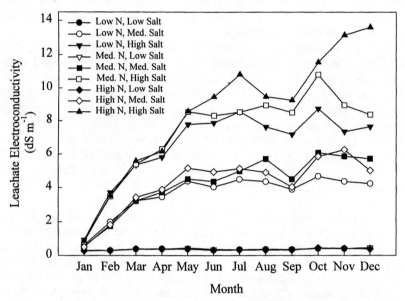

Figure 1. Electroconductivity of the leachate over time as a function of irrigation water salinity.

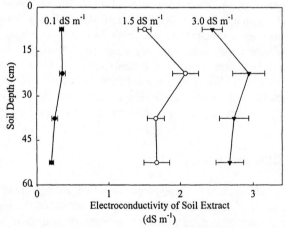

Figure 2. Soil salinity profiles at the end of the experiment as a function of irrigation water salinity. Error bars are standard deviation.

with depth or that the high leaching fraction resulted in rapid mixing of irrigation with soil water and obscured fractional water uptake by the root system near the soil surface. This is in contrast to Devitt (*19*) who reported increasing salinity with soil depth and greater water uptake near the soil surface by bermudagrass irrigated with saline water. These differences are likely due to the use of smaller columns (60 vs 120 cm) and the shorter period of salinization (11 vs 24 months) in the present study.

Irrigation was scheduled to provide a 30% leaching fraction, both to control soil salinity levels and to promote NO_3 leaching. This represents an excessive amount of water under most conditions, and one that could be difficult to maintain on fine-textured soils. However, high leaching fractions are an important strategy in managing saline waters and 30% would not be uncommon given the salinity levels used in this study. There was considerable fluctuation in the actual leaching fractions, expressed as monthly averages, during the first several months of the experiment (Figure 3), due in part to the inaccuracy and low resolution of the vacuum gauge used to adjust the applied tension. This problem was corrected by replacing the gauge with a mercury manometer. Over the course of the experiment, actual leaching fractions were reasonably close to 30% for seven of ten months. Leaching fractions for two of the remaining three months were approximately twice the 30% target, which would increase the potential for NO_3 leaching.

Total monthly irrigation data for the April through September period were used to examine the effects of N and salinity on irrigation requirement (data not shown). High salinity reduced irrigation by 9% (P < 0.001) compared to the low salt treatments, whereas high N increased irrigation 10-14% (P < 0.001) relative to low-N treatments. Similar effects of N rate (10) and salinity (11) have been reported for water use by bermudagrass turf.

Average monthly NO_3-N concentrations in the leachate ranged from < 0.1 to 2.9 mg N L^{-1} (Figure 4). The highest NO_3 concentrations occurred from March through May, whereas consistently low values were found from June through November. Since the tall fescue was growing slowly during late winter/early spring (data not shown), the high values could be due to a lower growth demand for N, or to the higher leaching fractions in March and May. Volume weighted averages for NO_3 concentration in the leachate were approximately 0.75-1.0 mg N L^{-1}. For comparison, the NO_3 concentration of the irrigation water ranged between 0.1 and 0.2 mg N L^{-1}. It should be noted that of the nearly 1000 samples analyzed during this study, none was above the critical level of 10 mg NO_3-N L^{-1}.

There were no obvious effects of either N application rate (Figure 4a) or salinity (Figure 4b) on NO_3 concentration of the leachate. Monthly average NO_3 concentrations were not affected by salinity, and were significant for N rate only for the March data. These results are consistent with previous research demonstrating the ability of turfgrass systems to immobilize applied N rapidly (*16*) but further suggest that the N uptake systems of turf roots and associated microorganisms were not appreciably impaired by the salinity levels used in this study.

Cumulative NO_3-N leached over the eleven month experiment amounted to approximately 10 mg N $column^{-1}$ (Figure 5). This represents approximately 2%, 1% and 0.7% of the applied N for the low, medium and high N rates, respectively. Much

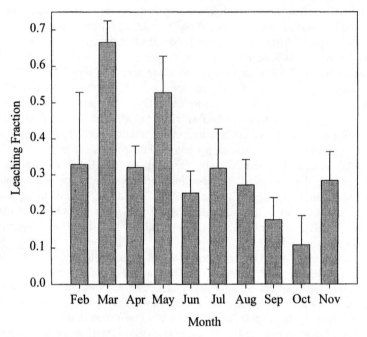

Figure 3. Monthly average leaching fraction, combined across all treatments. Error bars are standard deviation.

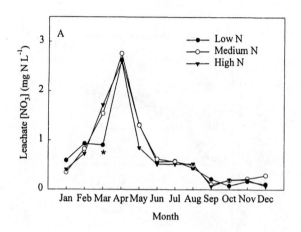

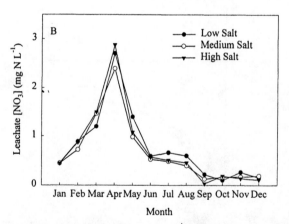

Figure 4. Volume weighted monthly average NO₃-N concentrations in the leachate as a function of N application rate (A) and salinity (B).

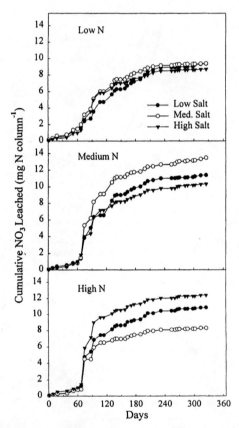

Figure 5. Cumulative NO₃-N leached as a function of N application rate and salinity.

of the total loss occurred during March through May, with very little additional loss recorded thereafter. As noted above, there was no significant effect of N rate on leaching when the data were expressed as NO_3-N concentration. However, when the data were expressed as a percentage of monthly-applied N, leaching losses decreased with increasing N rate during eight of the twelve months (Figure 6). This result is certainly counterintuitive and underscores the impact of data expression on interpretation.

Clipping dry weight and percent N in the tissue (data not shown) were used to calculate the amount of N partitioned to leaf tissue and removed in clippings (Figure 7). Nitrogen removal increased with increasing N application rate, but there was no effect of salinity. The slopes of curves were used to estimate average daily N allocation to leaf tissue. This can be compared to the average daily N addition rate (monthly rate/30 days) to estimate long-term N uptake efficiency (Table II). Nitrogen recovery in clippings was 93% for the low N rate, but declined to 79% and 75% at the medium and high N rates, respectively. These values, especially at the low rate, indicate efficient absorption of applied N by the turfgrasses, and suggest

Table II. Long-term and short-term N uptake efficiency for tall fescue at three N application rates and three salinity levels.

N Rate	Salt Level	N Removed in Clippings	Long-Term Uptake Efficiency	Short-Term [15]N Allocation
kg N ha^{-1}	dS m^{-1}	kg N ha^{-1} day^{-1}	% of applied N	
25	0.1	0.79 a	95 a*	24 e
	1.5	0.77 a	92 a	20 f
	3.0	0.78 a	93 a	20 f
50	0.1	1.25 b	75 b	32 cd
	1.5	1.42 b	85 b	48 a
	3.0	1.30 b	78 b	29 de
75	0.1	1.85 c	74 c	39 b
	1.5	1.90 c	76 c	26 e
	3.0	1.88 c	75 c	36 bc

* Data values in a column followed by the same letter are not significantly different at $P = 0.05$

that over the long term, a mature turf allocates most acquired N to new leaf tissues. It should be considered that this research was performed on a relatively young turf system with clippings removed, and in which very little soil organic N is likely to have accumulated. Consequently, mineralization of organic N would supply little N to this turf. If mineralization of soil organic N contributed significantly to the

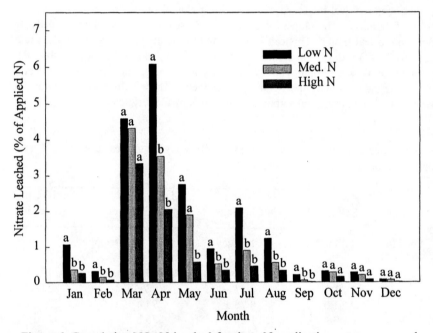

Figure 6. Cumulative NO$_3$-N leached for three N application rates, expressed as a percentage of applied N. Within a given month, values marked with the same letter are not significantly different at $P = 0.05$.

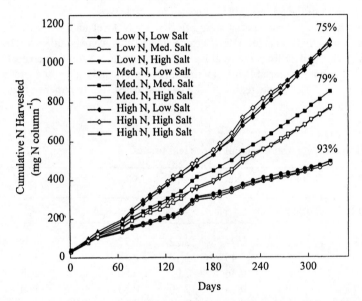

Figure 7. Cumulative N removed in clippings as a function of N application rate and salinity. The percent values associated with the curves represent the ratio of N harvested vs. N applied for the three rates of N.

nutrition of the turf, as might be the case in an old turf system, it is possible that additional applied N would not be absorbed as efficiently. Under such conditions, NO_3 leaching might be higher than reported here.

Short-term allocation of recently-absorbed N to new leaves was examined by applying ^{15}N-labeled fertilizer to the columns in September and determining ^{15}N content in clippings produced during the subsequent month (Table II). Tall fescue allocated from 22% to 37% of the applied N to leaves, with allocation to new leaves, expressed either as absolute amount or as percent of applied, increasing with N application rate. Based on the long term allocation pattern discussed above, it is apparent that short term allocation is more accurately interpreted as N cycling, since ultimately most of the absorbed N is allocated to new leaves.

Tissue ion concentrations were affected by N level and salinity, but there were no interactions (Table III). Because the results were similar for the two harvests, only the data from the final harvest are presented. Salinity increased tissue [Na] and [Cl] but decreased [Mg]. Increasing N significantly increased tissue [Na], [K], [Mg] and [Cl], but decreased [Ca]. In most cases, the treatment effects were relatively small.

Table III. Ion concentrations in tall fescue leaf tissue.

N Rate	Salt Level	Na	K	Ca	Mg	Cl
kg N ha[-1]	dS m[-1]		% of dry weight			
25	0.1	0.24	2.22	0.72	0.30	1.14
	1.5	0.45	2.14	0.78	0.20	1.41
	3.0	0.56	2.02	0.86	0.16	1.59
50	0.1	0.24	2.70	0.53	0.29	1.28
	1.5	0.45	2.46	0.62	0.22	1.56
	3.0	0.56	2.41	0.70	0.18	1.60
75	0.1	0.26	2.76	0.52	0.31	1.43
	1.5	0.56	2.46	0.65	0.24	2.04
	3.0	0.71	2.47	0.68	0.20	1.99
Significance						
Nitrogen		*	***	**	***	***
Salinity		***	ns	ns	***	***
N X S		ns	ns	ns	ns	ns
LSD$_{0.05}$		0.11	0.23	0.10	0.05	0.39

*,**,*** Significant at the 0.05, 0.01 and 0.001 levels, respectively.

Root distribution was determined with soil depth in 15 cm increments. Approximately 80% of the total root mass was in the top 15 cm and 20 % at 15-30 cm; no roots were recovered beneath 30 cm. Root mass increased with increasing salinity, but decreased with N rate (data not shown). Recent work with creeping bentgrass turf has shown that nitrate leaching is at least partially governed by rooting depth (20), with deeper root systems being more efficient at absorbing N. This may explain the lack of salt effect in NO_3 leaching in this study, since rooting depth was unaffected. However, the fact that there was also no effect of N rate on NO_3 leaching is somewhat perplexing. As noted above, N uptake efficiency decreased with N rate, which might logically result in greater leaching losses. This conclusion is based on N harvested in leaf tissue and does not consider N storage in roots, shoots and soil organic matter. In fact, roots plus shoots from the high-N treatment contained 223 kg N ha^{-1}, compared to 158 kg N ha^{-1} in the roots plus shoots from the low-N treatment (data not shown). This increase in plant biomass N at higher N rates accounts for much of the apparent reduction in N uptake efficiency, indicating that long-term patterns of N allocation to leaves may not accurately estimate N uptake efficiency. It also suggests that tall fescue root systems are highly effective at N absorption across the range of N application rates commonly used by turf managers.

CONCLUSION

The results of this study indicate that irrigation of tall fescue with moderately saline water should not increase NO_3 leaching, as long as adequate irrigation prevents excessive salt accumulation in the rootzone. Measures of growth, such as clipping production and N allocation to leaf growth, were also not affected by salinity. This indicates that the grass was under relatively low stress, which probably explains the lack of salt effect on NO_3 leaching. If salinity stress were greater, or if multiple stresses were imposed, it is possible that N use efficiency would decline and NO_3 leaching increase.

LITERATURE CITED

1. Devitt, D.A. *Agron. J.* **1989**. *81*:893-901
2. Letey, J., J.W. Blair, D. Devitt, L.J. Lund, and P. Nash. *Hilgardia.* **1977**. *45*:289-318.
3. Aslam, M., R.C. Huffaker, and D.W. Rains. *Plant Physiol.* **1984**. *76*:321-325.
4. Pessarakli, M. and T.C. Tucker. *Soil Sci. Soc. Am. J.* **1988**. *52*:698-700.
5. Ward, M.R., M. Aslam, and R.C. Huffaker. *Plant Physiol* 1986. *80*:520-524.
6. Brown, K.W., J.C. Thomas, and R.L. Duble. *Agron. J.* 1982. *74*:947-950.
7. Snyder, G.H., E.O. Burt, and J.M Davidson. In *Proceedings of the 4th International Turfgrass Research Conference;* Sheard, R.W., Ed.; University of Guelph, Guelph, Ontario. 1981. pp 313-324.

8. Snyder, G.H., B.J. Augustin, and J.M. Davidson. *Agron. J.* **1984.** *76*:964-969.
9. Morton, T.G., A.J. Gold, and W.M. Sullivan. *J. Environ. Qual.* **1988.** *17*:124-130.
10. De Nobili, M., S. Santi, and C. Mondini. *Fert. Res.* **1992.** *33*:71-79.
11. Cohen, S.Z., S. Nickerson, R. Maxey, A. Dupuy Jr., and J.A. Senita. *Ground Water Monitor. Rev.* **1990.** *10*:160-173.
12. Rieke, P.E. and B.G. Ellis. In *Proceedings of the 2nd International Turfgrass Research Conference*; Roberts, E.C., Ed.; Amer. Soc. of Agron., Madison, WI 1974. pp 120-130.
13. Starr, J.L. and H.C. DeRoo. *Crop Sci.* **1981.** *21*:531-535.
14. Mancino, C.F. and J. Troll. *HortSci.* **1990.** *25*:194-196.
15. Gold, A.J., W.R. DeRagon, W.M. Sullivan, and J.L. Lemunyon. *J. Soil and Water Conserv.* **1990.** *45*:305-310.
16. Bowman, D.C., J.L. Paul, W.B. Davis, and S.H. Nelson. *J. Amer. Soc. Hort. Sci.* **1989.** *114*:229-233.
17. Carlson, R.M. *Anal. Chem.* **1986.** *58*:1590-1591.
18. Bowman, D.C., J.L. Paul, and R.M. Carlson. *Commun. Soil Sci. Plant Anal.* **1988.** *19*:205-213.
19. Devitt, D.A., J. Letey, L.J. Lund, and J.W. Blair. *J. Environ. Qual.* **1976.** *5*:283-288.
20. Bowman, D.C., D.A. Devitt, M.C. Engelke, and T.W. Rufty. *Crop Sci.* **1998.** *38*:1633-1639.

BEST CHEMICAL MANAGEMENT PRACTICES

Chapter 11

KTURF: Pesticide and Nitrogen Leaching Model

Steven K. Starrett[1] and Shelli K. Starrett[2]

Departments of [1]Civil Engineering and [2]Electrical and Computer Engineering,
Kansas State University, Manhattan, KS 66506

The objective of this research was to develop two computer models that accurately predict pesticide and nitrogen leaching through turfgrass areas using Artificial Neural Network (ANN) modeling techniques. After much investigation, the inputs used to train the pesticide ANN model were reduced to: pesticide solubility, K_{oc}, time after application, and the irrigation application practice. Similarly, nitrogen form, percentage of soil that is sand, time after application, and the irrigation application practice were the inputs used to train the nitrogen ANN model. Most of the training and testing data were collected from a leaching study that used 50-cm turfgrass-covered undisturbed soil columns. On average, the model's predictions were within about 4% of testing case values. The ANN approach proved to be a feasible modeling technique. An interactive World Wide Web (www) site has been developed where these models can be used online.

Concentrated efforts have been made in recent years to describe water and pesticide movement in agricultural fields. Consequently, a variety of models have been proposed that differ in their conceptual approach and degree of complexity. These models are influenced by the environment, training, and biases of their developers. PRZM/RUSTIC, EXAMS, SWRRB, CREAMS/GLEAMS, SURFACE, HSPF/STREAM, VLEACH, PESTAN, AGNPS, and GUS (*1*) are some of the available models that attempt to predict the movement of pesticides in the environment or address the likelihood of pesticides contaminating drinking water supplies. Unfortunately, as the degree of modeling sophistication has increased so has the complexity of determining the required inputs. A high level of knowledge is required in order to understand and implement these models in any predictive technique. Also, these models were developed for the agricultural industry using soil and plant conditions from cropland. This makes it difficult for the models to simulate pesticide transport on golf courses or managed turfgrass areas since the turfgrass conditions are very different crop conditions.

Artificial Neural Networks (ANNs) have been used in a wide range of applications, including metro trains, cameras, camcorders, vacuum cleaners, automobiles, washing machines (2), and transport engineering (3). Artificial Neural Networks are also being used in environmental-engineering-related fields including the prediction of the hydraulic conductivity of compacted clays (4), identification of soil compaction characteristics, determination of aquifer parameters (5), identification of probable failure modes in underground openings (6), and liquefaction potential assessment (7).

The major difference between traditional modeling techniques and ANN techniques is that in the traditional method, the developer creates mathematical relationships between inputs and outputs, in contrast, the ANN program learns the relationships based on the training data set. Traditional models are built starting with a mathematical expression describing the relationships between the variables under consideration. The success of the model depends on how well the theoretical equations match the real system behavior. ANN techniques are powerful because they directly utilize data measured in the real system to build the model. ANN's allow the modeler the flexibility to choose which parameters will be used as input to the model, and thus inputs are chosen that are easily measurable in realistic situations. The ANN technique forms the functional relationship between the given input data and desired output data without needing a formal theory or mathematical model. The ANN training process constructs a new mathematical model specific to the problem of interest. Both techniques have advantages and disadvantages.

The objective of this work was to develop computer models that accurately predicted pesticide and nitrogen leaching when applied to turfgrass areas.

Materials and Methods

Artificial Neural Network Methodology. Generally speaking, ANN's are designed to mimic the learning processes achieved in the brain. During the training process, a large number of training data sets (pairs of questions and correct answers) are presented to the network. The network "learns" to predict the correct answers to the training data questions through a repeated mathematical process. Once the network is producing satisfactory answers to the training data questions, the training is complete. Unlike biological brains, most ANN's cease to learn once the training process is complete. After training, the ANN simply predicts an answer when presented with a new input set (question). The testing of an ANN involves presenting it with data sets with which it was not trained and determining the accuracy of its answers to these new questions. All of this is done using mathematical methods and numerical calculations.

The network topology (how the ANN is put together), the form of the learning rules, and the activation functions (part of the mathematical internal functioning of the ANN) can all be variable. This leads to a wide variety of network types such as competitive learning, the Hopfield network, and the back-propagation network *(8-10)*. Back-propagation networks are currently being used by most neural-network application developers *(3-7, 11-24)*.

A step-by-step description of a back-propagation learning algorithm is given in numerous publications *(4, 19, 25-28)*. A brief description of the ANN structure and training process follows.

The processing units (often called neurons or nodes) in a back-propagating ANN are arranged in layers. Most ANNs have an input layer, an output layer and a minimum of one hidden layer (Figure 1). Each processing unit performs a certain mathematical function (Figure 2) and then passes on its answer (output) to other units. The layers in an ANN are interconnected with weighted communication connections. The signal coming over the connection is multiplied by the weight assigned to that connection before it is used by the receiving unit. Within a unit, the entering signals are summed, a bias is added, and the result is passed through the transfer function (f(x)) of that unit (Figure 2). The output of the unit is the answer obtained from the transfer function, and this answer is passed on to the next processing layer (or is the final answer if the unit is in the output layer). Transfer functions can be linear or non-linear. If linear functions are used in both the hidden layer and the output layer that the network is compressed to a single layer.

The presence of the hidden units permits the ANN to learn the relationships between the inputs to the outputs. In the ANN, the development of an output answer takes place in a feed-forward manner, from the input layer through the hidden layer to the output layer (bottom to top in Figure 1). The process of learning is achieved by back-propagating the errors (difference between the correct answer and the ANN's guess) at the output layer back through the hidden layer(s) to the input layer. No communication is permitted between the processing units within a layer; however, the processing units in each layer send their output to the processing units in the next layer. It is the presence of the hidden units that allows the ANN to represent and compute the complicated associations between the input and output patterns.

The connections between the various nodes in the three-layer paradigm represent the most important part of the computational ANN. Associated with each connection is a numerical value representing the strength or the weight of that connection. Associated with each node (other than the input nodes) is a value representing the bias. The connection strengths and biases are determined during the process of training the ANN. At the beginning of the training process, the connection strengths and biases are assigned random values.

A set of data points (inputs and corresponding outputs) is used to train the network. Each training situation is presented to the network being trained and the

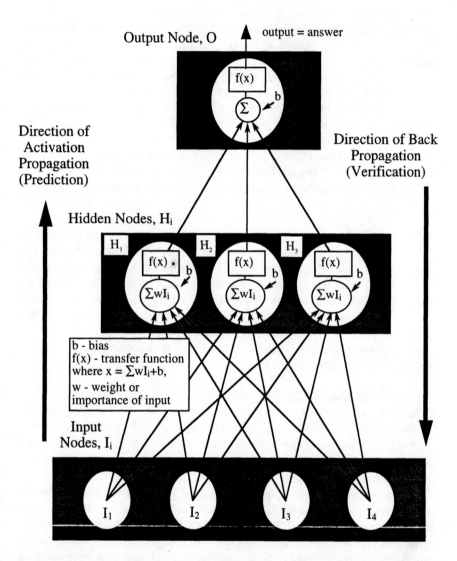

Figure 1. Example of an Artificial Neural Network (ANN) with back-propagation (Reproduced with permission from ref. 35. Copyright Marcel Dekker, Inc.).

network yields a predicted output. The error between the network prediction and the desired output is used to adjust the weights and biases of the network according to the chosen learning rule (backpropagation algorithm in this case). The training sets are presented repeatedly until the network predictions match the desired values within a given error tolerance. The choice of training data is critical to the range of applicability of the ANN model. The training data should encompass the same (or larger) range of conditions as those under which the model will be used. In other words, in order to accurately predict leaching of a pesticide with solubility of 69 ppm, 69 ppm should be within the range of the training data solubility inputs.

When the iterative training process has converged, the weights and biases learned are stored for later use by the ANN. As a result, the ANN will be able to utilize the stored weights and biases to quickly produce responses to new sets of inputs.

Development of ANN Models. The training and testing sets of data were formed by collecting leaching results from research articles. Most of the data used to train the ANN was collected from research conducted by Dr. Nick Christians, Dr. Al Austin, (Iowa State University) and Steven K. Starrett that investigated the transport of pesticides and nutrients applied to turfgrass (*29-33*). The research conducted at Iowa State University used undisturbed, turfgrass covered soil columns 50 cm in depth. The soil columns were at field capacity when the pesticides and nutrients were applied. Some general characteristics of the seven pesticides used are presented in Table I.

The Neural Network Toolbox of the computer software package MATLAB® (The MathWorks, Inc. 24 Prime Park Way, Natick, MA 01760-9779. (508) 653-1415) was used for this work. The hidden-layer transfer function used was a tanh-sigmoid function ($f(x) = \tanh(x)$), and the output-layer transfer function was a linear function (Figure 1). The test cases were randomly removed from each data set; therefore, the test cases were previously unseen by the ANN.

For comparison purposes, 1st and 2nd order polynomial regression models were developed using Mathematica® software (Wolfram Research, Inc. 100 Trade Center Drive, Champaign, Illinois 61820-7237. (217) 398-0700). This program determined the best-fit curve for each specified order of regression. To compare the ANN and the regression results, the same input data sets were used for training and testing.

Results and Discussion

Pesticide Model. The inputs for the ANN model consisted of pesticide solubility, pesticide soil:water partitioning coefficient (Koc), time after application, and the irrigation application practice. One hundred sixty-eight sets of data were used to

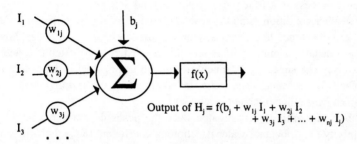

Hidden Node H_j

Output of $H_j = f(b_j + w_{1j} I_1 + w_{2j} I_2 + w_{3j} I_3 + ... + w_{nj} I_j)$

Figure 2. The mathematics inside a neuron (Reproduced with permission from ref. 35. Copyright Marcel Dekker, Inc.).

Table I. Properties of pesticides.

Pesticide	Water Solubility (mg L^{-1})	K_{OW}[1]	K_{OC}[2]	Field Half-Life (days)
metalaxyl	8400.0	50.0	50	70
isazofos	69.0	1000.0	100	34
chlorpyrifos	2.0	1000000.0	6070	30
pendimethalin	0.3	1500000.0	5000	90
2,4-D	796,000.0	0.4	20	10
MCPP	0.3	0.3	20	21
dicamba	400,000.0	0.2	2	14

[1] K_{ow} = octanol:water partitioning coefficient.
[2] K_{oc} = soil:water partitioning coefficient normalized by the organic carbon fraction of soil.
(Reprinted with permission from ref. *35*. Copyright Marcel Dekker, Inc.).

train the ANN. The network was given 1200 training epochs (training cycles) to learn the appropriate weights and biases. After training, a simulation of 21 test cases was run to test the ANN's accuracy (Table II). The trained ANN was used to predict the percentage of applied pesticide that leached through 50 cm of turfgrass covered soil for test cases (Table II) to which it had not been exposed previously. The regression equation predictions for the test cases were also determined and compared with the ANN predictions (Table III, Figure 3).

Predicting the complex pesticide transport process is truly a challenge. The transport process is a function of pesticide and soil characteristics, vegetation type, weather conditions, and the type and number of microbes present. Models have been developed to theoretically predict the leaching of pesticides; however, there are usually a number of inputs that are extremely difficult to determine accurately. The success of a model developed using ANN is dependent on the data used for training. Thus, the development of the training database that contains a wide range of conditions is critical.

The inputs initially investigated were the pesticide's water solubility, soil adsorptivity (K_{oc}), octanol-water partitioning coefficient (K_{ow}), half-life and rate of application values; the soil's organic matter, bulk density, and pH; the cumulative irrigation values, the irrigation application practice (heavy or light applications), and the time after the pesticide application that the leachate samples were collected. The importance of the various inputs was investigated, and it was found that many of these inputs had no affect on the predicted outputs. Possible explanations of why these inputs did not affect the output are: the input was not independent of other inputs used, variations in the input were orders of magnitude smaller than changes in output, or that the input value was not accurate. After determining the importance of each input listed above, the number of inputs used to train the ANN was reduced to four which were: pesticide solubility, K_{oc}, time after application, and the irrigation application practice. It is important to note that inputs used were parameters that are simple to determine. This increased the usability of the model for turfgrass managers since complex inputs, such as unsaturated hydraulic conductivity at varying moisture contents, are difficult to determine.

The training data were from soil at a moisture content of field capacity when the pesticides were applied; therefore, under greatly different conditions the ANN's estimated pesticide leaching could be inaccurate. If conditions were wetter than field capacity it would be expected that more pesticide would leach, and similarly if conditions were drier, less would leach.

By a trial and error process, the optimum number of hidden layers and hidden nodes was determined to be one and three, respectively. Many different transfer functions were investigated (logarithmic, sigmoidal, linear) by comparing predicted output values to the corresponding measured test cases. A tanh-sigmoidal function was determined to be superior for the hidden-node transfer function and a linear function was superior for the output-node transfer function.

Table II. Test case inputs.

Test Case	Pesticide Solubility (ppm)	Pesticide Koc (mg kg^{-1})	Irrigation Index[1]	Time (days) after pesticide application
1	69	100	1	12.5
2	69	100	1	22.9
3	69	100	0	8.3
4	69	100	0	22.9
5	2	6070	1	4.2
6	2	6070	1	20.8
7	2	6070	0	8.3
8	2	6070	0	20.8
9	8400	50	1	22.9
10	8400	50	0	12.5
11	0.3	5000	1	4.2
12	0.3	5000	1	20.8
13	0.3	5000	0	4.2
14	796,000	20	1	0.1
15	796,000	20	1	27.1
16	796,000	20	0	22.9
17	400,000	2	1	0.1
18	400,000	2	1	22.9
19	400,000	2	0	20.8
20	660,000	20	1	8.3
21	660,000	20	0	16.7

[1]infrequent irrigation regime (one 2.54 cm application per week) = 1, frequent irrigation regime (four 0.63 cm applications per week) = 0
(Reprinted with permission from ref. *35*. Copyright Marcel Dekker, Inc.).

Table III. Error of Artificial Neural Networks (ANN) and regression equations in predicting pesticide leaching for test cases.

Method	SSE
ANN	17.4
1st order regression	528.4
2nd order regression	522.3

SSE = sum of the squared error between measured and predicted values
(Reprinted with permission from ref. *35*. Copyright Marcel Dekker, Inc.).

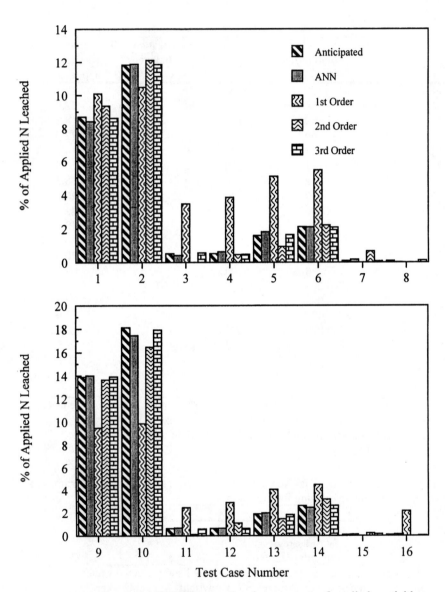

Figure 3. Measured and KTURF's predicted percentage of applied pesticide leached for different test cases (Reproduced with permission from ref. 35 Copyright Marcel Dekker, Inc.).

The ANN model was able predict within a few percentage points high leaching situations and low leaching situations. Test cases 17 and 18 are high leaching situations that the ANN did a good job of predicting, however, the regression models underestimated by about 60%. In general the ANN had a much lower error than the regression models (Table III).

Nitrogen Model. After extensive testing, the best architecture for the nitrogen leaching ANN model was determined to be one hidden layer, three hidden nodes, and training 1000 epochs. Again the tanh-sigmoid and linear transfer functions were used for the hidden and output layers, respectively. The model inputs used were irrigation rate, time after application, nitrogen form and percent sand of soil. The 16 testing points from each data set were input (Table IV) into the trained ANN model and a corresponding output prediction was produced (Figure 4).

The Sum of the Squared Error (SSE) between the anticipated values (from measured data and technical publications) and the predicted values (produced by the model) was calculated to be 0.3. Given that errors in the input parameters can easily occur, such as determining the percentage of sand in the soil or the irrigation rate, SSE of 0.3 demonstrates virtually no difference between the ANN predicted and the anticipated values.

As stated earlier, for comparison, a statistical regression model was developed using 1st, 2nd, and 3rd order equations in terms of four input variables; irrigation rate; time; nitrogen form; and soil parameter. The equations were calculated by *Mathematica®*.

The 16 test sets (Table IV) were plugged into these equations and outputs were calculated. The linear regression did not fit well with the anticipated output, SSE = 170. However, the 2nd order equation performed better with a SSE = 6.0, and the 3rd order equation, achieved a SSE = 0.1, comparable to that of the ANN model (Figure 4). A 4th order equation was also developed, but it showed no improvement over the 3rd order equation. The ANN and 3rd degree models display a high degree of accuracy when comparing the anticipated and the model predicted values.

The authors feel that the accuracy of the regression model in the present study was due to the simplicity of the data set and should not be considered the norm (*19, 26, 38*).

Conclusions

Artificial Neural Networks (ANN) proved to be a feasible modeling technique for predicting pesticide and nitrogen leaching. The ANN was much better at prediction pesticide leaching than the 1st or 2nd order regression equations. Apparently the relationships between the inputs and output were too complex to

TABLE IV. Data used in testing nitrogen leaching models.

Test Case	Irrigation Rate[1]	Time (days)	Nitrogen Form[2]	Percent Sand	Anticipated leachate (%)
1	1	2.1	1	65	8.70
2	1	16.7	1	65	11.82
3	0	4.2	1	65	0.52
4	0	18.8	1	65	0.52
5	1	6.3	0	65	1.60
6	1	20.8	0	65	2.12
7	0	8.3	0	65	0.09
8	0	22.9	0	65	0.09
9	1	14.6	1	100	13.88
10	1	29.2	1	100	18.10
11	0	2.1	1	100	0.64
12	0	18.8	1	100	0.64
13	1	4.2	0	100	·1.87
14	1	20.8	0	100	2.59
15	0	6.3	0	100	0.11
16	0	22.9	0	100	0.11

[1] 1 - infrequent irrigation, 2.54-cm applications
0 - frequent irrigation, 0.83-cm applications

[2] 1 - liquid
0 - granular

(Reprinted with permission from ref. 39 Copyright 1997. Marcel Dekker, Inc.).

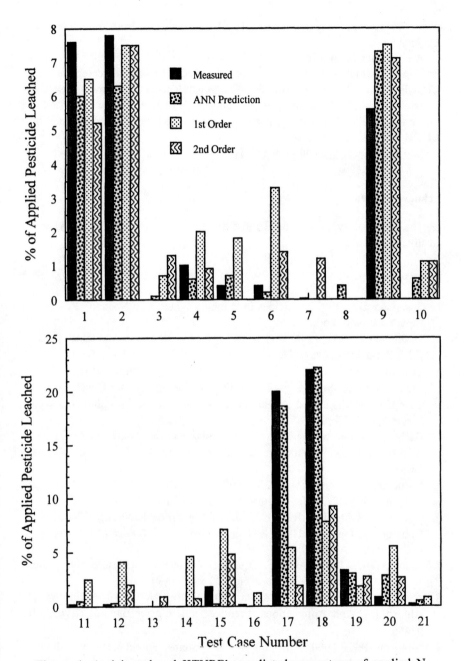

Figure 4. Anticipated and KTURF's predicted percentage of applied N leached for different test cases (See Table 1 for inputs) (Reproduced with permission from ref. 39 Copyright 1997. Marcel Dekker, Inc.).

be predicted by simple regression and the ANN was able to "learn" these complex relations.

An interactive World Wide Web (www) site called KTURF has been developed at which these ANN models can be used. Used as an assessment tool, KTURF can help to reduce pesticide and nitrogen leaching by allowing users to experiment with different management schemes. Practices can be optimized to reduce the likelihood of pesticide and nitrogen leaching beyond the rootzone. The KTURF site is located at:

> http://www.eece.ksu.edu/~starret/KTURF/

Acknowledgments

The financial support of the United States Golf Association (USGA) was greatly appreciated.

References

1. Gustafson, D.I., 1993, Pesticides In Drinking Water, Van Nostrand Reinhold, New York, New York.
2. Kartalopoulos, S.V., 1996, Understanding Neural Networks And Fuzzy Logic, IEEE Press, Piscataway, NJ.
3. Ghaboussi, J.; Sidarta, D.J.; and Lade, P.V., 1994, Neural Network Based Modeling in Geomechanics, Proceedings of the eighth Int'l Conference on Computer Methods and Advances in Geomechanics, Morgantown, West Virginia, pp. 153-164.
4. Basheer I.A. and Najjar Y.M., 1994, Designing And Analyzing Fixed-Bed Adsorption Systems With Artificial Neural Networks, *J. of Env. Systems*, 24(3):291-312.
5. Abdul Aziz, A.R. and Wong, K.V. 1992, A Neural-Network Approach To The Determination Of Aquifer Parameters, *Ground Water*, 30(2):164-166.
6. Lee, C. and Sterling, R., 1992, Identifying Probable Failure Modes For Underground Openings Using A Neural Network, *Int. J. Rock Mech. Min. Sci. and Geomech*. Abstr. 29:49-67.
7. Goh, A.T.C., 1994, Seismic Liquefaction Potential Assessed By Neural Networks. *J. of Geotech. Engineering*, 120:1467-1480.
8. Fausett, L., 1994, Fundamentals Of Neural Networks Architecture, Algorithms, And Applications, Prentice Hall, Upper Saddle River, NJ 07458.
9. Simpson, P.K., 1990, Artificial Neural Systems: Foundations, Paradigms, Applications And Implementations, New York: Pegamon Press.10. Werbos, P.J., 1994, The Roots Of Backpropagation From Ordered Derivatives To Neural Networks And Political Forecasting, John Wiley. New York.

11. Chao, L.C. and Skibniewski, M.J., 1994, Estimating Construction Productivity: Neural-Network-Based Approach, *Journal of Computing in Civil Engineering*, 8(2):234-251.

12. Flood, I. and Kartam N., 1994, Neural Networks In Civil Engineering. I: Principles And Understanding, " *Journal of Computing in Civil Engineering*, 8(2):131-148.

13. Flood, I. and Kartam N., 1994, Neural Networks In Civil Engineering. II: Principles And Understanding, *Journal of Computing in Civil Engineering*, 8(2):149-162.

14. Gagarin, N.; Flood, I., and Albert, P., 1994, Computing Truck Attributes With Artificial Neural Networks. *J. Computing in Civil Engineering.* 8(2):131-148.

15. Ghaboussi, J.; Garret Jr., J.H. and Wu, X., 1991, Knowledge-Based Modeling Of Material Behavior With Neural Networks, *Journal of Engineering Mechanics*, 117(1):132-153.

16. Garrett, J.H.; Ghaboussi, J. and Wu X., 1992a, Neural Networks, Chapter 6 in *Expert Systems For Civil Engineers*, Robert Allen (Ed.), pp. 104-143.

17. Garrett, J.H.; Ranjitham, S. and Eheart J.W., 1992a, Application Of Neural Network To Groundwater Remediation, Chapter 11 in *Expert Systems for Civil Engineers*, Robert Allen (Ed.), pp. 259-269.

18. Karunanithi, N.; Gernney W.J.; Whitley, D. and Bovee K., 1994, Neural Networks For River Flow Prediction, *Journal of Computing in Civil Engineering*, 8(2):201-221.

19. Najjar, Y.M. and Basheer, I.A. 1996, Utilizing Computational Neural Networks For Evaluating The Permeability Of Compacted Clay Liners, *Geotech. and Geol. Engg. Int. J.* 14(3):193-212.

20. Penumadu, D., 1993, Strain Rate Effects In Pressuremeter Testing And Neural Network Approach For Soil Modeling, Ph.D. Dissertation, Georgia Institute of Technology.

21. Rizzo, D.M. and Dougherty, D.E., 1994, Characterization Of Aquifer Properties Using Artificial Networks: Neural Kriging, *Water Resources Research*, 30(2):483-497.

22. Rogers, J.L., 1994, Simulating Structural Analysis With Neural Network, *Journal of Computing in Civil Engineering*, 8(2):252-265.

23. Rogers, L.L., and Dowula, F.U., 1994, Optimization Of Groundwater Remediation Using Artificial Neural Networks With Parallel Solute Transport Modeling, *Water Resources Research*, 30(2):457-481.

24. Wang, L. and Feng, X., 1993, Comprehensive Classification Of Rock Stability, Blastability And Drillability Based On Neural Networks, *Proc. 34th U.S. Symposium on Rock Mechanics*, Vol II, Madison, Wisconsin, 2: 741-744.

25. Medsker, L. R. 1994, Hybrid Neural Network And Expert Systems, Kluwer Academic Publishers, Boston.

26. Hecht-Nielsen, R., 1990, Neurocomputing, Addison-Wesley Publishing Company, Inc. Reading, MA 01867

27. Hertz, J., Krogh, A. and Palmer, R.G., 1991, Introduction To The Theory Of Neural Computation, Addison Wesley, Redwood, CA.

28. Basheer, I.A. and Najjar, Y.M., 1995, Estimating Hydraulic Conductivity Of Compacted Clay Liners, *J. Geotech. Engg*, 120:675-676.

29. Starrett, S.K.; Christians, N.E., and Austin, T.A., 1995, Fate Of Amended Urea In Turfgrass Biosystems, *Commun. Soil Sci. and Plant Anal.*, 26:1595-1606.

30. Starrett, S.K.; Christians, N.E., and Austin, T.Al, 1995, Fate Of Nitrogen Applied To Turfgrass Covered Soil Columns, *J. of Irrigation and Drainage Eng.*, 121(6):390-395

31. Starrett, S.K.; Christians, N.E., and Austin, T.Al, 1996, Comparing Dispersivities And Soil Chloride Concentrations Of Turfgrass Covered Undisturbed And Disturbed Soil Columns, *J. of Hydrology*, 180:21-29.

32. Starrett, S.K.; Christians, N.E., and Austin, T.Al, 1996, Movement Of Pesticides Under Two Irrigation Regimes Applied To Turfgrass, *J. Environ. Qual.*, 25:566-571.

33. Starrett, S.K.; Christians, N.E., and Austin, T.Al, 199X, Movement Of Trimec Under Two Irrigation Regimes Applied To Turfgrass, under review

34. Wauchope, R.D.; Buttler, T.M.; Hornsby, A.G.; Augustijn-Beckers, P.W.M., and Burt, J.P., 1992, The SCS/ARS/CES Pesticide Properties Database For Environmental Decision-Making, *In* G. W. Ware, (ed.) Reviews of environmental contamination and toxicology, Springer-Verlag, New York. N.Y. 123:1-35.

35. Starrett, Steven K.; Starrett, Shelli K.; Najjar, Y.M.; Hill, J.C. and G. L. Adams, 199X, Modeling Pesticide Leaching Using Artificial Neural Networks, *Commun. Soil Sci. and Plant Anal.*, in press

36. Brown, K.W.; Duble, R.L. and Thomas, J.C.. 1977. Influence of management and season on fate of N applied to golf greens. Agronomy J. 69:667-671.

37. Synder, G.H.; Augustin, B.M., and Davidson, J.M.. 1984. Moisture sensor-controlled irrigation for reducing nitrogen leaching in bermudagrass turf. J. Agronomy. 76:964-968.

38. Masters, T. 1993. Practical neural network recipes in C++. Academic Press Inc. San Diego, CA 92101 pp. 173-199.

39. Starrett, Steven K.; Starrett, Shelli K., and Adams, G. L.. 1997, Using Artificial Neural Networks and Regression to Predict Percentage of Applied Nitrogen Leached Under Turfgrass, *Commun. Soil Sci. and Plant Anal.*, 28:497-507.

Chapter 12

Calibration and Validation of Runoff and Leaching Models for Turf Pesticides, and Comparison with Monitoring Results

T. E. Durborow[1], N. L. Barnes[1], S. Z. Cohen[1], G. L. Horst[2], and A. E. Smith[3]

[1]Environmental and Turf Services, Inc., 11141 Georgia Avenue, Suite 208, Wheaton, MD 20902
[2]Department of Horticulture, University of Nebraska, Lincoln, NE 68583
[3]Department of Crop and Soil Sciences, University of Georgia, Athens, GA 30602

Computer models that simulate pesticide leaching and runoff were originally developed with a focus on agriculture. Due to distinct botanical and agronomic differences, there are questions about the applications of these models to turf. We evaluated the U.S. EPA's PRZM model and the USDA's GLEAMS model using data from turf test plot studies funded by the U.S. Golf Association. The plots simulated golf greens and fairways in Georgia, Nebraska, and Pennsylvania. The GLEAMS model performed surprisingly well in the Georgia evaluation: 11 of 12 runoff events were predicted moderately to very well for hydrology (CV=43% using all data) and pesticides. The PRZM3.0 evaluation of the Penn State runoff study produced less favorable results, but the differences may be more related to the nature of the field study and not the model's performance. The PRZM2.3 and 3.12 predictions of percolate volumes ranged from poor to good for the Georgia study (CV=59%) and excellent for the Nebraska study (CV=18% with no calibration). Predictions of pesticide leachate ranged from poor to good, but PRZM tended to over predict pesticide mass. Tentative conclusions are that PRZM and GLEAMS are useful for turf runoff predictions. PRZM is conservative for leachate predictions of more soluble pesticides.

There is a large difference between perceived risks and actual risks attributed to turf management at golf courses. The results from field or watershed-scale golf course water quality monitoring studies are encouraging but limited (Cohen, et al. *J. Environ.*

Qual., in press). One tool that is available to extrapolate results to different points in time and space, e.g., other golf courses, is computer simulation modeling.

The concept of using computer simulation models at the field and watershed scales to predict turf chemical impacts is being established *(1)*. However, pesticide transport models have not been thoroughly evaluated for application to turf in a calibration/validation analysis. That is the purpose of this study.

The approach we have taken is to apply these models to the recent U.S. Golf Association (USGA)-sponsored environmental research results. The relevant research in this area has all been done on small scale test plots (micro plots). Therefore we are evaluating the models using data from carefully controlled environments. The ultimate goal is to provide guidance for using these models to predict turf chemicals leaching to ground water and running off to surface water at the field and watershed scale locations throughout the U.S.

The basic concept is to calibrate the models against the results using site-specific data, and vary key inputs within acceptable ranges to achieve a reasonable match of predicted vs. observed results. First, we calibrate the water output, then the pesticide output. If the calibration is successful using reasonable assumptions and site-specific data for input parameters (e.g., wilting point, soil half life), then the study helps validate the model for this application. If predicted and observed results are consistently and significantly different, or if the results can only be made consistent if unreasonable and unexplainable assumptions are made about certain input parameters, then use of the model (as constructed) for this application may not be valid.

Methods

Two studies conducted at the University of Georgia, Georgia Experiment Station in Griffin, GA, *(2, 3)* were simulated. A greens-turf leaching study and a fairway-turf runoff study were conducted in 1992-1994, supported by the USGA Greens Section Environmental Research Program. The models used to simulate these field studies were PRZM (EPA's Pesticide Root Zone Model) and GLEAMS (USDA's Ground Water Loading Effects of Agricultural Management Systems model). We ultimately used PRZM version 2.3 for our leaching assessment. PRZM version 2.0, which was the current version at the time we initiated the work, was known to overestimate pesticide runoff, which was one reason why GLEAMS was chosen at that time to assess the Georgia runoff study.

PRZM was then calibrated against the results of a fairway-turf leaching study conducted by Horst, et al. *(4, 5)* at U. of NE, and PRZM was also calibrated against data generated in a fairway-turf runoff study by Linde, Borger, and Watschke at Penn State. GLEAMS was not selected for the second phase for two reasons: the runoff problem with PRZM was resolved in the development of releases 2.3 and 3.x, and EPA now prefers to use PRZM for pesticide risk assessments. We used EPA's more recent version of PRZM, version 3.0, for our runoff assessment and PRZM version 3.12, released in March 1998, for our leaching assessment.

The information presented in the reports prepared by the researchers for the USGA did not have the extensive detail needed for model calibration/validation studies. Therefore a questionnaire was developed that was completed by the U. of GA, U. of NE, and Penn State researchers. The questionnaire pointed out several data gaps for our work, which resulted in additional experimental work by all three institutions and ourselves. There was an extensive amount of follow up to obtain as much complete and correct site-specific input parameters as possible in an attempt to minimize the amount of parameter estimation.

Model Calibration/Validation Overview

Model calibration is the process of developing and fine-tuning a set of input parameters to a known set of conditions to generate predictions that are comparable to what was obtained from that known set of conditions. The ASTM (6) definition is "a test of a model with known input and output information that is used to adjust or estimate factors for which data are not available." Successfully calibrating a model therefore demonstrates its ability to simulate a specific set of conditions, and allows one to extrapolate to other points in space or time.

Model validation, on the other hand, is the process of applying the model to a wide variety of conditions and evaluating its performance compared to the results of the variable settings. The ASTM (6) definition is "a comparison of model results with numerical data independently derived from experiments or observations of the environment." Successful validation of a model demonstrates its ability to simulate a wide variety of conditions.

Our study is an attempt to calibrate and partially validate - or invalidate - these models for turf applications. We gauge the usefulness of the calibrated PRZM and GLEAMS models for simulating the test conditions of the turf research studies. We also do a partial validation study by evaluating performance of the models under a variety of turf conditions.

The approach taken for the calibration of the models with the data obtained from the university research studies involved two steps. Step one was to calibrate the hydrology component of the models. It was necessary to enter all available measured field data into the model input files and then enter reasonable approximations for the model parameters that cannot be specifically measured (e.g., runoff curve number). The model runoff predictions were compared with each study's measured percolate and runoff water losses and the more uncertain model parameters were then fine-tuned (or calibrated) to obtain a better fit of predicted runoff to the measured results. When this was done to an acceptable level, it demonstrated that the soil hydraulics are properly represented. Only then could the pesticide component be properly calibrated. The pesticide calibration was step two. There was more uncertainty in the pesticide parameters for the field studies. With the exception of the U. of NE study, the only pesticide parameters known for these specific plots was the target application rates. The environmental fate parameters for the Georgia and Pennsylvania studies were therefore all initially selected based on data obtained from literature and other sources, not necessarily generated under conditions applicable to the site specific settings. Site specific pesticide environmental fate parameters were determined for the U. of NE study.

Test Plot Descriptions

U. of Georgia Leaching Plots. The test plots are described in Smith and Tillotson *(2)*. We modeled pesticides and water movement through the greenhouse lysimeter plots.

The greenhouse plots were developed as small greens with rooting mixtures of 85:15 sand:sphagnum peat moss for the North plots and 80:20 for the South plots. The 85:15 mixture supports the growth of bentgrass and the 80:20 bermudagrass. We used the 85:15 mixture for the calibration/validation study.

U. of Nebraska Leaching Plots. The field research at the University of Nebraska was conducted in conjunction with Iowa State University. The test plots are described in detail in Horst and Christians *(4)* and Horst, et al. *(5, 7)*. The focus of the calibration/validation work targeted only the University of Nebraska (U. of NE) results. The U. of NE conducted their field research at the John Seaton Anderson Turfgrass Research Facility near Mead, NE. The soil was the Sharpsburg silty clay loam. Field plots were excavated and transferred to a greenhouse for the leaching study. We modeled pesticide and water movement through the greenhouse lysimeter plots.

U. of Georgia Runoff Plots. The turf runoff study was conducted at the Georgia Experiment Station in Griffin, GA by Albert Smith and David Bridges of the University of Georgia and is described in Smith *(3)*.

The study consisted of 12 bermudagrass turf plots, each equipped with runoff collection, measuring, and subsampling devices. The plots received natural rainfall and irrigation. The plots were intensively irrigated on four subsequent days following each of the pesticide applications. The rate of application was approximately 1.2"/hr. Runoff was measured and flow proportioned subsamples were collected using a tipping bucket device. The subsamples were then analyzed for pesticide residues and nutrients.

Penn State Runoff Plots. The turf runoff study was conducted at the Landscape Research Center on campus at Penn State University in University Park, PA. It is described in some detail in Linde, et al. *(8)*, Linde, et al. *(9)*, and Linde *(10)*.

The study consisted of bentgrass and ryegrass plots maintained to typical golf course fairway conditions. Runoff events occurred due to simulated and natural rainfall. Four pesticides and fertilizers were applied to three replicate plots of each turf type over the course of the study and runoff and leachate samples were collected at various times following applications and analyzed for the pesticides and nutrients. We conducted our computer model calibration study based on data obtained from the ryegrass plots.

Model Calibration Exercises

U. of Georgia Leaching

Hydrology. An initial set of input parameters was developed based on the questionnaire provided to the field study researchers. Initial daily irrigation and minimum/maximum temperature data were provided for the greenhouse plots. Plot geometry and physical properties for the soil profile were provided based on a single thickness (32.5 cm). The first run of the model used the initial data provided, but it was quickly apparent that PRZM2.3 would not run properly using a single uniform soil horizon of 32.5 cm (13 in).

A request was made to the U. of GA research team to discretize the soil column into smaller horizons and provide physical soil parameters for those horizons. A parallel plot that did not receive pesticide treatment was sacrificed for analysis to represent the other greenhouse plots. The top 6 inches of that plot was analyzed in 2 inch increments to provide physical parameters for the 85:15 plots. Specific parameters included soil particle size distribution, moisture content at field capacity, saturated hydraulic conductivity, organic matter, and bulk density. Wilting point was the only key parameter that was not analyzed at that time. It was measured, at a later date, for the whole lysimeter core. Only one set of soil characteristics was given to represent all of the 85:15 plots, and this was apparently based on a test plot that was not part of the modeled data set. Therefore only one value was used for wilting points throughout the four horizons modeled. The original valued provided for field capacity was used to represent the last horizon.

Pesticides. Three different pesticides were applied in treatments to the greenhouse plots made on June 29, 1992. Only one pesticide was applied in treatments November 5 & 20, 1991 to greenhouse lysimeter plots. Six different pesticides were applied in treatments to the outdoor mini-green plots made on July 3 & 17, 1992 and June - September, 1993. Dr. Smith provided application rates and formulations for these pesticides. Some environmental fate data were available for four pesticides in the 1991-1993 Summary report. Some of these values were used in the modeling and the remaining values were obtained from published literature and past experience.

Only the June 29, 1992 and November 5 & 20, 1991 application scenarios were modeled due to incomplete data for the other treatments.

Calibration of the pesticide component involved adjustment of K_{oc} and half-life ($t\frac{1}{2}$) for the pesticides 2,4-D, dicamba, and MCPP (mecoprop). These environmental fate parameters seem to have the most significant effect on the PRZM2.3 model output. The values selected for environmental fate parameters are given in the following.

U. of Nebraska Leaching

Hydrology. An initial set of input parameters was used based on a questionnaire submitted to Dr. Horst. A log of daily temperature and irrigation was provided as

Table I. Application Rates and Environmental Fate Data

	2,4-D DMA salt	Dicamba DMA salt	Mecoprop DMA salt
applic. rate[‡] (kg/ha)	0.56 0.28	0.067	0.56 1.4
foliar t½ (days)	6	9	50*
soil t½ (days)	6	9	9/12[††]
water t½ (days)	53	90	12
Henry's Law Const.	5.6e-9	3.3e-9	5.0e-7
K_{oc}	20	8[††]/35	20

[‡] These are the rates targeted by the U. of GA researchers.
[††] Initial value.
* This was used as input, but it is likely incorrect. Knisel et al. *(19)* indicate a t½ of 10 days, and Horst (see Table II) indicates 7 days.

well as daily evapotranspiration values. Moisture potentials were needed for each segment of the soil cores used as PRZM3.12 input parameters. Therefore field replicates were taken adjacent to the original cores, divided into eight corresponding segments, and sent for moisture potentials analyses. The first two segments were 1 inch thick (2.5 cm) each, the third segment was 2 inches (5 cm) thick, and the remaining segments were 4 inches (10 cm) thick to 24 inches deep (60 cm). Analyses consisted of field capacity, wilting point, bulk density, and organic matter from the four corresponding field replicates. The values used in the model were the mean values for each horizon. Prior to removal of the representative replicate cores they were irrigated for 2 days and allowed to drain for 24 hours.

The original field capacity for the last two horizons had to be adjusted because of the models preset particle density (P.D.) used in initial moisture saturation, i.e., field capacity cannot exceed initial saturation ($\theta = [1 - B.D. \div P.D.]$). The values used were within the range of actual measured field capacity for the four replicates cores.

Pesticides. Seven different pesticides were applied to the closed system turf plots on September 22, 1992. However, only four were modeled. Application rates were provided for each pesticide. An important pesticide parameter in the PRZM model is soil half life (t½). Field dissipation (DT_{50}) values were given for 2,4-D, dicamba, and mecoprop by Dr. Horst in verdure, thatch, and soil. A weighted average was then used for the first horizon from the thatch and soil DT_{50}'s. Horizon 2 & 3 input used the soil value and the remaining horizons (4-8) used one third of the soil value. Soil t½ values for isazofos and chlorpyrifos were taken from Horst, et al. *(5)*. The values selected for the four modeled pesticides are given on the following table.

Table II. Application Rates and Environmental Fate Data

	2,4-D	isazofos	chlorpyrifos	mecoprop
applic. rate (kg/ha)	1.14	2.24	1.12	0.61
foliar t½ (days)	7	0.0	0.0	7
soil t½ (days)	11[†]/18	35	30	8[†]/9
water t½ (days)	18	35	30	9
Henry's Law Const.#	0.0	0.0	0.0	0.0
K_{oc}	20	155	6070	20

[†] 1st horizon (approximate value)
Zero values were input for the Henry's law constant due to the fact that all-inclusive DT_{50} values were used for the verdure.

U. of Georgia Runoff

Hydrology. An initial set of input parameters was developed based on the information provided by Dr. Smith on the questionnaires. Daily rainfall and average temperature data were provided for June, July, and December, 1994 in addition to the amounts of irrigation applied on the specified days following each of the pesticide applications. Using a stochastic weather generator program, daily rainfall and temperature data were produced for the remainder of the year, since GLEAMS requires complete data as a continuous, daily time step model. The plot geometry and slope data were provided and the soil properties were measured. The University of Georgia collected samples from the study site and analyzed them for specific parameters of the model for each of the horizons of the soil profile, including soil particle size distribution, moisture contents at field capacity and wilting point, saturated hydraulic conductivity, and bulk density (which was used to estimate porosity). Five measurements for each parameter and each horizon were made. The mean values for field capacity (1/10th bar), wilting point (15 bar), and bulk density were used. Saturated hydraulic conductivity measurements were done in laboratory conditions on small diameter cores, rather than in the field, and highly variable results were obtained (CV = 47% to 157%). The minimum values of Ksat were used for each horizon, since the runoff produced by the simulated rainfall on the plots was relatively high, i.e., runoff was equal to 27% to 49% of rainfall during the 2" events.

Organic matter content was not measured in these plots. Values from the sample data provided with the GLEAMS model for the Cecil sandy loam soil profile were used. An organic carbon content of 10% for the thatch/soil layer at the surface of the profile was selected based on an evaluation of available literature and an assumption that not all of the organic matter in thatch is readily available for pesticide sorption (11-17). This parameter is not significant to the hydrology simulation, but does have implications in the next step of calibration, pesticides, due to the fact that organic matter can retard pesticide movement.

Crop parameters were formulated to represent a dense stand of turf (leaf area index = 3.0 to 4.0 throughout the year) with an active growing period between mid-March to early-November.

The initial runoff curve number used to begin the hydrology calibration study was selected from Table 2-2a, runoff curve numbers for urban areas, in TR-55 (18). The value selected was for 'open space,' intended to represent lawns, parks, golf courses, cemeteries, etc. Good conditions were assumed for the turf plots, i.e., grass cover >75%, and the hydrologic soil group for Cecil sandy loam is B, according to Exhibit A-1 in TR-55. Therefore a curve number of 61 was used initially in the calibration of the hydrology component, to represent the apparent conditions of the turf plots.

Measurement of sediment erosion was not part of the study design and therefore no field measurements were available for these parameters.

The runoff curve number was thus the focus of subsequent calibration iterations for the hydrology component of the GLEAMS model. Uncertainties in other sensitive hydrology parameters were reduced with the input of site-specific data in the soil profile description.

Pesticides. Eight pesticides were applied in treatments made on June 13, July 18, and December 5. Dr. Smith provided application rates and formulations for these pesticides. Oryzalin was not analyzed in the runoff collected from the plots and site-specific environmental fate data were not available for any of the pesticides. Initial values for the environmental fate parameters were taken directly from the pesticide data Table P-2 in Knisel, et al. (19). These values are based primarily on Wauchope, et al. (20) as well as other sources.

The dissipation half life data in this table are based almost exclusively on results from studies that focus on agricultural or bare ground scenarios. This, of course, is usually not relevant to turf applications. However, it probably would be SOP for people or companies who have minimal experience with computer models, turf, or both. Therefore we felt this would be a good first step. We then modified the input data based on a combination of sources including the open literature, EPA documents and data bases, data sheets and personal communications with the manufacturers of the products, and our experience in this field. Appropriate values for these parameters were selected from studies pertinent to the conditions of the site, i.e., soil type, climate, crop, etc., to the extent that the data were available.

Calibration of the pesticide component was more complicated than the hydrology calibration. There were ultimately seven pesticides, all with different environmental fate properties. Further there were uncertainties with all of the environmental fate parameters and the distribution of the pesticides in the turf/thatch/soil profile upon application, since no actual field data were available. The properties that were given the most attention in this calibration were K_{oc}, the foliar and soil half lives, and the distribution of the applied chemicals between the turf and the thatch/soil. Water solubility is a sensitive parameter, but there is more certainty in the published values for this parameter than the others. Some consideration was also given to the uncertainties in the potential for pesticide uptake into the plant and the washoff coefficient that describes the potential for rainfall and irrigation to flush the pesticides off of the turf canopy. The initial values selected from the GLEAMS User Manual for these parameters for each of the pesticides are given below in Table III. Pesticide input parameters selected for advanced calibration runs of the GLEAMS model are given in Table IV. Plant uptake of the pesticides was assumed to be very high (>15%) for the water soluble chemicals 2,4-D, dicamba, and mecoprop, moderate for the dinitroaniline herbicides benefin and pendimethalin (21), and minimal for chlorpyrifos and chlorothalonil. Initial distribution of the pesticide applications was assumed to be primarily in the turf canopy (90%) with a small fraction filtering through to the thatch/soil interface (5%) and the remainder lost to volitalization and drift.

Penn State Runoff

Hydrology. An initial set of input parameters was devleoped based on the information provided by Dr. Linde on the questionnaires and information gleaned from Harrison (34), Harrison, et al. (35), Linde, et al. (8), Watschke, et al. (unpublished report), Linde, et al.(9),Linde, et al.(10), the PRZM user manuals, USDA, NRCS web sites, and personal communications with the researchers. Daily rainfall and average temperature measurements for the Penn State region spanning the study period were provided by the researchers. The amount of simulated rainfall by irrigation for each of the study events was input into the daily weather file. ETS staff visited the study plots and collected soil samples for analysis of bulk density, texture, organic matter content, field capacity and wilting point.

The initial runoff curve number used to begin the hydrology calibration study was selected from Table 2-2a, runoff curve numbers for urban areas, in TR-55 (18). The value selected was for 'open space', intended to represent lawns, parks, golf courses, cemeteries, etc. Good conditions were assumed for the turf plots, i.e., grass cover >75%, and the hydrologic soil group for the Hagerstown silt loam is C, according for Exhibit A-1 in TR-55. Therefore a curve number of 74 was used initially in the calibration of the hydrology component, to represent the apparent conditions of the turf plots

The runoff curve number was thus the focus of subsequent calibration iterations for the hydrology component of the GLEAMS model. Uncertainties in other sensitive hydrology parameters were reduced with the input of site-specific data in the soil profile description.

Thirty-seven runoff events were recorded between May 1992 and October 1995. Five of these occurred due to natural precipitation. Irrigation was applied at a constant rate of 152 mm/hr from 1992 to 1993 and at 139.5 mm/hr in 1994 and

Table III. Environmental Fate Parameters for Initial Calibration Runs Selected from the GLEAMS User Manual (Based Primarily on Wauchope, et al. (20)).

		2,4-D dma salt	dicamba dma salt	mecoprop dma salt	benefin	pendimethalin	chlorpyrifos	chlorothalonil
applic. rate[1]	(kg/ha)	2.25	0.56	1.68	1.68	168	1.12	9.52
water sol'y	(mg/L)	796,000	400,000	660,000	0.1	0.28	0.4	0.6
foliar t½	(days)	9.0	9.0	10	10	30	3.3	5.0
soil t½	(days)	10	14		30	90	30	30
Koc	(ml/g)	20	2	20	9,000	5,000	6,070	1,380
washoff fraction		0.45	0.65	0.95	0.20	0.40	0.65	0.50

[1]This is the rate targeted by the U. of Ga. researchers

Table IV. Pesticide Input Parameters Selected for Advanced Calibration Runs of the GLEAMS Model

		2,4-D dma salt	dicamba dma salt	mecoprop dma salt	benefin	pendimethalin	chlorpyrifos	chlorothalonil
applic. rate[1]	(kg/ha)	2.25	0.56	1.68	1.68	168	1.12	9.52
water sol'y	(mg/L)	796,000[2]	720,000[2]	660,000[2]	0.1[3]	0.5[4]	2.0[5]	1.2[6]
foliar t½	(days)	4[7]	7[8]	3	2	2[9]	6[10]	3[11]
soil t½	(days)	6[12]	9[2]	9[2]	21[3]	34[4]	10[13]	13[6]
Koc	(ml/g)	109[14]	32[15]	32[15]	11,000[3]	15,100[16]	13,400[10]	2,680[6] recalc'd
washoff fraction		0.65	0.85	0.95[2]	0.20[2]	0.45[2]	0.10[9]	0.50[2]

[1]Targeted application rate.
[2]Taken from Knisel, et al. (19) and/or Wauchope, et al. (20)
[3]Taken from EPA, 1989a (22)
[4]Taken from EPA, 1991 (23)
[5]Taken from EPA, 1992 (24)
[6]Taken from EPA, 1989b (25)
[7]Taken from Altom and Stritzke (26) & Smith, et al.(27)
[8]Taken from EPA, 1989c (28)
[9]Taken from Hurto and Prinster (29)
[10]Taken from Racke (30)
[11]Taken from Pitts, et al. (31)
[12]Taken from Smith (32)
[13]Taken from Horst et al. (5)
[14]Taken from Rao and Davidson (33)
[15]Apparent GLEAMS limit actual=2.0
[16]Taken from Spieszalski, et al. (12)

1995. The plots were not irrigated for consistent time periods however, throughout the study. The plots were irrigated until runoff began and then the irrigation was turned off. Further, the plots were pretreated prior to all but two of the scheduled runoff events in 1994-1995 with an unspecified amount of irrigation water in an attempt to standardize the soil moisture levels for each event. We observed a distinct relationship between the magnitude of the runoff curve number and the success with which we were able to fit certain of the events. That is, low curve numbers (60-70) produced good results for the events where irrigation was left on for longer periods of time (25-75 min) but poor results for the events where irrigation was of short duration (5-20 min). Conversely, we observed good results with high curve numbers (80+) relative to short duration irrigation events and poor results relative to the long duration events. This is discussed in more detail later in this paper.

Pesticides. Four pesticides were applied between May 1992 and September 1993 and runoff sampling and pesticide analysis occurred on 16 dates from June 1992 to October 1993.

Rather than start with environmental fate values from agriculture-based sources as we did with the U. of GA simulations, we selected more appropriate values for these parameters pertinent to the conditions of the site, i.e., soil type, climate, crop, etc., to the extent that the data were available. Again, we relied on a combination of sources including the open literature, EPA documents and data bases, data sheets and personal communications with the manufacturers of the products, and our experience in this field.

The key parameters varied in an attempt to calibrate the predictions against observed results were K_{oc}, and the foliar and soil half lives. PRZM3 supports many options for the distribution of the chemical in the soil profile, however, options for defining the distribution in the plant canopy is limited. The penetration of the chemical application through the canopy to the soil surface was simulated according to the PRZM model's linear distribution model. That is, the amount that gets through to the soil surface is a linear function of the density of the crop canopy, which in this case was assumed to be a maximum of 80% coverage at maturity.

Sensitivity Analysis

Leaching Studies. During the process of attempting to calibrate the model to the conditions of the studies, it became apparent that certain parameters exhibited greater influence on the water and pesticide leaching predictions. These more sensitive parameters required special care in selection and interpretation.

The input parameter that demonstrated the most influence on the outcome of the leachate volume predictions was the average mean temperature. (Evapotranspiration increases as temperature increases, thereby reducing percolate volume.) Maximum and minimum temperatures were provided by Dr. Smith for the U. of GA greenhouse plots. Therefore a daily average was used based on the max and min values provided. A calibration run showed that an increase in temperature of 7°C reduced the leachate volume by approximately 44% and reduced the pesticide concentration by approximately 40%. Daily average temperatures were provided for U. of NE by Dr. Horst and used as input.

The temperature parameter was varied to show sensitivity to the model but could not be altered because values were actually measured. [We assumed that the average temperature equals the mean of the maximum and minimum greenhouse temperatures. If this assumption is wrong, it could have a significant influence on percolate water predictions.] Based on experience with other projects, field capacity is a sensitive parameter and can significantly change results when varied. However, field capacity is a measurable parameter and was therefore used as a site-specific input parameter to reduce uncertainty in the results.

Past experience indicates that the degradation rate constant(s) (k) is the most critical pesticide chemistry parameter for leaching assessments. This is because it occurs as an exponential function in the model algorithms. It also has a tendency to vary significantly depending on site conditions. Similarly, the soil-water distribution coefficient (K_d) or its soil organic carbon analog (K_{oc}) can significantly impact pesticide leaching assessments as well.

Runoff Studies. The one input parameter that demonstrated the most influence on the outcome of the runoff volume predictions was the runoff curve number. This parameter is also probably the most uncertain with respect to modeling turf scenarios. An increase in the runoff curve number of about 21% during one of the initial set of calibration runs in the Georgia study resulted in increases in the runoff volumes predicted from the irrigation/rainfall events following the pesticide applications of greater than 100%. The predictions of runoff were also particularly sensitive to the parameters that describe the hydraulic characteristics of the soil profile. These include the saturated hydraulic conductivity, field capacity and wilting point, and porosity.

Virtually all of the parameters that describe pesticide fate are sensitive parameters. The most significant parameter, aside from the application rate, that affected the magnitude of the pesticide results was K_{oc}. The pesticides with K_{oc}s in the thousands range showed runoff concentrations generally in the tens of parts per billion or less. The phenoxy-herbicides, with K_{oc}s less than 100, demonstrated higher runoff losses, generally in the hundreds of parts per billion. Water solubility also appeared to significantly impact the results, but these values are to a large extent correlated to K_{oc}. The distribution of the pesticides between the turf canopy and the thatch/soil surface was also important in determining the resulting runoff concentrations.

The soil and foliar pesticide degradation half lives were very important parameters for simulating the overall declines observed in the field plots of the pesticide concentrations in runoff on the days following the applications. With only a few exceptions, the observed concentrations in runoff, on the days subsequent to each of the dates when pesticides were applied, became successively smaller. For the less mobile pesticides, those with low water solubilities and high K_{oc}s, we interpreted the generally nonlinear decline in these concentrations to indicate that some process, other than equilibrium partitioning of the pesticides into runoff water, was taking place in the turf canopy, thatch, and on the soil surface that contributed to reducing the availability of the pesticides to runoff. The half lives entered for the pesticides, particularly the foliar half lives and the half lives at the top of the soil profile (in this case the thatch layer), showed a significant impact in the shape of the curves of the pesticide concentrations in runoff plotted over time from the dates of applications.

The organic carbon content of the soil surface, in this case the thatch layer, was both uncertain and significant. The uncertainties arise because actual measurements from the field study were not available and the extent to which the organic matter comprising thatch is available as sorption sites is not well understood. It is significant because it impacts the partitioning of the pesticides between the thatch and the runoff water.

Results and Discussion

Leachate Studies

University of Georgia. The PRZM2 prediction of leachate volumes ranged from poor to fair (CV=59%±34%). In most cases, but not all, PRZM2 tended to under-predict rather than over-predict leachate volumes. This may have been due to the fact that we did not have complete soil characteristics for the actual plots modeled, or it may be due to an evapotranspiration method inappropriate for turfgrass, or some other reason.

The pesticide predictions matched the observed concentrations more often than expected based on the inconsistent prediction of water flux. The best match was found with the herbicide 2,4-D, followed by dicamba (assuming a K_{oc} of 35) and mecoprop (assuming a K_{oc} of 20 and a soil t½ of 9 days). PRZM2 consistently and significantly over-predicted mecoprop leachate concentrations.

The model was first run with given data and compared with the actual field collected data. Each of three pesticides were applied to separate plots. However, only one leachate volume is obtained by the model output for all three sets of pesticide plots. There are observed leachate volumes for 2,4-D plots, dicamba plots, and mecoprop plots. All three different leachate volumes were compared with the same predicted leachate volume from PRZM2. These comparisons are shown on Figure 1.

Figure 1 shows the leachate volume for all study plots and is within the range of the collected volume for six of the eight collection periods. To obtain a better fit to the leachate volume collected in the greenhouse the field capacity was reduced 20%. Although field capacity was analyzed from a turf plot, it was measured for only one unused turf plot to represent all greenhouse and mini-green plots. PRZM2 output showed no change in the leachate volume for all but two collection periods. The August 17 collection showed only a 20% increase in the leachate volumes and 8/24 showed less than a 1% increase. No change occurred for the 7/20 and 8/3 collection periods where no leachate volumes were produced in PRZM. Therefore original field capacity values were used in the remaining calibration of the pesticide component, since it had no significant affect on the PRZM leachate volume.

Calibration of the pesticide component resulted in changing the most sensitive parameters for those pesticides where the parameter was the most uncertain (i.e. K_{oc} and t½ for dicamba and mecoprop, respectively). The initial environmental fate

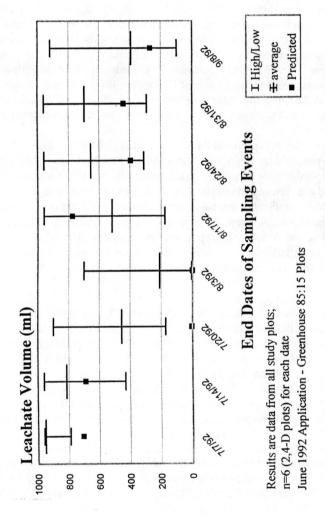

Figure 1. PRZM2.3 Simulation of U. of GA Leaching Plots -- Hydrology Comparison

values used for 2,4-D produced good concentration results from PRZM2 when compared with concentrations of analytical results from the collected leachate. Therefore no further calibration was warranted for 2,4-D (Figure 2). The initial K_{oc} value of 8 for dicamba produced higher results in all but the two collection periods where no leachate volume in PRZM2 was produced. Therefore the K_{oc} was increased. It was changed to 35 based initially on a water solubility of 6500 ppm and use in Kenaga's regression equation (36). Increasing the K_{oc} to 35 lowered the predicted concentration of dicamba to virtually non-detect levels. This produced a good-fit with the U. of GA results. However, comparison of the observed versus the predicted concentrations using a K_{oc} of 35 shows that PRZM (2.3) still over predicts the leachate mass (Figure 3). (The K_{oc} value of 35 exceeds the USDA/ARS Pesticide properties database recommended value of 13 (range = 13-21). However, ours is a dynamic turf system, more relevant than the static, equilibrium soil system run in standardized lab tests. Basically, more sorption in the system occurs than might be predicted).

We varied the mecoprop soil t½ from the initial 12 days to 9 days. The first four collection periods show a very good match with the analytical data, while the last four collection periods show the predicted concentration to steadily increase until the last time period and then appears to stabilize. Decreasing t½ to nine days decreases the predicted concentration but still has the same curve affect in the latter collection periods as the initial run. A review of literature in our files shows that nine days is a more appropriate soil t½ for mecoprop in southern turfgrass. Therefore, nine days was used in the simulation. Comparison of observed and predicted shows that PRZM2 consistently over predicted the leachate mass except for two dates (Figure 4).

University of Nebraska. Although PRZM3.12 generally over predicted the daily percolate volumes, the predictions matched very well with the observed volumes averaged on a weekly basis (Figure 5). The CV was 18%±19% without any calibration. Most of the key data parameters were provided as measured values by the researchers. We obtained measured daily evapotranspiration data for the field study from the researchers. However, there is no option for entering observed evapotranspiration data in PRZM3.12. Instead pan evaporation data are entered or calculated by the model using temperature data. Therefore we defaulted to the model's simulation of ET based on the measured daily temperatures.

The pesticide predictions in our PRZM3.12 runs significantly over-predicted the amount of pesticide mass leached when compared to the observed field study (see one example in Figure 6), with one notable exception. The predicted leachate mass of the low-mobility pesticide chlorpyrifos (see Figure 7) matched very closely with that observed in the U. of NE field study.

Runoff Studies

University of Georgia. The GLEAMS model performed surprisingly well considering the special considerations of turf compared with agriculture, for which (agriculture) the model was developed. The model predicted runoff water fairly well (Figure 8).

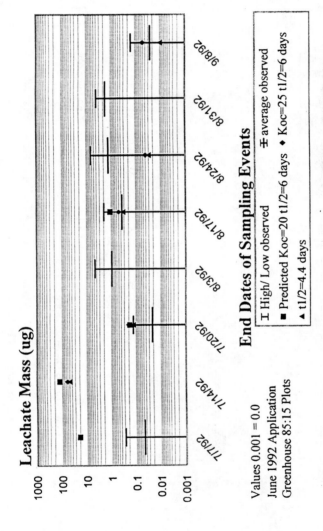

Figure 2. PRZM2.3 Simulation of U. of GA Leaching Plots -- Pesticide
Leachate Results - 2,4-D

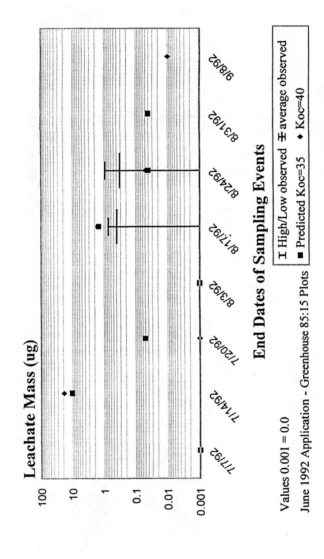

Figure 3. PRZM2.3 Simulation of U. of GA Leaching Plots -- Pesticide Leachate Results - Dicamba

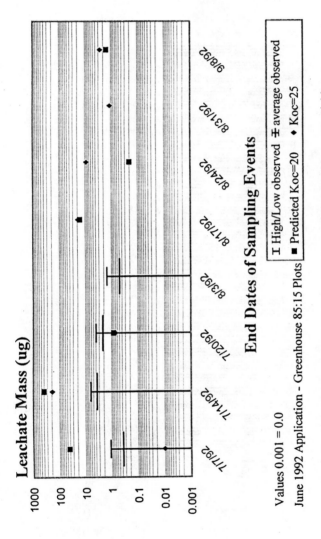

Figure 4. PRZM2.3 Simulation of U. of GA Leaching Plots -- Pesticide Leachate Results - Mecoprop

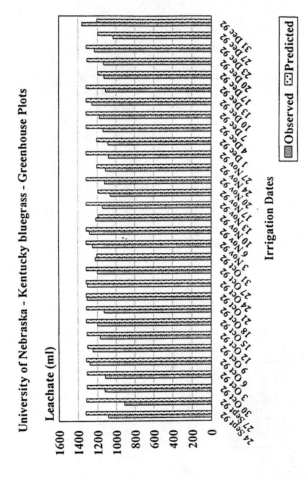

Figure 5. PRZM3.12 Leaching Simulation Modeling Project

Observed leachate is the sum of irrigation collected on and after the day of irrigation application

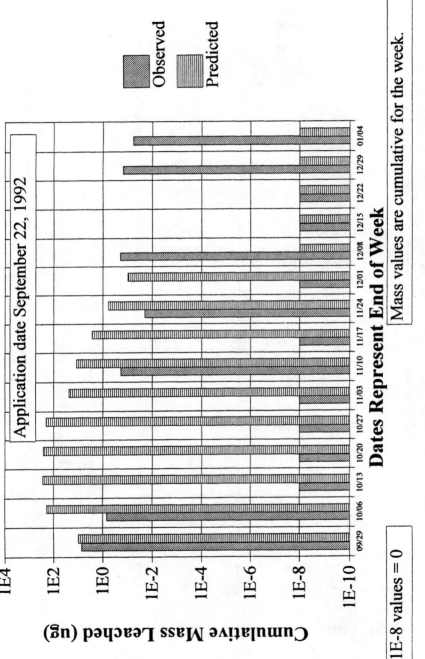

Figure 6. Univ. of Nebraska Greenhouse -- Mecoprop - Observed vs. Predicted

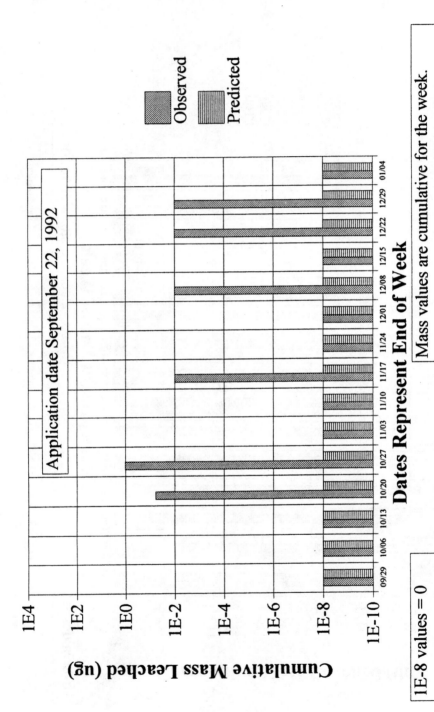

Figure 7. Univ. of Nebraska Greenhouse -- Chlorpyrifos - Observed vs. Predicted

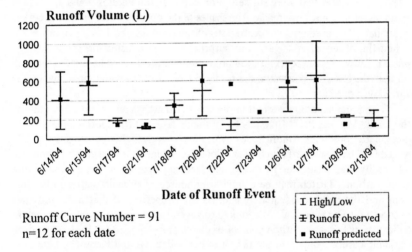

Runoff Volume (L)

Date of Runoff Event

Runoff Curve Number = 91
n=12 for each date

⊥ High/Low
Ŧ Runoff observed
■ Runoff predicted

Figure 8. GLEAMS Simulation of U. of GA. Runoff Plots -- Calibration of
Surface Runoff

The average difference between observed runoff and predicted runoff, using all 12 data points, was 7% as a percent of rainfall and 43% in terms of total runoff volume. If only one runoff event is excluded (7/22/94) the numbers improve to 5.7% and 22%, respectively. The calibrated runoff curve number selected (91) was higher than initially anticipated for turf. In hindsight, the high curve number seems reasonable. The soil comprising the runoff plots was heavily compacted when they were constructed and measured hydraulic conductivity rates confirm the slow downward movement of water more representative of HSG C or D than the recommended HSG B. Further, the intense rainfall events occurred over a very short duration relative to the 24 hour period implicit in the runoff curve number method and the daily time step of the model.

The pesticide predictions also compared well with observed concentrations of the seven pesticides that were studied. For example, see Figures 9 and 10. There were no discernible negative trends. The shapes of the observed and predicted curves were similar. The predicted concentrations of the troublesome July 22 runoff event matched the observed concentrations surprisingly well considering the hydrology mismatch. However, the concentration predictions of the Trimec herbicides were all lower than the observed concentrations for the 6 Dec 94 event when, presumably, the turfgrass was dormant.

The organic carbon content in the soil and thatch is an important parameter in the GLEAMS and PRZM models for pesticide fate and transport simulation. In particular, the predicted runoff losses of pesticides with high K_{oc}s are significantly impacted by the organic carbon content. Soils that maintain low organic carbon will more readily lose pesticides to runoff and leaching than soils that exhibit higher organic carbon. Thatch may be a significant source of organic matter in the turf system and the need to adequately address the properties of thatch in computer simulation modeling is critical. We looked to several studies to give us some insight into the potential organic carbon content of thatch (11, 13-17, 37). In these studies, the organic carbon content in the thatch of bermudagrass and Kentucky bluegrass ranged from 16 to 43% and the soil in the turf root zone immediately below the thatch was generally elevated compared to soil without turf. The extent to which the high organic carbon is actually available for pesticide sorption is not certain. The upper limit of the range recommended in the GLEAMS data entry preprocessor for organic carbon content is 10%.

A runoff curve number of 91 appeared to produce the best fit of predicted runoff to observed runoff in the Georgia runoff simulation. This value was much higher than expected, particularly for turf and a presumed hydrologic group B soil. A curve number of 91 is more representative of a poor stand of grass on a hydrologic group D soil, e.g., a clay soil.

This result would appear to significantly contradict the long-held assumption that turfgrass retards runoff flow better than most other land covers. However, it should be noted that the runoff curve number approach was originally derived from watershed-scale data for use in predicting watershed-scale impacts (38). Our results support the contention of Harrison, et al. (35) and Rallison and Miller (38) that the use of established runoff curve numbers may not be appropriate to evaluate microplot results such as these. The CN of 91 was also used by Wauchope et al. (39) in a lawn

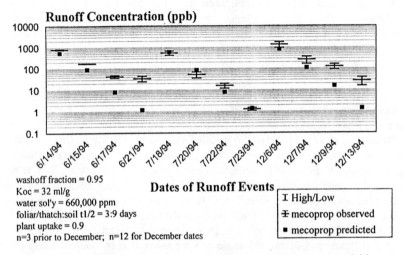

washoff fraction = 0.95
Koc = 32 ml/g
water sol'y = 660,000 ppm
foliar/thatch:soil t1/2 = 3:9 days
plant uptake = 0.9
n=3 prior to December; n=12 for December dates

Figure 9. GLEAMS Simulation of U. of GA Runoff Plots -- Pesticide Runoff Results - Mecoprop: Soil Surface % O.C. = 10.0

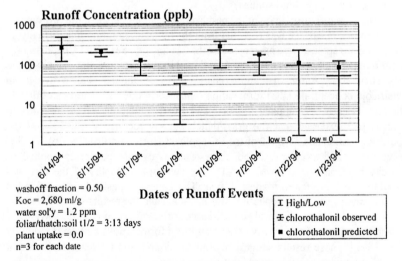

washoff fraction = 0.50
Koc = 2,680 ml/g
water sol'y = 1.2 ppm
foliar/thatch:soil t1/2 = 3:13 days
plant uptake = 0.0
n=3 for each date

Figure 10. GLEAMS Simulation of U. of GA Runoff Plots -- Pesticide Runoff Results - Chlorothalonil: Soil Surface % O.C. = 10.0

turf study of sulfometuron-methyl in a sandy soil. They also instrumented intense rainfall.

Further, the GLEAMS (and PRZM) models run on a 24 hour time step. Runoff calculations assume rainfall throughout the day and do not allow for consideration of short duration, intense rainfall events as those produced artificially during the U. of GA and PSU studies. A 2" rainfall event occurring over a 1.75 hour period would produce more runoff than one occurring over a 24 hour period on sandy loam soil. We were reasonably able to compensate for this with the higher runoff curve number.

The foliar dissipation of pesticides from turf has been the subject of numerous studies including Altom and Stritzke *(26)*, Cowell, et al. *(40)*, Hurto and Prinster *(29)*, Kuhr and Tashiro *(41)*, Pitts, et al. *(31)*, Sears and Chapman *(42)*, Smith, et al. *(27)*, and Stahnke, et al. *(43)*. Most of the studies ETS reviewed indicated that pesticide dissipation from turf is more rapid than in bare soil or agricultural settings. Dissipation from the foliage in particular can be very rapid. For example, Pitts, et al. *(31)* treated Kentucky bluegrass with four fungicides including chlorothalonil with soil half lives reported in Wauchope, et al. *(20)* to be between 26 and 70 days and foliar half lives between 5 and 30 days. They found that the foliar half lives ranged from 1.1 to 4.4 days. Selecting shorter half lives in the foliage and thatch for the GLEAMS and PRZM input, consistent with these and the other studies cited in Table IV, appeared to produce much better fits of the predicted pesticide concentrations in runoff to the observed concentrations, when compared to the initial modeling runs using typical parameters suggested in Wauchope, et al. *(20)* based primarily on agricultural and bare ground studies.

The potential distribution of the pesticide applications between the turf canopy and the thatch/soil surface was revisited from the initial assumption that virtually all of the pesticides applied were intercepted by the turf foliage. Sears and Chapman *(42)* examined the fate of four insecticides, including chlorpyrifos, on Kentucky bluegrass turf. Their study found that approximately 97% of applied chlorpyrifos in an emulsifiable concentrate (EC) formulation remained in the turf canopy and thatch. Kuhr and Tashiro *(41)* applied granular and EC formulations of chlorpyrifos to bluegrass and observed that the grass/thatch initially retained about 30% of the granular application and 57-59% of the EC applied. Hurto and Prinster *(29)* studied the dissipation of total and dislodgeable residues of five pesticides applied to Kentucky bluegrass (*Poa pratensis* L). The density and quality of the turf was equated to a well-maintained residential lawn. The amount of pesticide retained in the upper turf canopy following application was measured and the results showed that the percent of applied petrochemical solvent-based formulations such as ECs averaged 15% and the percent of applied dry or aqueous based formulations averaged a slightly higher 28%. These results suggest that applications made to dense turf are not necessarily completely intercepted by the turf canopy.

The K_{oc} is a sensitive parameter in the simulation pesticide fate and transport with GLEAMS. The differences between the predicted results of the phenoxy herbicides with low K_{oc}s (demonstrating low potentials to sorb to organic carbon in the soil) and the pesticides with similar application rates but much higher K_{oc}s (benefin, pendimethalin, and chlorpyrifos) were generally in the range of a factor of

100 for the Georgia study. It became apparent that there are some limitations on the ranges of certain parameters in GLEAMS. The lowest K_{oc} that could be used for the highly water soluble dimethylamine salt species of the phenoxy-herbicides dicamba and mecoprop was 32. As the K_{oc}s were reduced below this value the concentrations were predicted to actually increase rather than decrease. This may indicate a problem with the model and should be investigated in future simulation work.

Pennsylvania State University. We were able to obtain a reasonable match of predicted runoff compared to observed runoff. However, we did so by adjusting the runoff curve number throughout the duration of the study period. This was necessary because we observed a strong dependency of the runoff curve number to the duration of the irrigation. The irrigation was applied at a consistent rate throughout the study, but the duration was variable. The U. of GA study runoff prediction matched well with one curve number, but the irrigation regime remained consistent throughout the study.

Initial calibration steps using a single CN for the duration of the study produced poor runoff water results compared to the observed runoff (Figure 11). The predicted runoff best fit the observed runoff when the curve number was varied throughout the study period according to this relationship (Figure 12). Although this worked well for the calibration exercise, this would not be the typical application of the model for predictive simulation. This does, however, suggest the possible use of probabilistic simulation techniques. The monte carlo simulation component of PRZM to vary the input range of the runoff curve number may not be an entirely appropriate means to overcome limitations in the model relative to variabilities in the environment, such as storm duration, but it could at least generate hypotheses for testing.

Our attempts to calibrate simulated pesticide runoff losses compared to observed losses were complicated by irregularities in the pesticide losses observed in the field study. For example, mecoprop was not detected in the event on June 20, but it was detected at 4.84 mg on June 24 and again at 40.2 mg on June 27 (see Figure 13). The runoff volume was actually greater on June 24 than it was on June 27 (Figure 12). This pattern may be impossible to mimic in PRZM, or any other model.

In general, K_{oc} and water solubility significantly govern the magnitude of predicted pesticide loss. The foliar and soil half-lives impact the shape of the curve of pesticide runoff mass plotted against time.

The pesticide calibration remains incomplete, but given seemingly anomalous data as it was presented to us, continuing the calibration exercise with PRZM, or any other model, would only yield marginal improvements in the results.

In hindsight, the Penn State study was a very good one for an analysis of the hydrology. The fact that much of the focus of the Penn State study related to the detailed hydrologic aspects of turf contributed significantly to this result. On the other hand, the pesticide results were not thoroughly analyzed when our study was initiated. Although wealthier than most in terms of data for a turf runoff study, the study was not an ideal candidate for our pesticide calibration efforts.

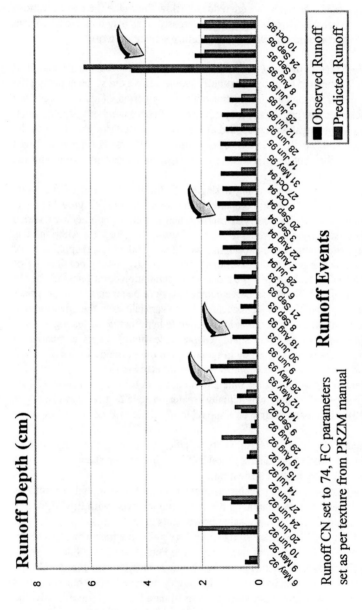

Figure 11. PRZM3.0 Simulation of PSU Runoff Plots -- Ryegrass Trials -
Surface Runoff Calibration - Early Steps

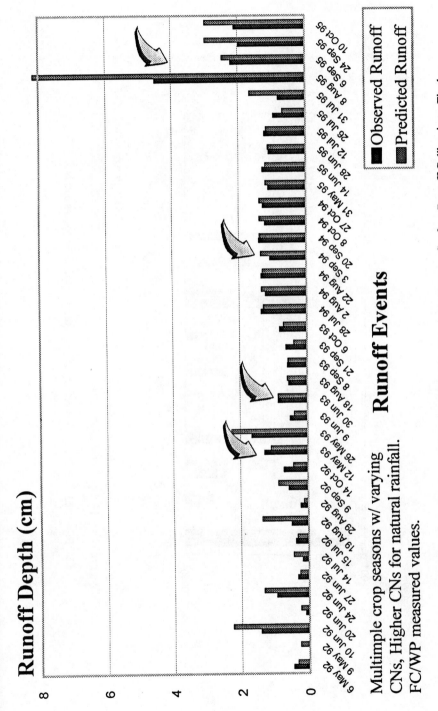

Figure 12. PRZM3.0 Simulation of PSU Runoff Plots -- Ryegrass Trials - Surface Runoff Calibration - Final

224

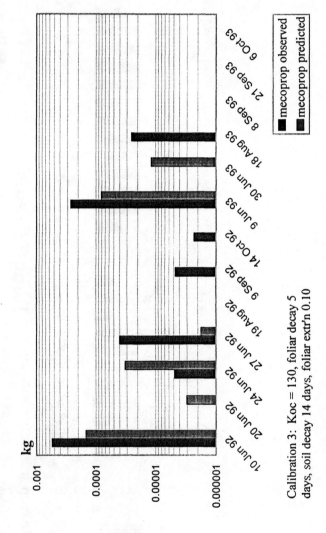

Calibration 3: Koc = 130, foliar decay 5
days, soil decay 14 days, foliar extr'n 0.10

Figure 13. PRZM3.0 Simulation of PSU Runoff Plots -- Ryegrass Trials -
Pesticide Calibration - Mecoprop

Conclusions and Recommendations

Conclusions. The GLEAMS model did a good job predicting runoff volumes and pesticide runoff at the Georgia site. With one exception, only a small amount of reasonable 'tweaking' of the input parameters was done to achieve a good fit of the data. The exception was the runoff curve number (CN) which had to be increased to 91. This can be explained by the fact that the CN is used by GLEAMS in 24 hr time steps, and the field studies used rainfall simulators to create short, intense 'storm' events.

Similarly, we were able to model the Penn State runoff depths successfully by adjusting the CN in accordance with the 'storm' duration. However, the pesticide modeling is problematic. The field results - as presented to us - may be impossible to predict with any model.

We were more successful modeling leachate volumes at U. of NE using PRZM3.12 than when we applied PRZM2.3 to the U. of GA data.

We generally over predicted pesticide leachate for the four water-soluble pesticides: isazofos, 2,4-D, dicamba, and mecoprop. Leachate predictions of the low-soluble chlorpyrifos matched observations very well.

PRZM and GLEAMS appear to be useful for turf runoff predictions. PRZM appears to be overly conservative for leachate predictions of more soluble pesticides.

Recommendations. 1. GLEAMS and PRZM3.x should each be applied to all data sets to improve the usefulness of this study. 2. The evapotranspiration algorithm in PRZM - Hammon's formula - should be carefully evaluated to determine its relevance to turf, particularly for dormant periods caused by the change in seasons or high temperatures. This could be done by integrating a critical literature review with the results of this study and the modeling evaluation of another independent turf leaching data set, e.g., Snyder and Cisar *(44)*. 3. Users should carefully consider all relevant hydraulic and chemical aspects of thatch before applying any solute transport model to the turf system. An assumption of 10% 'effective' organic carbon in thatch (the portion available for pesticide sorption) appears reasonable based on literature available in 1998. However, more research is needed on the effectiveness of thatch organic matter on pesticide sorption relative to soil organic matter. (Ten percent may be too conservative.) 4. Researchers studying turf test plots should obtain organic matter, field capacity, wilting point, and bulk density data for each depth studied if there is a reasonable chance the results will be used to calibrate or evaluate models.

Acknowledgments

We gratefully acknowledge the USGA's Greens Section Research Committee, who funded most of this work. We also thank the Penn State researchers D. Linde, J. Borger, and T. Watschke for sharing their data.

226

Literature Cited

1. Cohen, S.Z.; Durborow, T.E.; Barnes, N.L. In *Fate and Significance of Pesticides in Urban Environments*; Racke, K.D.; Leslie, A.R., Eds.; ACS Symposium Series 522; American Chemical Society: Washington, DC, 1993; pp. 214-227.

2. Smith, A.E.; Tillotson, W.R. In *Pesticides in Urban Environments, Fate and Significance*; Racke, K.D.; Leslie, A.R., Eds.; ACS Symposium Series 522; American Chemical Society: Washington, DC, 1993, pp. 168-181.

3. Smith, A.E. *USGA Green Section Record*, 1995, *33*(1), 13-14.

4. Horst, G.L.; Christians, N. Spring 1994. Pesticide and Fertilizer Fate in Turfgrasses Managed Under Golf Course Conditions in the Midwestern Region - University of Nebraska - Lincoln. *USGA Progress Report*.

5. Horst, G.L.; Shea, P.J.; Christians, N.; Miller, D.R.; Stuefer-Powell, C.; Starrett, S.K. *Crop Sci.* 1996, *36*, 362-370.

6. *Standard Practice for Evaluating Environmental Fate Models of Chemicals*; American Society of Testing and Materials: Philadelphia, PA, 1984.

7. Horst, G.L.; Powers, W.L.; Miller, D.R.; Shea, P.J.; Wicklund, E.A. *Crop Sci.* 1994, *34*, 292-295.

8. Linde, D.T.; Watschke, T.L.; Borger, J.A. In *Science and Golf II: Proceedings of the World Scientific Congress of Golf*; Cochran, A.J.; Farrally, M.R., Eds.; E & FN Spon: London, UK, 1994, pp 489-496.

9. Linde, D.T.; Watschke, T.L.; Jarrett, A.R.; Borger, J.A. *Agron. J.* 1995, *87*, 176-182.

10. Linde, D.T. 1996. Runoff, Erosion, and Nutrient Transport From Creeping Bentgrass and Perennial Ryegrass Turfs. *Doctorate thesis submission at Pennsylvania State University Department of Agronomy*.

11. Niemczyk, H.D.; Krause, A.A. *J. Environ. Sci. Health* 1994, *B29*(3), 507-539.

12. Spieszalski, W.W.; Niemczyk, H.D.; Shetlar, S. *J. Environ. Sci. Health* 1994, *B29*(6), 1137-1152.

13. Carrow, R.N.; B.J. Johnson. 1992. Final Report: 1989-1991, Influence on Soil and Plant Characteristics of Ringer Fertilizers. *Unpublished report from Univ. of Georgia, Agronomy Dept*.

14. Carrow, R.N.; Johnson, B.J.; Burns, R.E. *Agron. J.* 1987, *79*, 524-530.

15. Hurto, K.A.; Turgeon, A.J.; Spomer, L.A. *Agron. J.* 1980, *72*, 165-167.

16. Lickfeldt, D.W.; Branham, B.E. *J. Env. Qual.* 1995, *24*, 980-985.

17. Dell, C.J.; Throssell, C.S.; Bischoff, M.; Turco, R.F., *J. Env. Qual.* 1994, *23*, 92-96.

18. USDA Soil Conservation Service, *Technical Release 55: Urban Hydrology for Small Watersheds*, 2nd Edition, USDA, Engineering Division: Washington, DC, 1986.

19. Knisel, K.A.; Davis, F.M.; Leonard, R.A. *GLEAMS Version 2.0 Part III: User Manual*; USDA Agricultural Research Service: Tifton, GA, 1992.

20. Wauchope, R.D.; Buttler, T.M.; Hornsby, A.G.; Augustine-Beckers, P.W.M.; Burt, J.P. In *Reviews of Environmental Contamination and Toxicology*; Ware, G.W., Ed.; Number 123; Springer-Verlag: New York, NY, 1992, pp 26-37.

21. Weber, J.B. *Weed Tech.* 1990, *4*, 394-406.

22. *Pesticide Environmental Fate One Line Summary - Benfluralin*; Environmental Protection Agency, Environmental Fate & Effects Division: Washington, DC, 1989a.

23. *Pesticide Environmental Fate One Line Summary - Pendimethalin*; Environmental Protection Agency, Environmental Fate & Effects Division: Washington, DC, 1991.

24. *Pesticide Environmental Fate One Line Summary - Chlorpyrifos*; Environmental Protection Agency, Environmental Fate & Effects Division: Washington, DC, 1992.

25. *Pesticide Environmental Fate One Line Summary - Chlorothalonil*; Environmental Protection Agency, Environmental Fate & Effects Division: Washington, DC, 1989b.

26. Altom, J.D.; Stritzke, J.F. *Weed Sci.* 1973, *21*(6), 556-560.

27. Smith, A.E.; Aubin, A.J.; Biederbeck, V.O. *J. Environ. Qual.* 1989, *18*, 299-302.

28. *Pesticide Environmental Fate One Line Summary - Dicamba*; Environmental Protection Agency, Environmental Fate & Effects Division: Washington, DC, 1989c.

29. Hurto, K.A; Prinster, M.G. In *Fate and Significance of Pesticides in Urban Environments*; Racke, K.D., Leslie, A.R., Eds.; ACS Symposium Series 522; American Chemical Society: Washington, DC, 1993; pp. 86-99.

30. Racke, K.D. In *Rev. Environ. Contam. Toxicol.*; Ware, G.W., Ed.; Springer-Verlag: New York, NY, 1993.

31. Pitts, C.; Turco, R.; Throssell, C (Purdue U.). In *USGA Environmental Research Program, Pesticide and Nutrient Fate*; 1995 Annual Project Reports; U.S. Golf Association: Far Hills, NJ, 1995, pp 114-123.

32. Smith, A.E. *Weed Res.* **1980**, *20*, 355-359.

33. Rao, P.S.C.; Davidson, J.M. In *Environmental Impact of Nonpoint Source Pollution*; Overcash, M.R.; Davidson, J.M., Eds.; Ann Arbor Science Publishers, Inc.: Ann Arbor, MI, 1980, pp 23-65.

34. Harrison, S.A. *Effects of Turfgrass Establishment Method and Management on the Quantity and Nutrient and Pesticide Content of Runoff and Leachate;* Thesis in Agronomy - Master of Science; Pennsylvania State University College of Agriculture: State College, PA, 1989.

35. Harrison, S.A.; Watschke, T.L.; Mumma, R.O.; Jarrett, A.R.; Hamilton, Jr., G.W. In *Pesticides in Urban Environments*; Racke, K.D.; Leslie, A.R., Eds; ACS Symposium Series 522; American Chemical Society: Washington, DC, 1993, Chapter 17.

36. Kenaga, E.E. *Ecotoxicol. Environ. Saf.* **1980**, *4*, 26-38.

37. Ledeboer, F.B; Skogley, C.R. *Agron. J.* **1967**, *59*, 320-323.

38. Rallison, R.E.; Miller, N. In *Proc. International Symposium on Rainfall - Runoff Modeling*; Singh, V.P., Ed.; Water Resources Publications: Littleton, CO, 1982; pp 353-364.

39. Wauchope, R.D.W.; Williams, R.G.; Marti, L. *J. Environ. Qual.* **1990**, *19*, 119-125.

40. Cowell, J.E.; Adams, S.A.; Kunstman, J.L.; Mueth, M.G. In *Pesticides in Urban Environments: Fate and Significance*; Racke, K.D.; Leslie, A.R., Eds.; ACS Symposium Series 522; American Chemical Society: Washington, DC, 1993, pp 100-112.

41. Kuhr, R.J.; Tashiro, H. *Bull. Environ. Contam. Toxicol.*, **1978**, *20*, 652-656.

42. Sears, M.K.; Chapman, R.A. *J. Econ. Entomol.* **1979**, *72*, 272-274.

43. Stahnke, G.K.; Shea, P.J.; Tupy, D.R.; Stougaard, R.N.; Shearman, R.C. *Weed Science*, **1991**, *39*, 97-103.

44. Snyder, G.H.; Cisar, J.L. (U. Fla.) In *USGA Environmental Research Program, Pesticide and Nutrient Fate*; 1995 Annual Project Reports; U.S. Golf Association: Far Hills, NJ, 1995, pp 54-83.

Chapter 13

Dicamba Transport in Turfgrass Thatch and Foliage

M. J. Carroll[1], R. L. Hill[1], S. Raturi[1], A. E. Herner[2], and E. Pfeil[2]

[1]Department of Natural Resources Sciences and Landscape Architecture,
University of Maryland, College Park, MD 20742
[2]Environmental Chemistry Laboratory, U. S. Department of Agriculture/NRI,
BARC-East, Beltsville, MD 20705

The retention of dicamba to turfgrass foliage and thatch, and the transport of dicamba in undisturbed columns of soil and soil+thatch were examined in three separate studies. The use of linear equilibrium (LEM) and two site non-equilibrium (2SNE) models to predict the transport of dicamba in the columns of soil and soil+thatch were also investigated. Rainfall that occurred during the first 8 hours following foliar application removed about 70% of the dicamba present on the foliage. Thereafter, dicamba became increasing resistant to washoff. The high concentration of organic matter present in columns containing a surface layer of thatch reduced dicamba leaching. Although dicamba sorption to thatch was higher than to soil, when normalized for the presence of organic matter (K_{oc}), thatch organic matter was a less effective sorbant of dicamba than was soil organic matter. The 2SNE model gave reasonable estimates of dicamba transport when the model retardation factor (R) was calculated using laboratory derived adsorption coefficients. The LEM model satisfactorily described dicamba transport only when R was curve-fitted. Use of column R's based on laboratory derived adsorption coefficients resulted in poor LEM estimates of dicamba transport.

Pesticides applied to mature turf move into the soil only after being washed off foliage and moving through turfgrass thatch. Any attempt to predict the movement of pesticides applied to turf requires that the retention characteristics of the pesticide to foliage and thatch be known. Pesticide movement from foliage to underlying porous media layers is usually modeled using foliar washoff algorithms (1). The use of foliar washoff algorithms requires that accurate estimates of the fraction of applied pesticide that is deposited on the foliage, and the fraction of pesticide removed from the foliage as a function of rainfall be known. In the case of the latter, the amount of time elapsed between pesticide application and the first rainfall event can significantly affect the fraction of pesticide transferred from foliage to the thatch layer (2). Thatch has a pore space distribution similar to that of a coarse sand and a chemical composition resembling a young organic soil (3). For these reasons it is more analogous to a porous media which supports plant growth than a surface mulch.

Pesticide transport in thatch and soil can be modeled using the convection-dispersion equation. Alternate forms of this equation can be used to describe instantaneous or kinetically-driven sorption, or alternately, two domain flow phenomena within the media. Regulatory agencies often use the linear equilibrium model (LEM) that assumes all pesticide sorption sites are identical and that sorption equilibrium occurs instantaneously between the pesticide in the bulk soil solution and the sorbed pesticide. Some users of the LEM account for the presence of thatch by simply increasing the overall organic matter content of the soil profile to account for the volume-averaged contribution of the thatch layer (4). Since the impact of the thatch layer on pesticide sorption and subsequent transport is not specifically considered when such as approach is used, it is felt the assumption of instantaneous sorption used in LEM's may lead to inaccurate predictions of pesticide transport in turfgrass that contains thatch.

Sorption of pesticides to a highly organic media such as thatch is presently believed to be a two stage process (5). The first stage consists of direct pesticide sorption to the external organic matter surface sites (class 1 sites), whereas the second stage consists of pesticide sorption to sites located within the pores or fissures of organic matter aggregates (class 2 sites). The latter process involves diffusive mass transfer of the pesticide which is highly dependent on the residence time of the solution containing the pesticide. A two-site non-equilibrium model (2SNE) which considers that transfer to some sorption sites is diffusion mediated may be more appropriate than a LEM model when predicting pesticide transport through a soil with a thatch surface layer (6).

Dicamba (3,6-dichloro-2-methoxy benzoic acid) is one of the most commonly used herbicides for postemergence control of broadleaf weeds in turf. It has a pKa of 1.95 (7) and is soluble in water (4500 mg L^{-1}). Due to its dissociated anionic character, it is minimally sorbed to most soils (8). Several studies have shown dicamba is readily transported below the rootzone of cool season turfgrass species when high amounts of rainfall or irrigation occur within two days of application (9,10). The high organic matter content of thatch allows this media to readily sorb non-polar compounds (11,12,13). Less is known, however about the sorptive properties of thatch for ionizable compounds like dicamba.

The objectives of our research were to comprehensively address through integrated studies: 1) the effect of residence time on the washoff of dicamba from turfgrass foliage; 2) dicamba sorption to turfgrass thatch and the soil; 3) the effect of thatch on the transport of dicamba through columns containing a surface layer of thatch and columns devoid of thatch and ; 4) the use of LEM and 2SNE models to predict the transport of dicamba through columns containing a surface layer of thatch and columns devoid of thatch.

Materials and Methods

Foliar washoff. A 10-month old stand of 'Southshore' creeping bentgrass (*Agrostis palustris* Huds.) was utilized for the foliar washoff study. Prior to the study, the turf was irrigated and fertilized as needed to sustain vigorous growth. The turf was mowed three times a week at 1.6 cm to promote the formation of a dense foliar canopy similar to that found on a golf course fairway.

The study area was divided into four strips and each strip subdivided into five 2.1- by 1.8-m plots. Individual strips were separated by 1.8-m alleys and the borders of adjacent plots within each strip were placed 2.1-m apart. Dicamba (Banvel 4S) was applied at 0.53 kg ai ha^{-1} in 393 L of water on 7 Aug, 1995. The strips were sprayed in a sequential fashion with one strip being sprayed every hour. All plots within a strip were sprayed within 5 minutes of one another. Spray was applied with a CO_2-pressurized sprayer equipped with six, 6501 Teejet nozzles. Nozzles were located 0.38 m above the turf surface and were spaced 0.27 m apart. All strips were sprayed in calm wind conditions between 800 and 1100 hr with a delivery pressure of 245 kPa. Stationary calibration of the six nozzles prior to spraying indicated individual nozzle delivery volume coefficient of variation was 3.4% for the sprayer operating conditions stated. The entire study area was mowed at 1.6 cm with a triplex mower 1 hr before spraying the first strip.

Individual plots were exposed to a 4.9 cm hr^{-1} rainstorm for a 40 minute duration at 1, 8, 24 and 72 hours after spraying the plots with dicamba. A rainfall simulator that produced drop size distributions and impact velocities similar to natural rainfall was used to apply the precipitation (*14*). Dicamba washoff was determined by collecting foliage from adjacent strips of turf within plots that were mowed immediately before and shortly after the simulated rainfall. In the case of the latter, the turf was allowed to dry for 1 hr before being mowed. The 'before' and 'after' foliage samples were collected using a 51-cm wide front throw mower set at 0.9 cm and samples placed into sealed zip-lock bags. The fresh weight of the foliage was determined immediately after harvest and the samples quickly placed into a freezer maintained at -20 °C. One plot within each of the four strips sprayed with dicamba was not subjected to simulated rainfall. Separate 'before' and 'after' foliage samples were collected from these plots within 20 minutes of spraying. The 'before' and 'after' samples from these plots were used to evaluate the uniformity of the dicamba spray application and to determine the initial levels of dicamba on the foliage.

Foliage samples were analyzed for dicamba using procedures and equipment specifications discussed elsewhere (*15*). Dicamba extracted from the foliage was analyzed by gradient HPLC-UV analysis. Extraction efficiency was determined by adding known amounts of analytical grade dicamba to unsprayed foliage samples. Dicamba freezer storage stability was evaluated by adding 1.0 mL of the field dicamba tank mix to 20 gm of unsprayed foliage. Mean foliar extraction recovery and freezer storage recovery levels for dicamba were 118% ± 4% (n =13) and 76 ± 14% (n = 4), respectively.

Thatch and soil sample collection. Soil and turfgrass thatch samples were collected from a three-year old stand of Meyer zoysiagrass (*Zoysia japonica* Steud.) at the University of Maryland Turfgrass Research and Education Facility. The zoysiagrass was established and maintained without the use of dicamba and contained a 1.5 to 2.0-cm thick thatch layer (*16*). The upper half of the thatch layer consisted of non-decomposed and partially decomposed plant biomass. The rest of the layer was comprised of well composed plant biomass intermixed with the underlying soil. The soil beneath the thatch layer was classified as a Sassafras loamy sand (fine loamy, mixed, mesic, Typic Hapludult). The soil contained 82% sand, 10 % silt, 8% clay and had a cation exchange capacity of 4.2 cmol kg^{-1}. Thatch and soil pH and organic matter content are presented in Table 1.

The thatch and soil used in the isotherm study were removed from the field using a 10-cm diameter golf green cup cutter. Ten centimeter deep cores containing a surface layer of thatch and mat were brought into the laboratory and the thatch and mat separated from the underlying soil using hand shears. The thatch+mat and soil were sieved by hand to pass through a 4-mm screen to homogenize each material. The thatch+mat and soil were then placed into separate plastic bags and the sealed samples stored at 4 °C until needed. In the leaching study, undisturbed soil, and soil plus thatch columns, 12-cm lgth. by 10-cm diam., were extracted from the surface 12 cm of the zoysiagrass site using a specially designed drop hammer-sleeve assembly. The columns containing soil only were obtained after removing all above ground thatch and foliage. The columns were transported to the laboratory immediately after collection.

Sorption isotherm. Sorption isotherms were determined using a mechanical vacuum extractor. This device controls the rate a which a solution moves through a column of thatch or soil. The columns are created by packing known amounts of media into syringe tube barrels. Since the sample is not shaken during the procedure little disruption of the media aggregates and organic matter occurs.Moreover, the flowing conditions used in this modified batch/flow technique better represent the physiochemical interactions that occur in the field. A detailed description of this technique is presented elsewhere (*17*). In this study, field-moist samples having oven dry weights equivalent to 6.7-g thatch, or 13-g soil, were added to syringe tube barrels after placing a 2.0-cm length by 2.5-cm diameter foam plug (Baxter Healthcare Corp., McGaw Park, IL, DiSPo PLUGS# T1385) into the bottom of the barrel. The thatch was gently tamped to a bulk density of 0.67 Mg m^{-3}, and the soil to a bulk density of 1.30 Mg m^{-3}, before placing a second foam plug over the top of the sample to ensure even distribution of the solution as it entered each sample. The presence of the foam plugs also prevented channeling between the syringe tube wall and the sample. The syringe tube barrels were 2.5-cm in diameter and held 60 mL of solution.

A combination of commercially formulated dicamba and ring-labeled ^{14}C dicamba were used in this study. The ring-labeled ^{14}C dicamba (98% purity) was obtained from Sigma Chemical Co., St. Louis, MO., and had a specific activity of 16.65×10^7 Bq mmol^{-1}. The commercial product of dicamba (Banvel) was formulated with 480 g L^{-1} dicamba. Separate stock solutions of commercially formulated dicamba and ^{14}C dicamba were created by dissolving each material in distilled-deionized water. Dicamba solution concentrations of 1, 4, 8, 10, and 20 mg L^{-1}, each of which contained 0.3 mg L^{-1}, or 2.31 $\times 10^5$ Bq L^{-1}, of ^{14}C dicamba, were made by adding appropriate amounts of each stock solution to known amounts of distilled-deionized water.

Sorption isotherms were determined by leaching 30 ml of 1, 4, 8, 10, and 20 mg dicamba L^{-1} through samples of thatch and soil for 24 hrs. One milliliter of the leachate from each sample was added to a 20-mL scintillation vial that contained 5 mL of Scintiverse II (Fisher Scientific Co., Fair Lawn, NJ) scintillation cocktail. The radioactivity of the resulting sample was determined by liquid scintillation counting using a Beckman LS 5800 series (Beckman Instruments, Inc., Fullerton, CA) liquid scintillation counter. Dicamba sorption to any material other than thatch or soil was taken into account by including syringe tube blanks. The blanks were identical to the syringe tubes containing thatch and soil except that they contained no thatch or soil.

The amount of dicamba sorbed at 24 hrs, x/m, (mg kg^{-1}) was determined from the formula $(C_i-C_e)V/M$, where C_i is the initial concentration of the solution added to a sample (mg L^{-1}); C_e, the dicamba concentration of the leachate (mg L^{-1}); V, the volume of solution added to thatch or soil; and M, the thatch or soil dry weight (kg). Dicamba sorption was fitted to the linear form of the Freundlich equation:

$$\log (x/m) = \log (K_f) + 1/n \log (c),$$ (1)

where x/m is the surface concentration of the sorbate per unit amount of sorbent (mg kg^{-1}), c is the sorbate equilibrium solution concentration (mg L^{-1}) and K_f and 1/n are constants that characterize the dicamba sorption capacity. Regression analyses were used to calculate 1/n and K_f.

Leaching. Columns collected in the field were brought to the laboratory and saturated from the bottom. The bottom end of each saturated column was trimmed and placed into a funnel containing a 12-µm pore diameter saturated, porous, stainless steel plate which was made vacuum tight. The column and funnel were then inserted into one port of a multi-port vacuum chamber. A null balance vacuum regulator was used to maintain a pressure of -10 Kpa within the vacuum chamber at the base of each column.

A 0.001 M CaCl$_2$ solution was applied to each column using a specially designed drop emitter (18) that uniformly distributed solution to the surface of each column. Leachate was collected in 400 mL sterile polyethylene cups located beneath the funnel of each column within the vacuum chamber. After steady-state flow conditions were achieved (0.85 cm hr^{-1}), 10 mL of 300 mg bromide L^{-1} was surface-applied to each column. Leachate was then collected every half an hour for the next 12 hours and the bromide concentration of the leachate determined by ion chromatography. After this initial leaching period, 10 mL of 40 mg dicamba L^{-1} was uniformly added to the surface of each column. The dicamba solution consisted of 2.31 x 10^5 Bq L^{-1}, of ^{14}C ring-labeled dicamba and 39.7 mg L^{-1} of commercially formulated dicamba. After adding dicamba, leaching solution inputs and the vacuum applied to the base of each column were discontinued for 24 hr to permit sorption of dicamba to the thatch and soil. During this time all columns were covered with plastic wrap to prevent volatile dicamba losses. After the 24 hr sorption period, the leaching process resumed. Sampling continued every half hour for 6 hr and then, once every one hour for the next 12 hours with dicamba in the leachate being determined using the liquid scintillation techniques described previously.

After collecting the last leachate sample, thatch layer thickness was measured before being separated from the soil, and the moisture contents of the thatch and soil layers gravimetrically determined. Sub-samples were stored in the freezer until the amount of dicamba remaining in each sample could be determined. Thawed thatch and soil sub-samples were first shaken for 2 hr in a 50:50 water and methanol solution. The resulting slurry was then subjected to vacuum filtration and the filtrate analyzed for ^{14}C. The amount of ^{14}C remaining in the sample was determined through combustion using a biological material oxidizer with the amount of ^{14}C evolved measured by liquid scintillation. The amount of ^{14}C removed by the water and methanol solution was considered to represent the easily extractable fraction of dicamba and dicamba metabolites present in the sample at the end of the leaching event. The amount removed by combustion was considered to be the tightly bound fraction of dicamba and dicamba metabolites remaining in the sample.

Estimating Transport Parameters from Breakthrough Curves (BTC's)

Theoretical background. The one dimensional convection dispersion equation (CDE) for steady-state transport of a solute through homogeneous soil is (19):

$$R*\delta C/\delta t = D*(\delta^2 C/\delta t^2) - v* (\delta C/\delta x) \tag{2}$$

Where C is solution-phase solute concentration ($\mu g \ cm^{-3}$), t is time (h), D is the hydrodynamic dispersion coefficient ($cm^2 \ h^{-1}$), R is the retardation factor (dimensionless), x is distance from solute origin (cm), and v is the average pore water velocity ($cm \ h^{-1}$). The R term reduces to one for non-reactive solutes and is greater than 1 when solute retention occurs. The retardation factor is defined as (20):

$$R = 1+[\rho K_f (1/n)c^{(1/n-1)} /\theta] \tag{3}$$

where ρ is the soil bulk density ($g \ cm^{-3}$) θ is the volumetric water content ($cm^3 \ cm^{-3}$), and K_f, c and $1/n$ are as previously defined.

The simplest approach is to assume that all pesticide sorption sites are identical and that equilibrium occurs instantaneously between the pesticide in the bulk soil solution and the pesticide adsorbed. This mathematical approach is called linear equilibrium sorption. Where bimodal porosity leads to two-region flow, or situations where the sorption process is controlled by two-site kinetic non-equilibrium sorption processes, non-equilibrium models may more accurately describe the transport of pesticides through soil. Chemical non-equilibrium models consider adsorption on some of the sorption sites to be instantaneous, while sorption on the remaining sites is governed by first-order kinetics (21). The two-site chemical non-equilibrium model (2SNE) conceptually divides the porous medium into two sorption sites: type-1 sites assume equilibrium sorption and type-2 sites assume sorption processes as a first-order kinetic reaction (6). In contrast, physical non-equilibrium is often modeled by using a two-region dual-porosity type formulation. The two-region transport model assumes the liquid phase can be partitioned into mobile (flowing) and immobile (stagnant) regions. Solute exchange between the two liquid regions is modeled as a first-order process. The concepts are different for both chemical and physical non-equilibrium CDE, however, they can be put into the same dimensionless form (2SNE) for conditions of linear adsorption and steady-state water flow (22):

$$\beta * R * \delta C_1/\delta t = 1/P * (\delta^2 C_1/\delta x_2) - \delta C_1/\delta x - \omega(C_1 - C_2) \tag{4}$$

$$(1-\beta) * R * \delta C_2/\delta t = \omega(C_1 - C_2) - \mu_2 C_2 + \gamma_2(x) \tag{5}$$

Where the subscripts 1 and 2 refer to equilibrium and non-equilibrium sites, respectively, β is a partitioning coefficient, ω is a dimensionless mass transfer coefficient, P is the Peclet number; and μ (h^{-1}) and γ (ug h^{-1}) define first-order decay and zero-order production terms, respectively, each represented in component contributions of both the liquid and solid phases.

Customarily β and ω are obtained by fitting solute BTC's to the non-equilibrium model using a non-linear least squares minimization technique (23). The values of β and ω obtained from the BTC's of non-interacting solutes can be used to evaluate the potential contributions from two-region flow. In the absence of two-region flow, β and ω may be used to evaluate the contributions from two-site kinetic non-equilibrium sorption (24). For interacting solutes, β represents the fraction of instantaneous solute retardation in two-site non-equilibrium model and ω the ratio of hydrodynamic residence time to characteristic time for sorption. They are equivalent to:

$$\beta = (\theta + f\rho K)/(\theta + \rho K) \tag{6}$$

$$\omega = k_2(1-\beta)RL/v \tag{7}$$

Where f is the fraction of equilibrium-type sorption sites, K is K_f when $1/n$ is unity, L is the length of transport (cm) and k_2 is the desorption rate constant (h^{-1}).

Model analysis. Results from the leaching experiment were used to evaluate the performance of the LEM and 2SNE models to predict dicamba transport. The models were compared for steady-state flux-averaged boundary conditions. Solute breakthrough curves were plotted for both bromide and dicamba using the relative concentrations (C/Co) versus the pore volumes of leachate. Convective transport parameters were estimated by a least squares minimization procedure (23) using the bromide breakthrough data. Actual mean pore water velocities were used and the bromide retardation factor, R, was assumed to be equal to 1. One and two domain flow forms of the convective dispersive equation were curve-fitted to the bromide leachate data. Values of the dispersion coefficient were used in subsequent dicamba simulations.

Dicamba model simulations used calculated mean pore water velocities, bromide-fitted dispersion coefficients, and retardation factors calculated from the column measured values of θ and ρ, and the K_f values determined in the isotherm study. Dicamba retardation factors for individual columns were calculated using thatch and soil K_f's in a volume-averaged approach where the relative volume of the thatch and soil layers were used as weighing factors in calculating a mean retardation coefficient for each column. The solution concentration of the dicamba added to the columns (40 mg dicamba L^{-1}) was also used to calculate the thatch retardation factors. The dimensionless partitioning coefficient (β), and the dimensionless rate coefficient (ω), were fitted for the 2SNE model.

Results and Discussion

Dicamba washoff. Figure 1 summarizes the washoff of dicamba from bentgrass foliage by providing the initial dicamba concentration present on the foliage as a function of residence time and the portion of pesticide remaining on the foliage after subjecting the turf to simulated rainfall. One of the 72-hr plots was inadvertently missed when spraying the strip with dicamba. Thus the means for the 72-hr residence time are based on three replicates. The percent of applied dicamba that was recovered from the foliage within 1 hour of application (T= 0) was 72% and is based on the measured delivery rate of the sprayer and the amount of Banvel placed in the sprayer tank. The retention fraction for the plots not receiving simulated rainfall (T = 0) was not significantly different than unity indicating that foliar interception of the applied dicamba was uniform over the area sampled. Foliar levels of dicamba declined 57% over the first 24 hours following the application of dicamba. Little decline in dicamba was observed over the next 48 hours. Rainfall that occurred during the first 8 hours following application removed about 70% of the dicamba present on the foliage. After this time dicamba became more resistant to washoff, however 44% of the dicamba present on the foliage could be removed with 3.3 cm of rain 72 hours after application. In a previous study, we found that most washoff of commercially formulated dicamba occurs during the initial 1.5 cm of rainfall (15).

Isotherms. Thatch had a much higher dicamba sorption capacity than did the underlying soil. Dicamba sorption to thatch (K_f = 0.82) was greater than to soil (K_f = 0.28) even though the pH of the thatch was higher than that of the soil (Table I). This indicates thatch organic matter had a greater effect on dicamba sorption than did pH for the thatch and soil examined here. Dell et al., (12) have suggested that lignin is the primary sorbent in turfgrass thatch. Although the structure of lignin is not fully known, lignin is composed of polymerized aromatic alcohols and phenolic acids. The alcoholic and acidic functional groups of these compounds likely served as the sites of dicamba sorption in the thatch (25). Entrapment, or physical absorption within cell wall structures of degrading cellulose-lignin fibrils, may also serve as a secondary means by which dicamba sorption to thatch takes place (26).

Using the isotherm K_f values and the thatch and soil organic matter contents, the calculated normalized sorption coefficients (K_{oc}) for the thatch and soil were 0.011 and 0.043 (m^3kg^{-1}), respectively. These values are greater than the dicamba K_{oc} (0.002) cited frequently in the literature (27). It is likely that our K_{oc} values were higher due to the low soil pH (5.3) and because we utilized commercially formulated dicamba for our isotherm solutions. Commercially formulated herbicides usually contain one or more ingredients that facilitate sorption to lipophilic substances such as plant leaves.

The slope (1/n) of the thatch isotherm was greater and significantly different from one whereas the slope of the soil isotherm was not significantly different from one (Table I). Isotherms having slopes greater than one are often found when the sorbant is a lipophilic surface and the sorbent is a hydrophilic solute (28). Thatch with its high organic matter content is generally considered to be a lipophilic media (12,13). Thus, the slope for our thatch isotherm appears to be consistent with the properties of dicamba which is highly polar.

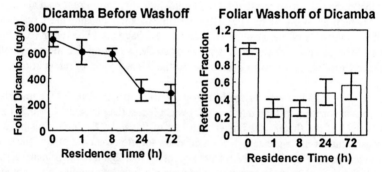

Figure 1. Dicamba on foliage prior to simulated rainfall and the fraction of dicamba retained on the foliage following a single 33 cm simulated rainfall event for four residence times.

Table I. Dicamba Freundlich isotherm sorption coefficients, pH and organic matter content for zoysiagrass thatch and the soil immediately below the thatch.

Media	pH	Organic Matter %	logKf	Kf $(mg^{1-1/n} kg^{-1} L^{1/n})$	l/n
Thatch	6.7	14.2	-0.088(±0.19)‡	0.82a§	1.26(±0.25)a
Soil	5.3	1.5	-0.546(±0.46)	0.28b	1.04(±0.56)a

‡Number in the parenthesis are the 95% confidence limits.
§Means within a column followed by the same letter are not significantly different
at the 0.05 level of probability using a two sample t-test.

Leaching. Mass recoveries of the ^{14}C dicamba are presented in Table II with total recoveries ranging near 100% when the standard errors of the estimates are considered. Soil columns experienced significantly greater leaching losses than those columns having zoysiagrass thatch layers. This reduced leaching in columns containing a surface layer of thatch compared to the columns devoid of thatch and can be attributed to the greater dicamba adsorptive capacity of the zoysiagrass thatch.

Table II. Mass balance of ^{14}C recovered from the column transport study.

Column ID	Easily Extractable	Tightly Bound	Leached	Total Recovery
	------------------------------- % -------------------------------			
Thatch*	13.11 (±6.15)§	30.64 (±7.56)	65.27 (±1.40)	109.33 (±13.45)
Soil only	2.87 (±1.77)	16.58 (±5.33)	82.77 (±0.41)	102.22 (±6.83)

* 3.4 cm surface layer zoysiagrass thatch + 9.75 cm soil
§ Values in parentheses indicate standard errors of the estimates.

Model evaluation and comparison. The physical properties and experimental averages for the soil and thatch+soil columns are presented in Table III. Although the mean pore water velocity for each column type was similar, the Darcian flux and soil water content were slightly greater for the thatch+soil columns. The presence of thatch lowered the mean bulk density of the thatch+soil columns, which in turn, increased the soil water content of these columns.

Mean pore water velocities (v) and hydrodynamic dispersion coefficients (D) based on the bromide leaching portion of the study are reported in Table IV. For the sake of brevity, the bromide leaching data has not been presented in this report although the flow phenomena observed has direct implications on subsequent dicamba leaching and warrants discussion. Relatively high pore water velocities may have enhanced the degree of preferential flow or the fraction of immobile water in some columns. The presence of observed macropores and root channels also suggest that preferential flow through macropores contributed to the two domain flow in some columns. The relationships between the D, v, and L are often described with Peclet numbers (P), where $P = (vL)/D$.

Peclet numbers for most columns ranged from 18 to 50 indicating that convective flow was dominating diffusion and that diffusion was minimal. Some columns exhibited asymmetrical tailing of the bromide tracer, indicating that there were significant effects of immobile water on transport. Peak bromide flow and asymmetrical tailing behavior were better described with a 2SNE model in two columns containing thatch+soil (ZT3, ZT4) and in two columns containing soil only (ZS1, ZS3). The LEM and 2SNE models performed equally well in describing bromide transport within the three remaining (ZT1, ZT2, ZS2) columns examined in this study. Thus there seemed to be minimal justification in using a 2SNE model to describe convective flow phenomena within these (ZT1, ZT2, ZS3) columns.

Table III. Physical properties and experimental conditions within columns used to examine dicamba leaching.

Column ID	Mean Pore Water Velocity	Darcy Flux	Soil Water Content	Bulk Density
	cm h^{-1}	cm h^{-1}	cm^{3}cm^{-3}	g cm^{-3}
Zoysiagrass Thatch + Soil (ZT)	3.77	1.08	0.29	1.27
Soil only (ZS)	3.77	0.97	0.26	1.61

Graphical comparisons of model estimations and measured values of dicamba transport are presented in Figure 2 for columns not exhibiting two-domain flow. The model transport parameters for all columns are presented in Table IV. In all columns the predicted dicamba BTC's based on the LEM model did not adequately describe the observed BTC's. If model evaluation is based on the coefficients of determination, the 2SNE model described dicamba transport fairly well with significantly improved fits compared to the LEM model suggesting two-site sorption. Since the leaching experiment was completed in a relatively short time period and the resulting dicamba recoveries were near 100% (Table II), no benefit was observed when considering production / decay components in the modeling efforts.

Poor fits were obtained from the LEM model when it was used to predict dicamba transport in the columns not exhibiting two-domain flow indicating either that two site sorption had occurred or that our dicamba sorption studies did not mimic sorption values that might be observed in thatch and soil columns subject to high flow regimes. To further examine the ability of the LEM model to predict dicamba transport we ran an addition simulation with this model. In this LEM simulation, we allowed the model to select the R that resulted in the best fit rather than fixing R to be the value we calculated using θ, ρ, and the laboratory determined K_f's. While significant improvements in the predictive capabilities occurred by allowing the LEM model to select R, the 2SNE utilizing R values based on laboratory-derived K_f's continued to have slightly higher coefficients of determination in the columns not exhibiting two-domain flow (ZT1, ZT2, ZS2). The inappropriateness of using the LEM model to obtain media K_f's in columns exhibiting two domain flow is readily apparent in columns ZT3 and ZS1. The R's obtained from the LEM simulations predicted negative dicamba adsorption in these two columns.

Non-equilibrium parameters (β and ω) for dicamba transport were optimized by fitting dicamba BTC's to the 2SNE model using independent estimates of R and v (Table IV). Because bromide exhibited no significant two domain flow in columns ZT1, ZT2 and ZS2; β and ω values obtained for the dicamba BTC's can be interpreted primarily as sorption related non-equilibrium parameters (*29, 30*). In this case, calculated values of f (fraction of sorbent for which sorption is instantaneous) and k2 (the desorption rate constant) using β and ω terms, respectively, may be interpreted as relating to sorption non-equilibrium without confounding effects from transport related non-equilibrium (*31*). Values of f were 0.08, 0.06 and 0.12 for columns ZT1, ZT2 and ZS2, respectively, indicating that a significant fraction of sorption sites do not participate in instantaneous retardation in these columns during dicamba transport. Values of k2 were 0.12, 0.10 and 0.10 for columns ZT1, ZT2, and ZS2, respectively, indicating there was little difference in dicamba desorption in the thatch+soil and soil only columns.

The values of f and k2 for the other columns (ZT3, ZT4, ZS1, and ZS3) cannot be interpreted solely in terms of two-site non-equilibrium sorption . Bromide BTCs under these column conditions demonstrated significant two domain flow; consequently, β and ω, hence f and k2, reflect contributions from both sorption related non-equilibrium and transport related non-equilibrium.

Conclusions

In the absence of rainfall, foliar levels of dicamba decline rapidly during the first 24 hours following application. Washoff of dicamba from turfgrass foliage is greatest during the first 8 hours following application then declines with increasing residence time. The amount of dicamba present on the foliage and extent to which residence time influences the washoff of dicamba from turfgrass foliage appear to justify the need to include a foliar washoff algorithm in models that are used to predict the fate of pesticides applied to turf.

Zoysiagrass thatch had a much higher sorptive capacity than the underlying soil. Dicamba sorption to thatch was higher than to soil even though the pH of the thatch was higher than the soil. This suggests that organic matter may influence dicamba adsorption to a greater degree than pH in soils that are moderately acidic. When normalized for the presence of organic matter (K_{oc}), thatch organic matter was a less effective sorbent of dicamba than was soil organic matter.

The presence of a surface layer of thatch significantly reduced dicamba leaching. When dicamba breakthrough curves were fitted to the LEM and 2SNE models, the latter did a better job predicting dicamba transport in the presence of two-domain flow. In absence of two domain flow, the LEM model could adequately describe dicamba transport only when the model was allowed to select a value of R that differed from what reality would tell us to be acceptable. Use of an R based on laboratory K_f's resulted in poor LEM estimates of dicamba transport. In the absence of two domain flow, the 2SNE model gave reasonable estimates of dicamba transport indicating that two site sorption may be occurring during the leaching process. These estimates were strengthened by the fact they were based on laboratory derived sorption parameters.

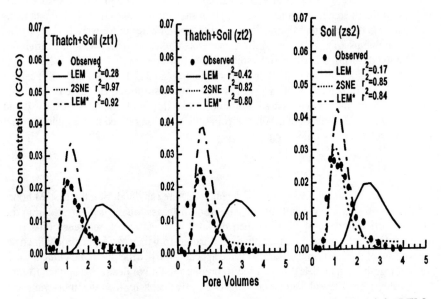

Figure 2. Dicamba breakthrough curves fitted with the LEM, 2SNE and the LEM model allowed to select R (LEM*), for colums possessing single domain flow.

Table IV. Estimated model transport parameters for dicamba breakthrough curves.

Column	Model	v cm h⁻¹	R	D cm² h⁻¹	pulse	β	ω	r²
ZT1	LEM	4.44	3.078	3.17	0.034			0.28
	2SNE	4.44	3.078	3.17	0.034	0.409	0.620	0.97
	LEM*	4.44	1.353	3.17	0.034			0.92
ZT2	LEM	4.05	3.172	2.737	0.0358			0.42
	2SNE	4.05	3.172	2.737	0.0358	0.383	0.629	0.82
	LEM*	4.05	1.296	2.737	0.0358			0.80
ZT3 ‡	LEM	3.548	2.565	0.9191	0.0286			0.19
	2SNE	3.548	2.565	0.9191	0.0286	0.39	0.657	0.89
	LEM*	3.548	0.959	0.9191	0.0286			0.95
ZT4 ‡	LEM	3.021	2.987	1.808	0.032			0.30
	2SNE	3.021	2.987	1.808	0.032	0.34	0.752	0.95
	LEM*	3.021	1.143	1.808	0.032			0.80
ZS1 ‡	LEM	3.77	2.757	2.364	0.0363			0.39
	2SNE	3.77	2.757	2.364	0.0363	0.37	0.259	0.95
	LEM*	3.77	0.996	2.364	0.0363			0.97
ZS2	LEM	4.03	2.77	2.28	0.0378			0.17
	2SNE	4.03	2.77	2.28	0.0378	0.436	0.493	0.85
	LEM*	4.03	1.30	2.28	0.0378			0.84
ZS3 ‡	LEM	3.498	2.713	1.8	0.0361			0.41
	2SNE	3.498	2.713	1.8	0.0361	0.37	0.458	0.88
	LEM*	3.498	1.034	1.8	0.0361			0.89

‡ Two domain flow present in column.
* LEM simulation where model was allowed to select a value for R.

Acknowledgments. The authors wish to express their thanks to the Maryland Turfgrass Council, the Maryland Agricultural Experiment Station and the United States Golf Association Greens Section Research for providing financial assistance to conduct this research

Literature Cited

1. Smith, C.N.; Carsel, R.F. *J. Environ. Sci. Hlth.* **1986**, B19, 323-342.
2. Willis, G.H.; McDowell, L.L.; Smith, S.; Southwick, L.M. *J Environ. Qual.* **1994**, 23, 96-100.
3. Hurto, K.A.; Turgeon, A.J.; Spomer, L.A. *Agron. J.* **1980**, 72,165-167.
4. Primi, P; Surgan, M.H.; Urban, T. *Ground Water Monit. Rev.* **1994** 14, 129-138.

242

5. Brusseau, M.L.; Rao, P.S.C. *Chemosphere*. **1989**, 18, 1691-1706.
6. van Genuchten, M.T.; Wagenet, R.J. *Soil Sci. Soc. Am J.* **1989**, 53:1303-1310.
7. Weber, J.B. In *Research Methods in Weed Science*; Editor Truelove, B., Southern Weed Science Soc. Auburn, Al, pp. 109-118.
8. Grover, R.; Smith, A.E. *Canadian J. Soil Sci.* **1974**, 54:179-186.
9. Gold, A.J.; Morton, T.G.; Sullivan, W.M.; McClory, J. *Water, Air and Soil Pollution*. **1988**, 37:121-129.
10. Watschke, T.L.; Mumma, R.O. *Final Report. U.S. Dept. of Interior, Geological Survey*, **1989**, p.64.
11. Hurto, K.A.; Turgeon, A.J. *Weed Sci*. **1979**, 27:141-146.
12. Dell, C.J.; Throssell, C.S.; Bischoff, M.; Turco, R.F. *J Environ. Qual*. **1994**, 23:92-96
13. Lickfeldt, D.W.; Branham, B.E. *J Environ. Qual*. **1995**, 24:890-985.
14. Meyer, L.D.; Harmon, W.C. *Trans. ASAE*. **1979**, 22:100-103.
15. Carroll, M.J.; Hill, R.L.; Pfeil, E.; Herner, A.E. *Weed Tech*. **1993**, 7:437-442.
16. Dernoeden, P.H.; Carroll, M.J. *Hort. Sci*. **1992**, 27:881-882.
17. Raturi, S.; Carroll, M.J.; Hill, R.L.; Pfeil, E.; Herner, A.E. *Inter. Turf. Soc. Res. J.* **1997**, 8:187-196.
18. Ogden, C.B., H.M. van Es, and R.R. Schindelbeck. *Soil. Sci. Soc. Am. J. 1997*, 61:1041-1043.
19. Lapidus, L; Amundson, N.R. *J. Phys. Chem*. 1952, 56:984-988.
20. Hashimoto, I.; Deshpande, K.B.; Thomas, H.C. *Ind. Eng. Chem Fundamen*. **1964**, 3:213-218.
21. Selim,H.M.; Davidson, J.M.; Mansell, R.S. *In Proc. Summer Comput. Simul.*Conf. *Wash DC*. Simul Counc.: LaJolla, CA, **1976**.
22. Nkedi-Kizza, P.; Bigger, J.; Selim, H.M.; van Genuchten, M. Th.; Wierenga, P. J.; Davidson, J.M.; Nielsen, D.R. *Water Resour. Res*. **1984**, 20:1123-1130.
23. Toride, N.; Leij, F.J.; van Genuchten, M. Th. *The CXTFIT code for estimating transport parameters from laboratory or field tracer experiments, Version 2*. Research Rpt. 137, U.S. Salinity Laboratory, USDA, ARS, Riverside, CA. **1995**.
24. Gaber, H.M.; Inskeep, W.P.; Comfort, S.D.; Wraith, J.M. *Soil Sci. Soc. Am. J.* **1995**, 59:60-67.
25. Dao, T.H. In *Allellochemicals: Roles in agriculture, forestry*; Waller G.R., Ed.; Sym. Series 330; ACS, Washington, DC, **1987**, p.358-370.
26. Dao, T.H.; *J Environ. Qual*. **1994**, 20:203-208.
27. Balogh, J.D; Walker, W.J. *Golf Course Management & Construction*. Lewis Publishers, Boca Raton, FL. **1992**.
28. Weber, J.B.: C.T. Miller. In *Reactions and movement of organic chemcials in soils;* Sawhney, B.L.; Brown, K. SSSA Spec. Publ. 22. ASA, CSSA, SSSA Madison, WI, **1989**, p. 305-334.
29. Parker, J.C.; van Genuchten, M.Th. *Determining transport parameters from laboratory and field tracer experiments*. Virginia Agric. Exper. Stat. Bulletin 84-3. Blacksburg, VA. **1984**.
30. Brusseau, M.L.; Jessup, R.E.; Rao, P.S.C. *Water Resour. Res*. **1989**, 25:1971-1988.
31. Brusseau M.L.; Jessup, R.E.; Rao, P.S.C. *Environ. Sci. Technol*. **1991**, 25:134-142.

Chapter 14

Monitoring Vadose-Zone Soil Water for Reducing Nitrogen Leaching on Golf Courses

G. H. Snyder[1] and J. L. Cisar[2]

[1]Everglades Research and Education Center, University of Florida/IFAS,
P.O. Box 8003, Belle Glade, FL 33430
[2]Fort Lauderdale Research and Education Center, 32005 College Avenue,
Fort Lauderdale, FL 33314

Nitrogen (N) fertilization is required to maintain high-quality turfgrass on golf courses. The nitrate form of nitrogen is very mobile in soils and is a contaminant in groundwater. Drinking water standards generally specify that nitrate-N should be below 10 mg N L^{-1}. Monitoring studies were conducted on vadose-zone water in newly-constructed sand-based golf greens at two courses (9 total greens) in Florida, and in six fairways on sand soil at one course. Lysimeters were used for monitoring in the greens, which were generally constructed according to United States Golf Association specifications. Ceramic cup water samplers placed at 0.5 m depth were used in the fairways. One of the golf courses was irrigated with reclaimed water that contained approximately 8 mg N L^{-1} in the nitrate form and monitored for groundwater N. Appreciable N leaching was observed in the golf greens. Nitrogen leaching occurred mostly in the nitrate form, except for a short period (30 to 45 days) following greens construction. During the period of time when the newly-planted grass was filling in (the grow-in period), N concentrations ranged from 20 to nearly 200 mg L^{-1}. After the greens were established, and when N fertilizations involved lower rates, less frequent applications, and greater use of controlled-release N sources, nitrogen concentrations generally were below 10 mg L^{-1}. Nitrogen in water samples taken from the fairways occurred mostly in the nitrate form, and generally was below 10 mg N L^{-1}. An on-going groundwater monitoring study at one of the golf courses revealed no increase in nitrates during the 7 months following course construction. Clearly, nitrate leaching can be substantial in sand-based golf greens, but it can be reduced considerably by modifying N fertilization rates, application frequency, and by utilizing controlled-release N sources. The fact that less N leaching was observed in fairways, and fairways constitute a considerably greater portion of the golf course area than greens, may partially explain why no increase of nitrate was observed in groundwater during the 7-months of the study.

Golf course construction has accelerated over the past several decades, both in the USA and throughout the world. Currently, there are approximately 16,000 golf courses in the USA, 932 more are under construction, and there are over 11,600 courses outside of the USA (National Golf Foundation, Jupiter, FL). Perusal of any issue of many golf course trade magazines or newspapers, such as Golf Course News (United Publications, Inc., Yarmouth, Maine), reveals numerous announcements of new golf courses being built worldwide. However, such perusal also indicates that some golf course developments have or are facing opposition from various organizations concerned with the effects of golf courses on the environment. For example, in just one issue of the previously-mentioned publication (Volume 10, No. 2, February 1998) one will find articles about the efforts to stop golf courses from being constructed by organizations ranging from the Environmental Coalition of Ventura County, on the west coast (p. 7), to actor Paul Newman's Newman's Own food products company, on the east coast (p. 10), and golf course designers who have set up departments within their companies to help clients deal with governmental permitting, including environmental permits (p. 49). The same issue also has an article by U.S. Environmental Protection Agency head Carol Browner discussing the importance of avoiding groundwater contamination by agrichemicals applied to golf courses (p. 19). Clearly, the golf course industry needs to minimize environmental impacts and document the degree to which this goal has been achieved.

There have, in fact, been a number of environmental impact studies conducted in association with obtaining permits for new golf courses, or on existing courses. Perhaps 85% of the monitoring studies have been conducted to obtain permits (Personal communication. 1994. Dr. M. M. Smart. The Turf Science Group, Cary, NC). Unfortunately, such studies are rarely published in scientific journals. These, and the published studies, generally have focused on groundwater monitoring. For example, Cohen et al. (1) monitored both nutrients and pesticides in groundwater near and below golf courses on Cape Cod. The United States Geological Survey monitored nutrients and pesticides in wells associated with Florida golf courses over a two-year period (2). Such studies have provided useful information, but groundwater studies have limitations.

Groundwater monitoring is expensive. Monitoring wells must be drilled to detailed and exacting specifications (*3*). Sampling is also highly structured and regulated (*4*). Interpretation of results depends on a detailed knowledge of the geology and hydrology of the area, which can be very expensive to obtain and is seldom as complete as is desired. Perhaps the greatest shortcoming of all is that groundwater contamination generally is the result of many years of mismanagement. Groundwater moves slowly and responds slowly to quality changes, influencing efforts to predict, monitor, and control groundwater pollution (*5*). Therefore, short-term management variables generally can not be studied and implemented by ground water monitoring. After contamination is identified in the groundwater, it may take many years to effect a correction.

The vadose zone is the unsaturated region in the soil above the water table. Since vadose-zone water is adsorbed to the soil particle surfaces, it cannot be recovered in wells. Grasses maintained with close mowing on golf courses generally have very shallow rootzones; less than 20 cm. If vadose-zone water is monitored below the rootzone, it generally can be assumed that nitrate-N in such water will be no more concentrated when it enters the groundwater, and may be less concentrated because of denitrification losses. Since the monitoring is conducted at shallow depths, changes in nutrient concentrations more closely reflect fertilizer management variables than can be identified by groundwater sampling. The opportunity exists to make management corrections before groundwater contamination occurs.

Vadose-zone soil water monitoring is not as structured and regulated as groundwater monitoring. There are a variety of techniques for sampling vadose-zone soil water. Each has advantages for certain situations, and some are inappropriate for certain soil conditions (*6*). For example, lysimeters (open top/closed bottom containers placed in the soil profile to collect percolate water) are appropriate when there is a textural discontinuity of finer textured soil overlying a coarser soil within the profile that can be reproduced in the lysimeter. However, in the absence of such a discontinuity, a lysimeter will underestimate percolation, and may alter nitrogen mineralization, nitrification, and denitrification by creating wetter conditions within the lysimeter than in the surrounding soil. In deep soils lacking textural discontinuities, soil water must be extracted from the vadose zone under vacuum, or nutrients must be adsorbed on some type of exchanger. The most commonly accepted method for sampling vadose-zone water in these soils involves the use of micro-pore surfaces through which the water is collected under vacuum without the entry of air. Although a variety of designs and materials have been used for this purpose, ceramic cup water samplers probably are used most often. Ceramics can influence the passage of cations such as calcium and magnesium, and react with phosphorus. Nitrate-N is less affected by the ceramic material. Some question still exists, however, as to what portion of the soil water is sampled, i.e., what is the relative proportion of the water that comes from micropores and macropores and is nitrate-N evenly distributed between these two pore sizes? This is less of a concern in coarse-textured soils and when nitrate-N concentrations are sufficiently great (> 1 mg L^{-1}). Of course, when only a portion of the soil water is sampled, a measurement of percolation must be made to quantify nutrient leaching. This problem does not exist when lysimeters can be used.

The authors have monitored pesticide leaching on a golf green constructed at an agricultural research center (7), and nutrient leaching on two golf courses. All three studies involved vadose-zone soil water. The latter two studies will be described herein.

Methods and Materials

Greens. Lysimeters were installed during the construction of three golf greens in July 1993, at a course in Palm Beach County, Florida, and in six greens from August through November 1997, at a course in Orange County, Florida. The lysimeters (35.6 cm inside diameter by 40.6 cm deep) were constructed entirely of stainless steel and were patterned after lysimeters previously described in detail (8). They contained an expanded metal rack covered with a perforated metal disk which together were held approximately 20 cm off the bottom of the lysimeter by a stainless steel stand. A stainless steel tube (0.6 cm outside diameter in Palm Beach County and 1.0 cm in Orange County) extended from the bottom of the lysimeter downhill to a collection point off the side or back of the green. A 0.6 cm diameter air-return line that entered the lysimeter just below the rack also extended to the sample collection point. Percolate water drained by gravity into a 10 L vessel placed in a 28 x 40 cm plastic valve box with removable cover.

The greens were built in general accordance with United States Golf Association (USGA) specifications for greens construction in which the rootzone mix directly overlies the gravel layer (9). For each lysimeter, after the subgrade was prepared, including the drainage system, the lysimeter was placed into the subgrade about three meters from the side or back edge of the green to a depth that positioned the metal rack at the same level as the top of the subgrade. Several centimeters of gravel provided by the greens construction contractor were water washed, sieved to exceed the size of the holes in the perforated metal, and placed on the rack in the lysimeter. The contractor then covered the subgrade with approximately 10 cm of gravel, placing gravel inside the lysimeter to the same level as was placed outside the lysimeter. Rootzone mix was spread to an apporximate depth of 30 cm over the gravel both inside and outside of the lysimeter. Therefore, after completion of the green, the rim of the lysimeter was approximately 20 cm deep and did not interfere with normal cultivation operations such as aerification and verticutting. The green was then grassed and maintained by the golf course without any further regard for the lysimeter. Percolate was collected periodically and, at the Orange County course, the volume of water was recorded . At both courses a 20-ml subsample was placed in a plastic scintillation vial and maintained frozen until analyzed for ammonium and nitrate nitrogen, and for orthophosphate. Analysis of the Palm Beach County was performed by automated colorimetry. Nitrogen in the Orange County samples was analyzed with an ammonium electrode, and phosphorus with a probe colorimeter. Total N leaching in the Orange County study was calculated as the product of the N concentration by the quantity of water

At each golf course, the greens were sprigged with 'Tifdwarf' bermudagrass (*Cynodon dactylon x C. Transvaalensis*) and fertilized and irrigated liberally to encourage the development of a solid cover of grass over the green. In late January, 1993, one of the greens at the Palm Beach course, and in late November, 1997, all of the greens at the Orange County course were seeded with cool-season grasses (*Poa trivialis and Agrostis tenuis*) to provide a good putting surface, since bermudagrass growth was slowed by cool weather. 'Tifway' bermudagrass was used on the fairways at the golf course in Orange County, which were overseeded with ryegrass (*Lolium perenne* L.) In late November. The Orange County course was irrigated with 'reclaimed' water (highly-treated sewage effluent) that contained approximately 8 mg NO_3-N and 2 mg PO_4-P L^{-1}. The application amount was approximately calculated from records provided by the agency that supplied the reclaimed water. However, the apportionment of this water on various entities of the golf course (greens, tees, fairways, roughs) could not be ascertained. The course in Palm Beach County was irrigated with well water.

Fairways.   Six fairways were sampled on the course in Orange County. Since the soil at this location was a deep sand with no textural variation in the upper portion of the profile (an Entisol), lysimeters could not be appropriately used. Consequently, vadose-zone water sampling was achieved with ceramic cup water samplers. The samplers were installed in November, 1997, after the fairways were constructed and grassed with 'Tifgreen' bermudagrass as the permanent cover, and overseeded in October. The ceramic cups (5 cm diameter by 5 cm long) were glued onto polyvinyl chloride (PVC) pipe approximately 10 cm long with a standard PVC cap glued in place. A nylon tube (0.3cm outside diameter) extended from the bottom of the ceramic cup through the cap for a distance of approximately 10 m to a valve box used for water collection, and a similar air return line extended from just below the closed top of the sampler to the valve box.. Stainless-steel fittings screwed into the PVC cap secured the nylon tubing. The samplers were placed 50 cm deep, approximately 8 meters into the fairways, and the collection point was located several meters into an adjacent rough. To collect soil water, the sample line was sealed off, and a vacuum was drawn on the air-return line. After sufficient time had elapsed for 10 to 20 ml of soil watet to enter the sampler (generally 0.5 to 2.0 hours, depending on soil moisture), the water was forced out the sample line by pumping air into the air-return line. The samples were placed into scintillation vials, and stored in a frozen condition until analysis. Fairway samples generally were taken weekly.

At each location, the golf courses were maintained by the golf superintendents without any consultation with the authors. Information about the fertilizations and irrigations were anecdotally received from the superintendent at the Palm Beach location. A calendar listing of fertilizations and fertilizers used was obtained from the superintendent at the Orange County location, but the information was not sufficiently detailed to identify specific greens or fairways that received fertilization, and no information relating to fertilizer spreader calibrations was received. A groundwater monitoring study was conducted by Woodard & Curran (Winter Haven, Florida) at the Orange County site for a decade before the golf course was constructed, and has continued to date.

Results and Discussion

Greens. Considerable nitrogen was found in the percolation from the greens during the 'grow-in' period following greens construction at the Palm Beach County site (Figure 1). Nitrate-N (NO_3 -N) often was over 20 mg L^{-1}, and ranged up to 80 mg through the fall of 1993 and winter of 1994. However, the concentration of nitrate-N generally decreased in January and February, 1994, and for greens 2 and 17 remained below 10 mg L^{-1} from late April through the end of the study in late June. Green 14 was the exception. The nitrate-N concentration in percolate from green 14 decreased from early January through mid-February. But then it abruptly increased and remained around 40 to 60 mg L^{-1} through March and April. When this was brought to the attention of the superintendent, he recalled that because green 14 received considerable shade, which in combination with the cool winter temperatures was unfavorable for bermudagrass growth, the green was overseeded and more heavily fertilized to encourage the overseeded grass. As the overseeded grass faded in May and the bermudagrass grew better, nitrate-N in percolate from green 14 decreased to below 10 mg L^{-1} by the last several weeks of the study.

Even higher nitrate-N concentrations were found percolate from greens at the Orange County golf course during the fall and winter grow-in/overseeding period. In the early part of the grow-in period, the nitrogen fertilizer often contained only ammonium-N (NH_4 -N). As is illustrated using the data from green 5 (Figure 2), most of the leached N initially was in the ammonium-N form. However, after 30 to 45 days, the concentration of ammonium-N decreased, and the concentration of nitrate-N increased. Apparently, nitrifying bacteria were not present and active in the newly-created sand green, but became established and active within a month after fertilization with ammonium-N began. Thereafter, virtually all of the percolate N was in the nitrate form. Again using data from green 5 for illustration, from late December 1997, through the end of the sampling period in May, 1998, nitrate-N generally was near or below 10 mg N L^{-1} (Figure

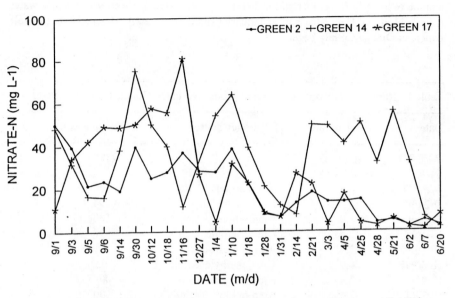

Figure 1. Nitrate-N in percolate water collected from three golf greens at the Palm Beach County site. Reproduced with permission from reference 10. Copyright 1997 Bedrock Information Systems, Inc.

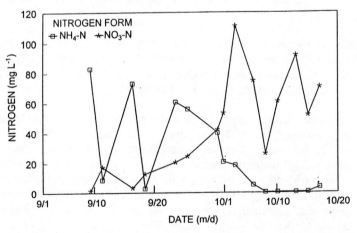

Figure 2. Ammonium and nitrate nitrogen in percolate water collected from Green 5 at the Orange County site shortly after construction and grassing of the green.

3). This decrease can be partially, but not entirely, explained by the fertilization practices at the course. Although the golf superintendent reported using a lower portion of water soluble N in early December and in February, there were applications of 100% water soluble N in late December and in January (Table 1). In early March, 1998, the golf course in Orange County began

Table 1. Reported fertilization of greens monitored in the Orange County Course.

Date	Nitrogen rate	Nitrogen source	Soluble nitrogen
(M/D/Y)	$(g\ m^2)$		(%)
9/22/97	4.9	Ammonium sulfate	100
9/24/97	4.9	Urea, methylene urea	94
9/26/97	4.9	Ammonium sulfate	100
10/1/97	2.4	Urea, methylene urea	98
10/16/97	2.4	Ammonium sulfate	100
10/30/97	4.9*	Urea, methylene urea	95
12/9/97	2.4	Urea, methylene urea	80
12/23/97	2.4	Ammonium sulfate	100
1/18/98	4.9	Ammonium sulfate	100
2/12/98	4.9	Coated urea	0
Total	39.0		

* Assumed rate.

responding to serious financial difficulties (11). According to the superintendent, virtually no fertilizer was used on greens or fairways from March through the end of May, 1998. This probably also contributed to the relatively low concentration of nitrate-N that was observed from March to the end of May (Fig. 3).

The total calculated nitrate-N leached in the 4 golf greens at the Orange County course that were established by October, 1997, ranged from less than 30 to over 100 g m^{-2} (Table 2). In fact, these values undercalculate the actual leaching somewhat since occasionally the 10L collection vessels were overflowing at the time of sampling. However, even with the undercalculation, the nitrogen leaching generally exceeded the amount of fertilizer N that was reported to have been applied (39.0 g m^{-2}, Table 1). Since the irrigation water contained N (approximately 8 mg L^{-1}), some of the leached N may have been derived from the irrigation water. The estimated total nitrate-N contribution

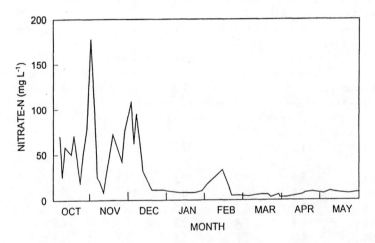

Figure 3. Nitrate-N in percolate water collected from Green 5 at the Orange County site during the monitoring study.

during the September 1997 through May, 1998 period is 8.5 g m^{-2} (data not presented). The calculation is based on the total amount of irrigation water used by the golf complex, including greens, tees, fairways, and roughs both on the course that was under construction at the time the study was initiated, and on an adjacent course and large driving range that was completed. It is possible that more than the "average" amount of water was used on the greens containing the lysimeters, which would have increased the N contribution. This would be consistent with water usage during a "grow-in" period. Nevertheless, if it is assumed that the lysimeters accurately assessed the quantity of percolate, total N leaching probably is greater than can be accounted for by the reported N fertilization and apparent irrigation.

Table 2. Monthly summary of Nitrate-N leached from greens at the Orange County Course.

	Green identification					
Month/Year	1	3	5	6	SW4	SE5
	- - - - - - - - - - - - - - - - (g m^{-2}) - - - - - - - - - - - - - - - - - -					
10/97	11.3	40.7	38.7	32.4	--	--
11/97	5.9	26.7	23.8	27.3	7.3	--
12/97	5.4	16.2	30.1	18.8	10.8	--
1/98	0.4	4.2	2.5	3.3	21.5	--
2/98	1.8	1.4	2.1	3.0	6.9	7.3
3/98	0.3	2.1	1.6	2.7	1.8	1.5
4/98	1.2	4.0	1.1	2.6	2.7	1.7
5/98	2.6	5.7	4.7	4.9	3.8	1.5
Total	28.9	101.0	104.6	95.0	54.8	12.0

Fairways Most of the nitrogen obtained in the fairway water samplers on the Orange County golf course was in the nitrate form. The concentration generally was below 10 mg L^{-1} (Table 3), and was similar to or less than the amount contained in the irrigation water. Although fairly high nitrate-N concentrations were observed in percolate from greens, especially during the grow-in period, greens constitute only 2 to 3 percent of the land area of a golf course. Consequently, nitrogen in percolate waters from fairways probably has a greater impact on groundwater nitrogen than that from greens. Through the summer of 1998, groundwater samples from wells in the vicinity of the golf course averaged at or below historic nitrate-N levels (< 4 mg N L^{-1}), according to a personal communication with David Macintyre, P.E., PB Water (Maitland, Florida). However, the groundwater monitoring wells are several meters into the water table. It is possible that there has been insufficient time for percolate from greens and fairways to reach the well points.

Table 3. Nitrate-N observed in percolate water 0.5 m below the soil surface in three fairways at the Orange County course, reported for illustration purposes one date in each month from February through March, 1998.

Date	Fairway		
	2	3	8
(M/D)	- - - - - - - - - - - - - - - (mg L^{-1}) - - - - - - - - - - - - - -		
2/24	9.7	4.7	7.2
3/18	4.1	4.4	3.9
4/21	7.8	9.3	6.2
5/12	-	8.1	8.1

Conclusion

High concentrations of nitrate-N were observed in percolate from golf greens during the grow-in period. After a complete grass cover was established, and with the use of less frequent N fertilization, lower rates of application, and sources of N that were less water soluble, the concentration of nitrates in percolate decreased to less than 10 mg N L^{-1}. Percolate water from fairways, which occupy far more land area than greens, generally contained less than 10 mg N L^{-1}. By monitoring vadose-zone N on a golf course, management practices can be modified to reduce N leaching before elevated concentrations of groundwater nitrates are observed.

Acknowledgment

The authors wish to express their appreciation to Mr. Mark Johnston for sample collection, and to Mr. David Rich for sample analyses. The assistance of Mr. Norman

Harrison and Mr. Jim Jeffers during equipment installation is gratefully acknowledged. We appreciate the information provided by Mr. Kevin Adams, Robert Sample and Chris Schnittker, and are in debt to Mr. Phil Ritson for facilitating our work on the golf course. The work was conducted under a grant from Woodard&Curran. WaterConserve II.

Literature Cited

1. Cohen, S.Z., S. Nickerson, R. Maxey, A. Dupuy Jr., and J. A. Senita. 1990. A ground water monitoring study for pesticides and nitrates associated with golf courses on Cape Cod. *Ground Water MR*, Winter. 160-173.
2. Swancar, Amy. 1996. Water quality, pesticide occurrence, and effects of irrigation with reclaimed water at golf courses in Florida. U. S. Geological Survey Water-Resources Report 95-4250. Tallahassee, FL. pp. 86.
3. USEPA. 1991. Environmental Compliance Branch Standard Operating Procudures and Quality Assurance Manual. USEPA, Region IV. Athens, GA.
4. ASTM. 1986. Standard Guide for Sampling Groundwater Monitoring Wells. Designation: D 4448 - 85a. American Society for Testing and Materials, Philadelphia, PA.
5. FAO. 1979. Groundwater Pollution. FAO Irrigation and Drainage paper 31. Food and Agricultural Organization of the United Nations, Rome. pg. vii.
6. Snyder, G. H. 1996. Nitrogen losses by leaching and runoff: methods and conclusions. Pg 427-432 in N. Ahmad (ed.) Nitrogen Economy in Tropical Soils. Kluwer Academic Publishers. Boston.
7. Cisar and Snyder. 1996. Mobility and persistence of pesticides applied to a USGA green III: Organophosphate recovery in clippings, thatch, soil, and percolate. *Crop Science 36*:1433-1438.
8. Cisar, J. L., and G. H. Snyder. 1993. Mobility and persistence of pesticides in a USGA-type green. I. Putting green facility for monitoring pesticides. *Int. Turfgrass Soc. Res. J. 7*:971-977.
9. USGA Green Section Staff. 1993. USGA recommendations for a method of putting green construction. *USGA Green Section Record*, March/April. pg. 1-3.
10. Snyder, G. H., and J. L. Cisar. 1997. An evaluation of nitrogen leaching in three golf course putting greens during establishment. Florida Turf Digest 14(2):14-16.
11. Jackson, J. 1998. Golf center lands in the rough. Orlando Sentinel, March 12.. Pg. B-1, B-6.

Chapter 15

Modeling Approaches for Assessment of Exposure from Volatilized Pesticides Applied to Turf

D. A. Haith[1], C. J. DiSante[1], G. R. Roy[2], and J. Marshall Clark[2]

[1]Agricultural and Biological Engineering, Cornell University, Ithaca, NY 14583
[2]Department of Entomology, University of Massachusetts, Amherst, MA 01003

Mathematical models can potentially be used to estimate volatilization of chemicals applied to turf. However, the complexity and limited testing of volatilization models restrict their general applicability. Alternatively, it may be possible to identify simple parameters or indices of the pesticide and environmental conditions which are associated with health hazards. Three indices were investigated, based on pesticide application rates, Henry constant, vapor pressure, organic carbon adsorption partition coefficient and wind speed. Regressions on these indices explained 70 – 90 % of the variation in measured vapor concentrations of 8 pesticides applied to turf. Classifications of the chemicals by an inhalation hazard quotient (HQ) were identical whether based on measured concentrations or concentrations estimated by regressions on any of the three indices.

Introduction

Some chemicals used for control of turfgrass pests have vapor pressures that suggest tendencies for volatilization. Moreover, since application typically leaves the pesticides exposed to air on grass surfaces rather than incorporated into soil, opportunities for gaseous losses may be greater than with comparable applications to agricultural crops. Measurements have confirmed that volatilization of pesticides does indeed occur, resulting in possible health hazards to turf users (1-4). However, generalizations are difficult. Some pesticides which are used on turf may not volatilize significantly, or may not be particularly hazardous or may only volatilize and threaten health under certain environmental conditions.

Hazard quotients (HQ) have been proposed for assessing the inhalation dangers from vaporized turf pesticides (*1,5*). The HQ is the estimated inhaled dose for a 70 kg adult divided by the chronic reference dose for the chemical,

$$HQ = I / Rfd \tag{1}$$

where I = daily inhaled pesticide dose (μg/kg-d) and Rfd = chronic reference dose for the pesticide (μg/kg-d). The inhaled dose is calculated from the concentration C (μg/m^3) of the chemical in the air and an assumed moderate breathing rate of 2.5 m^3/h over a four hour period for a 70 kg adult:

$$I = 4(2.5) \, C / 70 = C / 7 \tag{2}$$

Since HQ $\geq$ 1 indicates a potentially unsafe condition, concentrations which exceed Rfd by more than a factor of seven can be considered hazardous. Given reliable estimates of vapor concentrations, we can identify the chemicals and environmental conditions that should be avoided in turfgrass management.

Experimental studies can provide some of these estimates, but are not likely to be practical for the full range of chemicals, weather and site conditions encountered in practice. For these purposes, it would be useful to have mathematical models capable of predicting pesticide volatilization losses from turf. The state-of-the art in modeling volatilization of pesticides from soil surfaces is represented by models by Jury *et al.* (*6*) and Yates (*7*). The Jury *et al.* model estimates the pesticide vapor flux at the soil surface. Yates combines essentially the same flux calculation with the theoretical profile shape model of atmospheric transport (*8*) to predict downwind vapor concentrations. Although either of these models could presumably be extended to turfgrass, they require a range of parameters that may be difficult to estimate in many field settings.

A simpler modeling approach is development of empirical relationships between vapor concentrations and basic indices of chemical volatilization. If the relationships have reasonable predictive power, and indices can be determined from readily available chemical properties and environmental data, the approach can provide an efficient way of identifying volatilization hazards with pesticide applications to turf.

Volatilization Indices

Simple Indices. The most obvious indicators of pesticide volatilization are vapor pressure (p_v, mPa) and Henry constant (h, dimensionless). The first measures the general tendency of a chemical to escape as a vapor from solution and the second is the ratio of gaseous concentration or density (g, μg/m^3) to dissolved (solution) concentration (d, μg/m^3) at equilibrium:

$$h = g / d \tag{3}$$

where the concentrations g and d are typically measured under saturation conditions. Vapor density can be computed from p_v by the ideal gas law:

$$g = (M\ p_v) / (R\ T) \tag{4}$$

in which M = molecular weight, R = gas constant $(8300\ mPa{-}m^3/mol{-}°K)$ and T = temperature $(°K)$.

Vapor pressure, solubility and hence Henry constant are significantly affected by temperature. The higher the temperature, the greater the fraction of pesticide molecules that have sufficient kinetic energy to escape as vapor. Experiments have demonstrated diurnal variation in volatilization rates (1,2). During midday, when temperatures are at their highest, the vapor pressure of the pesticide increases and volatility is greatest.

Temperature effects on p_v can be determined from following procedure described as "method 2" in Lyman et al. (9):

$$\ln (p_v / p_{v0}) = [\ \Delta\ H_{v0}/(\Delta Z_b\ R\ T_0)]\ \{\ 1 - [3-2(T/T_0)]^m\ (T_0/T)$$

$$- 2m[3-2(T/T_0)]^{m-1}\ \ln(T/T_0)\} \tag{5}$$

where p_v = vapor pressure at temperature T, p_{v0} = vapor pressure at temperature T_0, ΔH_{v0} = heat of vaporization (cal/mol) at T_0, R = gas constant $(1.987\ cal/mol{-}°K$ in this equation), ΔZ_b = compressibility factor at boiling point (dimensionless), and m is a constant.

The compressibility factor equals one for ideal gases, but is assumed to be 0.97 in the examples given in Lyman et al. Heat of vaporization is approximated from the ideal gas law, and for the above dimensions,

$$\Delta\ H_{v0} / R\ T_0 = K_f\ [\ 22.84 + \ln (T_0 / p_{v0})] \tag{6}$$

with K_f is a constant, with mean value = 1.06 for a range of organic compounds.

The constant m is 0.19 for liquid chemicals, and may be either 0.36, 0.8 or 1.19 for solids, depending on boiling point. However, for the range of field temperatures of interest to pesticide volatilization $(280 - 310\ °K)$, equation 5 is relatively insensitive to m, and a value of m = 0.8 can be used with minimal loss of accuracy.

The Arrhenius equation can be used to model temperature dependence of the Henry constant:

$$h = h_0\ \beta^{\ T - T_0} \tag{7}$$

In this equation, h and h_0 are Henry constants for temperatures T and T_0, respectively, and β is a constant, which can be determined from the general observation that h typically doubles with a 10°C temperature rise (*10*); or $\beta = 1.072$

Combination Indices. Vapor pressure and the Henry constant are basic chemical properties that should influence pesticide volatilization. They are sensitive to environmental conditions through their dependence on temperature. However, there are other factors that mediate these influences in a turfgrass setting. Adsorption to organic matter limits volatilization opportunities, air movement removes the vapor from the site and moisture increases the total mass of dissolved chemical (*11*). A more general index which includes these phenomena can be obtained from an equilibrium mass balance of a turfgrass system to which the pesticide has been applied. The system consists of a grass and thatch layer with thickness z (cm) measured from the soil surface to the grass tips. The system consists of water, air and vegetation solids.

The pesticide may be in three phases: a vapor in the air, a solid adsorbed to the vegetation, or a liquid dissolved in water within or on the vegetation. The three phases are described by the dissolved and gaseous concentrations (d and g, as before) and a = concentration of chemical adsorbed to grass solids (μg/g). If α is the fraction of the layer occupied by the grass and thatch, we can develop expressions for A, D, and G = the total pesticide in adsorbed, dissolved and gaseous forms (g/ha), respectively, at equilibrium:

$$A = 100 \, a \, \alpha \, \rho \, z \tag{8}$$

$$D = 100 \, d \, \alpha \, \theta \, z \tag{9}$$

$$G = 100 \, g \, (1-\alpha) \, z \tag{10}$$

in which ρ = bulk density of the vegetation (g/cm^3) and θ = volumetric moisture content of the vegetation (cm^3/cm^3).

Under equilibrium conditions, dissolved and adsorbed concentrations can be related by a linear partitioning,

$$a = k \, d \tag{11}$$

where k is the adsorption partition coefficient (cm^3/g) . Combining equations 3 and 11 with 8 and 10, the total pesticide P (g/ha), is given by

$$P = A + C + G = [k \, \rho \, \alpha + \alpha \, \theta + h \, (1-\alpha)] \, 100 \, d \, z \tag{12}$$

Solving equation 12 for d and substituting in equation 10, the gaseous pesticide mass is

$$G = h \, d \, (1-\alpha) \, 100 \, z \quad = \quad h \, (1-\alpha) \, P \, / \, [k \, \rho \, \alpha + \alpha \, \theta + h \, (1-\alpha)]$$

$$= \ h \ P \ / \ [k \ \rho \ (\alpha/(1-\alpha)) + (\alpha/(1-\alpha)) \ \theta + h] \tag{13}$$

If we assume that roughly equal portions of vegetation and air, then $1-\alpha \approx \alpha$. Also, partition coefficients are typically estimated from k_{oc}, the organic carbon partition coefficient (cm^3/g), and f_{oc}, the organic carbon fraction of the solid medium:

$$k = k_{oc} \ f_{oc} \tag{14}$$

With these substitutions,

$$G = h \ P \ / \ [k_{oc} \ f_{oc} \ \rho + \theta + h \ ] \tag{15}$$

It should be recognized that equation 15 is a very simplified, lumped-parameter approach to the turf system. For example, adsorption is conceptualized as the same linear equilibrium process, whether in foliage, thatch or soil. The reality is likely more complex, with different adsorption mechanisms and associated parameters in each component. Similarly, dissolved chemical is unrealistically considered proportionally distributed to the moisture in the foliage, thatch and soil. However, the equation is not actually intended as an accurate model of pesticide behavior. Rather, as an approximation of chemical processes, it's purpose is to provide the basis for an indicator of gaseous losses.

The equation can be further simplified by examining the magnitudes of its parameters. Typical thatch bulk densities are 0.26 to 0.63 g/cm^3 (*12*) and organic matter contents of the grass and thatch range from 44% to 85% (*13*), corresponding to f_{oc} values of roughly 0.3 – 0.6. Typical values for k_{oc} are $10 – 10^4$. With these values, $k_{oc} \ f_{oc} \ \rho$ might range from 0.3(0.3)(10) = 0.9 to 0.6(0.6)10^4 = 3600. Moisture content θ might range from 0.1 for very dry conditions to 0.8-0.9 at saturation. Since Henry constant values are generally < 10^{-3}, it is clear that h can be neglected in the denominator of equation 15. Also, except in cases of very weakly adsorbed chemicals and very wet conditions, θ can also be neglected. The equilibrium volatilized pesticide within the grass/thatch layer reduces to

$$G = h \ P \ / \ (k_{oc} \ f_{oc} \ \rho) \tag{16}$$

Equation 16 provides an estimate of a total chemical mass available for volatilization loss from the grass/thatch system, and it is reasonable to presume that conditions which favor large values for G also produce large concentrations of inhaled vapors. However, air movement would dilute concentrations, suggesting that equation 16 should be divided by wind velocity. With this addition, and noting that f_{oc} and ρ will be constants for a particular turf system, a plausible indicator of pesticide vapor concentrations is

$$G \ / \ V \approx (h \ P) \ 10^6 \ / \ (k_{oc} \ V) \tag{17}$$

With V = wind speed (m/s), and 10^6 is added as a scaling factor.

This equation provides a simple, and intuitively appealing index for volatilized pesticide concentrations, suggesting increased volatilization with pesticide application and Henry constant, and reduced volatilization when the chemical is strongly adsorbed. In addition, temperature effects are included through their influence on h (equation 7).

Experimental Data

Volatilization indices were tested using data from on-going field turf experiments at the University of Massachusetts, Amherst. The 0.2-ha plots had well-established creeping bentgrass maintained at 1.3 cm height. Experimental design and sampling methods are described in Murphy *et al.* (*1*). Testing data covered 20 weeks during 1996 and 1997. Chlorpyrifos, diazinon, ethoprop, isazofos, and isofenphos were applied in weeks 1, 4, 7 and 12, and bendiocarb, carbaryl, and trichlorfon were applied in weeks 3, 6, 9 and 13. Ethoprop and isofenphos were also applied in weeks 16, 18 and 20. All chemicals were applied as sprays. Sampling periods were at various times between 8:00 AM and 7:00 PM on days 1, 2, 3, 5, and 7 each week the pesticide was applied. Day 1 sampling usually occurred four times a day. On Days 2 and 3, three samples were taken and one sample was taken on days 5 and 7. Sampling data included concentrations at 0.7 m height, surface and air temperatures, solar radiation and wind speed for each period. Measured concentrations generally fell to very low levels after the second day following application. Sampling periods were free of precipitation except for the first period on day 1 of week 12 and most periods within the first two days of week 13.

Table I. Chemical Properties for Eight Turf Pesticides

Chemical	Molecular Weight	Rdf (µg/kg-d)	Vapor Pressure @ 20°C (mPa)	Henry Constant @ 20°C	Organic Carbon Partition Coefficient (cm^3/g)
Bendiocarb	223.2	5.0	2.75	$9.03\ 10^{-7}$	385
Carbaryl	201.2	14.0	0.03	$1.86\ 10^{-8}$	288
Chlorpyrifos	350.6	3.0	1.60	$1.93\ 10^{-4}$	9930
Diazinon	304.3	0.09	7.35	$1.53\ 10^{-5}$	1520
Ethoprop	242.3	0.015	26.61	$3.78\ 10^{-6}$	104
Isazofos	313.7	0.02	7.45	$5.71\ 10^{-6}$	155
Isofenphos	345.4	0.5	0.22	$1.73\ 10^{-6}$	777
Trichlorfon	257.4	2.0	0.29	$2.53\ 10^{-10}$	15

Characteristics of the eight chemicals used in the study are given in Table I. Vapor pressures were taken from Tomlin (*14*) and adjusted to a common temperature using equations 4 and 5. Henry constants were determined by equation 3 from vapor pressures and solubilities (saturated concentrations) from Tomlin (*14*). Pressures were adjusted to the solubility datum temperature and converted to densities using equation

4. Partition coefficients were obtained from the USDA Pesticide Properties Database (http://www.arsusda.gov/rsml/ppdb2.html), and in most cases are the means of values from several sources. Reference doses were taken from Office of Pesticide Programs (5).

Analyses

Volatilization indices were compared with the maximum vapor concentrations measured during the two days following pesticide application. Thirty-seven concentrations were available, each corresponding to the maximum vapor concentration measured in the two days following application of a pesticide in a specific week. Henry constants and vapor pressures were determined for the surface temperatures measured during the periods of maximum concentration. Wind speeds from these same periods were used in equation 17. Data for these comparisons are listed in Table II. Indices were compared with concentrations using simple linear regression analyses.

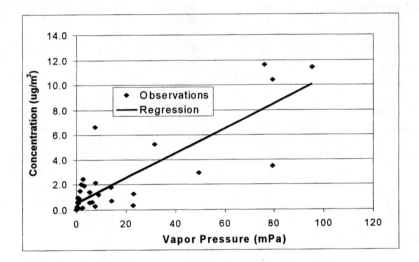

Figure 1. Maximum Vapor Concentration *vs.* Vapor Pressure.

Simple Indices. Two simple indices investigated were vapor pressure (p_v) and Henry constant (h). Regressions between concentrations and Henry constant were not significant. However, concentrations were strongly related to vapor pressure as shown in Figure 1. The associated regression equation,

$$C = 0.523 + 0.100\, p_v \tag{18}$$

has a standard error of 1.71 and $R^2 = 0.706$. It is obvious from Figure 1 that the observed concentrations are sometimes poorly predicted by the regression. However, it is interesting to note that for this range of chemicals, applications and environmental

Table II. Maximum Pesticide Vapor Concentrations and Associated Data for Two Days following Application to Turf.

Pesticide	Week	Wind Speed (m/s)	Surface Temperature (°C)	Application (g/ha)	Maximum 2-day Vapor Concentration ($\mu g/m^3$)
Bendiocarb	3	2.6	26.7	5832	0.58
Bendiocarb	6	2.3	30.2	4666	0.29
Bendiocarb	9	1.7	17.7	4666	0.13
Bendiocarb[a]	13	1.2	18.7	5832	0.18
Carbaryl	3	2.9	29.9	2803	0.05
Carbaryl	6	1.6	27.1	2243	0.04
Carbaryl	9	1.8	15.8	2243	0.01
Carbaryl[a]	13	1.2	18.7	2803	0.02
Chlorpyrifos	1	2.4	24.4	6997	2.42
Chlorpyrifos	4	2.5	32.1	7674	1.39
Chlorpyrifos	7	2.8	26.7	8577	1.89
Chlorpyrifos	12	3.0	18.4	9460	1.49
Diazinon	1	1.7	20.4	4204	2.15
Diazinon	4	2.5	32.1	4605	1.22
Diazinon	7	2.8	26.7	5146	1.78
Diazinon[a]	12	1.2	7.5	5676	2.00
Ethoprop	1	1.1	21.8	6327	5.24
Ethoprop	4	2.5	32.1	6907	3.44
Ethoprop	7	2.8	26.7	7719	2.91
Ethoprop[a]	12	1.2	7.5	8514	6.63
Ethoprop	16	1.3	31.6	10256	11.60
Ethoprop	18	1.7	32.1	7137	10.36
Ethoprop	20	1.2	34.3	6883	11.40
Isazofos	1	1.1	21.8	2099	1.21
Isazofos	4	2.5	32.1	2302	0.35
Isazofos	7	2.8	26.7	2573	0.72
Isazofos	12	3.0	18.4	2838	0.62
Isofenphos	4	1.5	29.2	2301	0.07
Isofenphos	7	2.8	26.7	2573	0.21
Isofenphos	12	1.6	22.6	2838	0.11
Isofenphos	16	1.3	31.6	3419	0.64
Isofenphos	18	1.7	32.1	2379	0.52
Isofenphos	20	1.2	34.3	2294	0.89
Trichlorfon	3	2.6	26.7	11484	0.96
Trichlorfon	6	1.1	23.0	9187	0.57
Trichlorfon	9	1.8	15.8	9187	0.23
Trichlorfon[a]	13	1.2	18.7	11484	0.26

[a] Rain during sampling period

conditions, vapor pressure alone explains more than 70% of the observed variation in concentrations.

Combination Indices. Figure 2 shows the relationship between concentrations and the G/V index given by equation 17. This index, which reflects the equilibrium volatilized mass in the turf divided by air velocity provides a better predictor of concentrations, with the regression

$$C = 0.384 + 0.019 \, G/V \tag{19}$$

with $R^2 = 0.875$ and standard error of 1.11. If data for rainy periods are excluded, a similar regression is obtained,

$$C = 0.332 + 0.019 \, G/V \tag{20}$$

With $R^2 = 0.914$ and standard error of 0.92.

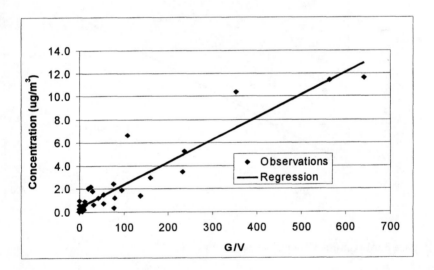

Figure 2. Maximum Vapor Concentration *vs.* G/V.

A similar index was also considered with vapor pressure replacing Henry constant,

$$G^*/V \approx (p_v \, P) \, 10^{-3} / (k_{oc} \, V) \tag{21}$$

and 10^{-3} as a scaling factor. The resulting regression is

$$C = 0.754 + 0.020 \, G^*/V \tag{22}$$

264

with $R^2 = 0.854$ and standard error = 1.20. As shown in Figure 3, this equation performs nearly as well as the regression on G/V, although prediction errors at low concentrations appear slightly larger

Hazard Assessment based on Volatilization Indices. Up to 90% of the observed variation in gaseous concentrations of these eight turf pesticides is explained by indices constructed from basic chemical properties and environmental variables. However, the scatter shown in Figures 1-3 suggest that concentrations predicted by the regression could have substantial errors. The asymptotic behaviors of G/V or G*/V are not reassuring, since extreme concentrations are obtained when air movement or adsorption become negligible (V or k_{oc} <<1). Certainly it would be unwise to extrapolate below the minimum velocities and partition coefficients given in Tables I and II.

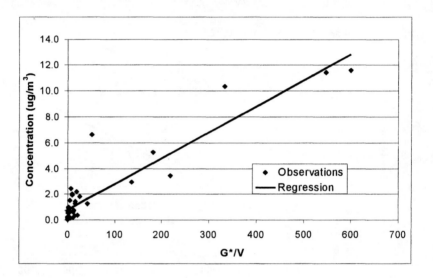

Figure 3. Maximum Vapor Concentration vs. G*/V.

Nevertheless, the regressions produce robust indicators of health hazards. Based on hazard quotients as given by equation 1, concern is not with the exact value of HQ, but rather with chemicals and conditions that are likely to produce values of HQ which approach one. Computed HQs based on measured concentrations and concentrations calculated by the regression equations are compared in Table III.

Although the regressions on volatilization indices often result in HQ values which are different than measured values, the differences are not large enough to produce misleading conclusions regarding hazards. Even the least accurate regression (equation 18, based solely on vapor pressure) clearly identifies the same hazardous chemicals (HQ > 1) as would be flagged by the measured concentrations. Diazinon, ethoprop and isazofos appear to have clear volatilization hazards, but bendiocarb, carbaryl, chlorpiryfos and isofenphos are relatively benign. Trichlorfon is a special case

Table III. **Comparison of Measured Hazard Quotients with Estimates based on Concentration Regression Equations for 8 Pesticides Applied to Turf.**

Pesticide	Week	Observed	Hazard Quotient (HQ) Estimated from p_v (Eq'n 18)	Estimated from G/V (Eq'n 19)	Estimated from G*/V (Eq'n 22)
Bendiocarb	3	0.02	0.03	0.02	0.03
Bendiocarb	6	0.01	0.04	0.02	0.03
Bendiocarb	9	0.00	0.02	0.01	0.02
Bendiocarb	13	0.01	0.02	0.02	0.03
Carbaryl	3	0.00	0.01	0.00	0.01
Carbaryl	6	0.00	0.01	0.00	0.01
Carbaryl	9	0.00	0.01	0.00	0.01
Carbaryl	13	0.00	0.01	0.00	0.01
Chlorpyrifos	1	0.12	0.04	0.09	0.04
Chlorpyrifos	4	0.07	0.05	0.15	0.05
Chlorpyrifos	7	0.09	0.04	0.11	0.05
Chlorpyrifos	12	0.07	0.03	0.07	0.04
Diazinon	1	3.41	2.04	1.40	1.80
Diazinon	4	1.94	4.49	1.94	2.55
Diazinon	7	2.83	3.07	1.52	2.02
Diazinon	12	3.17	1.14	1.23	1.49
Ethoprop	1	49.90	35.01	47.54	41.95
Ethoprop	4	32.76	80.43	46.79	49.14
Ethoprop	7	27.71	52.08	33.23	33.32
Ethoprop	12	63.14	12.10	23.69	17.35
Ethoprop	16	110.48	77.28	122.62	121.99
Ethoprop	18	98.67	80.43	69.20	70.94
Ethoprop	20	108.57	95.70	108.01	111.93
Isazofos	1	8.64	10.08	13.81	7.82
Isazofos	4	2.50	20.08	13.67	8.41
Isazofos	7	5.14	13.81	10.23	7.25
Isazofos	12	4.43	8.26	7.07	6.25
Isofenphos	4	0.02	0.17	0.15	0.22
Isofenphos	7	0.06	0.16	0.13	0.22
Isofenphos	12	0.03	0.16	0.14	0.22
Isofenphos	16	0.18	0.17	0.18	0.23
Isofenphos	18	0.15	0.17	0.15	0.22
Isofenphos	20	0.25	0.18	0.17	0.23
Trichlorfon	3	0.07	0.04	0.03	0.06
Trichlorfon	6	0.04	0.04	0.03	0.06
Trichlorfon	9	0.02	0.04	0.03	0.06
Trichlorfon	13	0.02	0.04	0.03	0.06

266

because although it appears safe, the metabolite dichlorvos is often present in much higher concentrations.

Conclusions

Gaseous loss of pesticide from turfgrass is influenced by chemical properties, environmental and site conditions and management practices. Estimates of likely concentrations inhaled by human turf users are required for assessment of possible health risks. Mechanistic mathematical models of the volatilization are possible vehicles for such estimates, but we are unaware of any successful attempts to apply such models to the realities of field conditions. The simpler empirical approach of volatilization indices may be a more feasible option. Measured pesticide concentrations can be directly related to basic chemical properties and environmental variables and these same relationships may be used in a screening process to identify hazardous situations.

Volatilization indices based on pesticide application rates, Henry constant, vapor pressure, organic carbon adsorption partition coefficient and wind speed show considerable promise as practical tools for hazard assessment. Regressions on three of these indices explained 70 – 90 % of the variation in measured air concentrations of 8 pesticides applied to turf. Classifications of the chemicals by an inhalation hazard quotient   (HQ) were identical whether based on measured concentrations or concentrations estimated by regressions on any of the three indices.

Although this research has demonstrated the feasibility of volatilization indices as pesticide management tools, it would be premature to use the actual regression equations presented here for that purpose. The regressions have not been tested with independent data sets and may not apply to chemicals with properties significantly different from the eight pesticides included in the study.

Acknowledgments

Support for this research was provided, in part, by the U.S. Golf Association,   Green Section Turfgrass and Environmental Research Program.

Literature Cited

1.   Murphy, K. C.; Cooper, R. J.; Clark, J. M. *Crop Sci.* **1996**, 36(6), 1446-1454.
2.   Murphy, K. C.; Cooper, R. J.; Clark, J. M. *Crop Sci.* **1996**, 36(6), 1455-1461.
3.   Taylor, A. W.; Glotfelty, D. E.; Turner, B. C.; Silver, R. E.; Freeman, H. P.; Weiss, A. *J. Agric. Food Chem.* **1977**, 25, 542-548.
4.   Turner, B. C.; Glotfelty, D. E.; Taylor, A. W. *J. Agric. Food Chem.* **1977**, 25, 548-550.
5.   Office of Pesticide Programs. Reference Dose Tracking Report; U.S. Environmental Protection Agency: Washington, DC, **1993**.

6. Jury, W. A.; Russo, D.; Streile, G.; El Abd, H. *Wat. Resources Res.* **1990**, 26(1), 13-20.
7. Yates, S. R. *J. Environ. Qual.* **1993**, 22(3), 481-486.
8. Wilson, J. D.; Thurtell, G. W.; Kidd, G. E.; Beauchamp, E. G. *Atmos. Environ.* **1982**, 16(8), 1861-1867.
9. *Handbook of Chemical Property Estimation Methods*; Lyman, W. J.; Reehl, W. F.; Rosenblatt, D. H., Eds.; McGraw-Hill: New York, NY, **1982**.
10. Montgomery, J. H. *Agrochemicals Desk Reference*; Lewis Publishers: Boca Raton, FL, **1993**..
11. Taylor, A. W.; Spencer, W. F. In: *Pesticides in the Soil Environment: Processes, Impacts, and Modeling*; Cheng, H. H., Ed.; Soil Science Society of America: Madison, WI, **1990**, 213-269.
12. Hurto, K. A.; Turgeon, A. J.; Spomer, L. A. *Agron. J.* **1980**, 72, 165-167.
13. Ledeboer, F. B.; Skogley, C. R. *Agron. J.* **1967**, 59, 320-323.
14. *The Pesticide Manual*; Tomlin, C., Ed.; British Crop Protection Council & The Royal Society of Chemistry: Cambridge, UK., **1994**.

Chapter 16

Best Management Practices to Reduce Pesticide and Nutrient Runoff from Turf

J. H. Baird[1], N. T. Basta[2], R. L. Huhnke[3], G. V. Johnson[2], M. E. Payton[4], D. E. Storm[3], C. A. Wilson[4], M. D. Smolen[3], D. L. Martin[5], and J. T. Cole[5]

[1]Department of Crop and Soil Sciences, Michigan State University, East Lansing, MI 48824
Departments of [2]Plant and Soil Sciences, [3]Biosystems and Agricultural Engineering, [4]Statistics, and [5]Horticulture and Landscape Architecture, Oklahoma State University, Stillwater, OK 74078

Vegetation buffer length (0, 1.2, 2.4, and 4.9 m), mowing height (1.3, 3.8, and 7.6 cm), and aerification were evaluated to reduce chemical surface runoff from bermudagrass [*Cynodon dactylon* (L.) Pers.] turf on a Kirkland silt loam (fine, mixed, thermic Udertic Paleustolls) with a 6% slope. Nitrogen, P, chlorpyrifos [o,o-diethyl o-(3,5,6-trichloro-2-pyridinyl) phosphorothioate], and the dimethylamine salts of 2,4-D (2,4-dichlorophenoxyacetic acid), mecoprop [2-(2-methyl-4-chlorophenoxy) propionic acid], and dicamba (3,6-dichloro-o-anisic acid) were applied at standard use rates on plots located upslope of buffers. A portable rainfall simulator applied precipitation rates of 51 or 64 mm h^{-1} for 75 to 140 min within 24 h after chemical application. Surface runoff losses expressed as percent of total applied ranged from 0.2 to 14% for 2,4-D, 0.2 to 15% for mecoprop, 0 to 7.8% for dicamba, 0.08 to 0.9% for chlorpyrifos, 0.3 to 6.3% for ammonium-N, 0.3 to 4.1% for nitrate-N, and 0 to 14% for phosphate-P. Greatest runoff losses occurred when heavy natural precipitation fell prior to initiation of the simulated rainfall experiments. All buffer treatments reduced chemical runoff compared to no buffer. The tallest buffer mowing height was most effective in reducing runoff. In most instances, aerification of turf in the buffer did not significantly reduce surface runoff.

Surface runoff is a primary means for potential loss of pesticides and nutrients from turf. On golf courses, surface water features are commonly found bordering the edge of highly maintained turf and thus have the potential to receive contaminants in runoff water. Since fairways comprise a large portion of golf course turf that receives regular applications of pesticides and nutrients, a study was initiated in 1995 at Oklahoma State University to develop management

© 2000 American Chemical Society

practices to reduce pesticide and nutrient runoff from bermudagrass turf
maintained under simulated fairway conditions.

Overland flow occurs when infiltration rate is exceeded by precipitation rate. Several factors affect surface loss of pesticides and nutrients including: 1) time interval between chemical application and precipitation event causing runoff; 2) amount and duration of precipitation event; 3) antecedent soil moisture; 4) slope; 5) amount and method of chemical application; 6) timing of chemical application in regard to plant uptake; 7) chemical properties; 8) rate of field degradation/transformation; 9) soil properties; and 10) vegetation type or density (1,2). Pesticides and nutrients are lost in runoff from croplands as dissolved, suspended, or sediment-bound particles (1,2). Runoff losses due to sediment are insignificant for most turf areas and, accordingly, turf has been shown to significantly reduce chemicals in surface runoff compared to tilled soils (3-5). In a 2-year study in Rhode Island on creeping bentgrass (*Agrostis palustris* Huds.) putting greens, surface runoff was recorded for only two natural precipitation events; one of these events occurred on frozen ground and the other involved wet soils which received 125 mm of precipitation within one week (6). Brown et al. (7) evaluated seasonal losses of nitrogen (N) in leachate and runoff from sand- and soil-based creeping bentgrass putting greens. Nitrate-nitrogen (NO_3-N) concentrations in surface runoff from soil-based greens exceeded 10 mg L^{-1} only once and all other samples were below 5 mg L^{-1}. Linde et al. (8) found that creeping bentgrass turf maintained under golf course fairway conditions reduced surface runoff when compared to perennial ryegrass (*Lolium perenne* L.). Few studies have been conducted regarding surface runoff of bermudagrass, the dominant turfgrass species used on golf course fairways and many areas in the southern United States.

Nitrogen and phosphorus (P) are important and commonly-applied essential nutrients required by the turfgrass plant for establishment, growth, and color. From an environmental standpoint, excessive losses of these nutrients into water resources can result in eutrophication. Aquatic problems associated with increased algal growth can result from total N and P concentrations as low as 1 mg L^{-1} and 25 µg L^{-1}, respectively (1,2). Nitrogen and P in surface runoff from turf is typically recovered in soluble forms as NO_3-N, ammonium-nitrogen (NH_4-N), and phosphate-phosphorus (PO_4-P), respectively (1). Morton et al. (6) studied the influence of overwatering and fertilization on N losses in runoff from Kentucky bluegrass (*Poa pratensis* L.) turf. Mean concentrations of NH_4-N and NO_3-N in runoff water ranged from 0.36 mg L^{-1} on overwatered, unfertilized turf to 4.02 mg L^{-1} on overwatered, fertilized turf. Gross et al. (3) found that runoff losses of N were significantly higher from fertilized tall fescue (*Festuca* arundinacea Schreb.)/Kentucky bluegrass turf when compared to unfertilized turf. No significant treatment differences were found with regard to runoff losses of all forms of P.

The herbicides 2,4-D (2,4-dichlorophenoxyacetic acid), mecoprop [2-(2-methyl-4-chlorophenoxy) propionic acid], and dicamba (3,6-dichloro-o-anisic acid) are commonly used alone or in combination on turf for selective control of many broadleaf weed species. Likewise, the insecticide chlorpyrifos [o,o-diethyl o-(3,5,6-trichloro-2-pyridinyl) phosphorothioate] is commonly used for control of surface- and root-feeding insects. Pesticides with water solubilites of greater than 10 mg L^{-1} are predominantly lost in the soluble form in surface runoff (9). The measured water solubilities of the dimethylamine salts of 2,4-D, mecoprop, and dicamba are 300,000 to 790,000 mg L^{-1}, 660,000 mg L^{-1}, and 850,000 mg L^{-1}, respectively; chlorpyrifos solubility is 2.0 to 4.8 mg L^{-1} (2). The runoff potential of pesticides is also greatly influenced by their adsorption to soil. The adsorption affinity of pesticides to soil can be expressed by the adsorption coefficient (K_{OC}). The K_{OC} values for 2,4-D, mecoprop, dicamba, and chlorpyrifos are 20, 20, 2, and 6,070 to 14,000, respectively (2). Larger K_{OC} values, like that of chlorpyrifos, reflect a stronger degree of adsorption to soil particles. Watschke and Mumma (10) studied the leaching and runoff characteristics of several pesticides including 2,4-D, dicamba, and chlorpyrifos applied to turf. Total losses in runoff were negligible for chlorpyrifos and less than 2% for the herbicides. White et al. (11) observed similar concentrations of 2,4-D lost in runoff from an agricultural watershed. A recent study in Georgia reported that 9, 13, and 14% of the applied 2,4-D, mecoprop, and dicamba, respectively, left the simulated bermudagrass fairways following a simulated rainfall event (12).

Vegetated waterways and no-till buffers have been shown to reduce surface runoff of chemicals in wetlands (13) and agricultural settings (14-16). Buffers help reduce surface runoff by: 1) increasing potential for infiltration; 2) reducing surface flow velocity to reduce both the erosive power and sediment carrying capacity of runoff water; 3) providing physical filtering of sediment or chemicals in solution; and 4) diluting applied chemicals (17). Dillaha et al. (18) observed reductions in surface loss of N and P from plots containing vegetation filter strips. The evidence from these and other studies suggests that vegetation provides an effective filter for reducing surface runoff. However, the influence of the buffer length, mowing height, and turf aerification practices on reducing nutrient and pesticide losses in surface runoff from turf has not been studied. Turfgrass aerification is a common cultivation practice used to help reduce thatch and soil compaction and to increase infiltration of water and oxygen. As a result, holes created by aerification may help reduce surface runoff of chemicals. While vegetation, including turf, may serve as an effective filter of chemicals, it is less effective when a heavy precipitation event closely follows the application of pesticides and fertilizer (2).

The objectives of this study were to: 1) evaluate the effects of vegetation buffers, specifically length, mowing height, and aerification, for reducing pesticide and nutrient runoff from simulated bermudagrass fairways; 2) determine the runoff potential of selected pesticides and nutrients including 2,4-D, dicamba, mecoprop, chlorpyrifos, NH_4-N, NO_3-N and PO_4-P; and 3) compare the runoff potential of different chlorpyrifos and nitrogen formulations. The pesticides and fertilizer were chosen for this study because of their widespread use by turf managers and potential for surface runoff based upon their physico-chemical properties.

Materials and Methods

Site Preparation. The experiments were conducted in July and repeated in August of 1995 and 1996 on a 1.2 ha sloped field in common bermudagrass located at the Oklahoma State University Agronomy Farm in Stillwater, Oklahoma. The soil is a Kirkland silt loam (fine, mixed, thermic Udertic Paleustolls). The turf was mowed to a 1.3 cm height three times a week and received fertilization and weed control to achieve golf course fairway turf conditions. Soil test results indicated that acidity, P, and potassium (K) levels were adequate for turf growth. Differential leveling techniques were used to define contour lines in order to determine suitable locations for eight rainfall simulator set-ups, each containing four plots. Plot locations were staked out to insure that plots were parallel to the slope and that they contained no significant surface depressions. The average slope of the plots was 6% (±0.6%).

Rainfall Simulator. A portable rainfall simulator was used to apply controlled precipitation simultaneously to four 1.8 m x 9.8 m plots. The rainfall simulator is based on the Nebraska rotating-boom design (19), and is capable of wetting a 15.2 m diameter area. The nozzles, located on a rotating boom 2.7 m above the ground, spray continuously and move in a circular pattern. The rainfall simulator boom was rotated at approximately 7 revolutions min^{-1}. A central alley, 3 m wide, allowed room for the simulator placement between plots 2 and 3 with at least a 1.5 m boom overhang at all corners to ensure uniform rainfall coverage (Figure 1). The rainfall simulator was set parallel to the land surface with the boom held at a constant height above the ground. Plot pairs were separated by 0.3 m. The rainfall simulator was calibrated prior to the experiments; however, three rain gauges were installed in the center alley, 2.4, 4, and 5.5 m from the boom center to measure delivered rainfall.

Each rainfall simulation experiment required 14,500 to 21,600 L of water, depending upon the rainfall intensity and length of the rainfall simulation event. A 19,000-L tanker was used for storage of water taken from a City of Stillwater fire hydrant. A 5.2 kW gasoline engine and pump, located at the tanker, pumped the water through a 5-cm high-pressure vinyl hose to the simulator. The pump provided a mast-head pressure set at 207 kPa.

Plot Construction. Plot borders were made using lengths of flexible 3.8 cm i.d. plastic discharge hose (Amazon Hose and Rubber Co., Chicago, IL) filled with masonry sand. Plot borders were laid onto the turf and a sand:bentonite clay (5:1 v/v) mix was used to seal the outside edge of the hose to eliminate runoff from flowing underneath the borders.

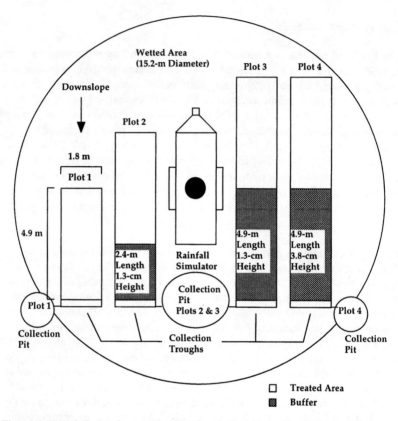

Figure 1. Rainfall simulator and example of plot layout. Ten booms with staggered nozzles not shown. (Reproduced with permission from ref. 25. Copyright 1997 American Society of Agronomy.)

275

August 1995 Simulated Rainfall Event. Treatments are shown in Tables I-VII. In August, an additional treatment containing no buffer was evaluated except that a 50% wettable powder (WP) formulation (DowElanco, Indianapolis, IN) was substituted for the granular formulation of chlorpyrifos and sulfur-coated urea (39% N) (Greenskote Trikote Poly SCU, Pursell Indust., Sylacauga, AL) for the urea formulation of N fertilizer. The treatment changes were made to evaluate the effects of formulation on chemical runoff. Fertilizer and insecticide were applied followed by irrigation as described in the July simulated rainfall event. Simulated rainfall was applied within 24 h of application of pesticides and nutrients. Based on the inordinately long times to start of runoff and low total runoff volumes in the July event, the decision was made to increase rainfall intensity to 64 mm h^{-1} for 75 min in the August event.

1996 Simulated Rainfall Experiments. Two experiments were conducted in July and repeated in August 1996 to evaluate the effects of buffer length and buffer mowing height on surface runoff of the applied chemicals. Simulated rainfall intensity was 64 mm h^{-1} for 75 min in both experiments. All other experimental procedures and rates of chemical application were the same as described for the 1995 experiments.

Buffer Length Experiment. The pesticides and fertilizer were applied to plots containing buffer lengths of 0, 1.2, 2.4, or 4.9 m. All buffers were mowed at 3.8 cm. The 50% WP and SCU formulations of chlorpyrifos and nitrogen, respectively, were applied in the buffer length experiment.

Buffer Mowing Height Experiment. The pesticides and fertilizer were applied to plots containing 4.9-m long buffers mowed at 1.3, 3.8, or 7.6 cm. The granular and urea formulations of chlorpyrifos and nitrogen, respectively, were applied in the buffer mowing height study.

Sampling. Simulated rainfall was applied onto two to four simulator set-ups per day. Start of surface runoff was recorded when a continuous trickle of water was first observed at the collection pit. Samples were collected at preset times after the start of runoff for individual plots using a nominal sampling schedule. Samples were taken more frequently early in the event during the rising portion of the hydrograph. Most plots were sampled 10 times during the simulated rainfall period. In addition to sample collection for chemical analysis, surface runoff volumes of 0.45 L, 1.0 L, or 5.0 L were collected depending upon the runoff rate. The time required to fill a calibrated container was used to calculate the runoff rate. The accuracy of the surface runoff rate was estimated to be ±3%. Discharge measurements, sample times, weather, precipitation, and plot conditions were recorded for each simulated rainfall event. A sample of the simulated rain water was taken directly from a simulator nozzle during all runs for chemical analysis. A single volume-weighted composite was prepared from runoff samples for each plot. In addition, a time series analysis was performed on a set of data from the July 1996 simulated rainfall event. Runoff samples collected from the non-buffered and 4.9 m buffer length treatments were analyzed individually for pesticide or nutrient concentration.

Pesticide Analysis. The following procedures were adopted from DiCorcia and Marchetti (*20*). All organic solvents were high performance liquid chromatography (HPLC) grade and deionized water was ≥ 15 MΩ. Runoff samples were vacuum filtered through 0.45 μ nylon membrane filters and refrigerated at 4 C until analyzed (<48 h). Filtered samples (100 mL) were passed through Carbopack B solid phase extraction (SPE) columns (Supelco, Inc.). Air was drawn through the column for 60 s to dry the SPE column. Acidic (2,4-D, dicamba, mecoprop) and base/neutral (chlorpyrifos) pesticides were adsorbed from the runoff water by the SPE column.

Chlorpyrifos was eluted first from the column with one 2 mL then two 1 mL volumes of a solution of 80:20 (v/v) methylene chloride (MeCl): methanol (MeOH). Column eluates were combined in a 4-mL HPLC vial and placed in a water bath at 35 C. Chlorpyrifos was concentrated by evaporation of the eluate to 1 mL under a gentle stream of ultra-high purity helium (He) gas. Eluate was reconstituted to 2 mL by volume with 50:50 MeOH:H$_2$O. Chlorpyrifos was determined by HPLC analysis as described below.

After chlorpyrifos was eluted, acidic pesticides were eluted from the SPE column by three 1 mL volumes of acidified 80:20 (v/v) MeCl:MeOH (acidified with 0.17% trifluoroacetic acid, TFA). Column eluates were combined in a 4 mL HPLC vial and placed in a water bath at 35 C under a stream of He until dry. The pesticide residue was redissolved in 2.0 mL of 50:50 (v/v) MeOH:H$_2$O. Acidic pesticides (2,4-D, dicamba, and mecoprop) were determined by HPLC analysis.

Pesticides were determined using a Model 500 HPLC (Dionex, Sunnyvale, CA). Components included a 50 μL injector, LC-18 HPLC column (Varian, Walnut Creek, CA), and UV-VIS detector (Dionex). Chlorpyrifos was analyzed by using a mobile phase flow rate of 1.5 mL min^{-1}. The mobile phase was 15% 1 mM sodium phosphate buffer (pH 6.7) and 85% acetonitrile (ACN). Chlorpyrifos had a retention time of 3.2 min and was measured at 230 nm. The detection limit was 0.1 μg L^{-1} chlorpyrifos in runoff water.

The mobile phase flow rate of 2,4-D, dicamba, and mecoprop was 1.5 mL min^{-1}. The acidic pesticides were separated using premixed MeOH:ACN (82:18, v/v) as an organic eluent and water acidified with TFA (0.17% TFA, v/v). The initial mobile phase was 50% organic eluent and 50% acidified water, which was linearly increased to 62% organic eluent over 15 min. Acidic pesticides were measured at 220 nm. Retention times of 5.7, 8.0, and 11.2 min were observed for dicamba, 2,4-D, and mecoprop, respectively. The detection limit was 0.1 μg L^{-1} of each acidic pesticide in runoff water samples.

Chlorpyrifos concentration in runoff water was also determined by enzyme-linked immunosorbent assay (ELISA) using the Chlorpyrifos RaPID

Assay and Ohmicron RPA-1 Photoanalyzer (Ohmicron Rapid Diagnostics, Newtown, PA). The detection limit was 0.1 µg L^{-1} for chlorpyrifos in runoff water samples.

Nutrient Analysis. Phosphate-phosphorus, NH$_4$-N, and NO$_3$-N were measured in filtered runoff samples. Phosphate-phosphorus was determined by the phosphomolybdate colorimetric procedure employed by Murphy and Riley (*21*). Ammonium-nitrogen and NO$_3$-N were determined by colorimetric methods using automated flow injection analysis (*22,23*). The detection limit was 0.01 mg L^{-1} of each nutrient in runoff water samples.

Experimental Design. In 1995, the experimental design was an unbalanced, randomized incomplete block. An unbalanced design was used because adjacent land was unsuitable to construct additional simulator set-ups needed to test the number of treatments in this study. Nevertheless, the design insured that important treatment comparisons were contained in the same simulator set-up (block) at least twice. Each treatment was replicated four times. Data were subjected to analysis of covariance (ANCOVA). The time to the start of runoff and total runoff volume for each plot were used as covariates in the analyses of concentration and mass, respectively, in order to account for differences in runoff that were due to differences in antecedent soil moisture or natural variation in soil properties. Contrasts were performed to make meaningful treatment comparisons at alpha levels 0.05 and 0.01. Due to limitations in sample size, treatment main effects were examined at alpha level 0.1 in order to provide more powerful comparisons. Instances in which the contrast was significant but the main effect was not significant were noted. In addition, soil moisture content of areas surrounding each plot was determined by gravimetric analysis prior to rainfall simulation.

In 1996, the experimental design was a randomized complete block. Each treatment was replicated four times. Data were subjected to ANCOVA. Soil moisture content of each plot was determined by gravimetric analysis prior to rainfall simulation and used as a covariate in the analyses of concentration and mass in order to account for differences in runoff that were due to differences in antecedent soil moisture or natural variation in soil properties. For the time series analysis, multiplicative models describing the observed behavior of runoff concentration over time for buffered and non-buffered plots were fitted on a log-transformed scale using linear mixed models (*24*). A spatial power covariance structure was used. Additional models for contaminant mass flow rates were fitted to evaluate the effect of buffers on total runoff mass.

Results and Discussion

Surface runoff losses of each chemical in all experiments conducted in 1995 and 1996 are summarized in Tables I-VII. Surface runoff losses expressed as percent of total applied ranged from 0.08 to 14% for 2,4-D (Table I), 0.2 to 15% for

Table I. Surface Runoff of ... ... on the 1995-96 Oklahoma State University Simulated Rainfall Experiments. Stillwater, OK.

Buffer (m)		Nitrification	Date Applied	Natural Rainfall[a] (mm)	Simulated Rainfall (mm)	Soil Moisture (%)	Time to Start of Runoff (min)	Runoff Depth (mm)	Concentration (µg L⁻¹)	Mass (µg)	Percent of Total Applied (%)
--	0	No	July 1995	0	83	17.0	56	7.2	314	30800	3.1
--	0	No	August 1995	165	80	18.4	15	63	174	97500	9.8
--	0	No	August 1995	165	80	17.7	17	56	166	86500	8.6
--	0	No	July 1995	19	80	17.7	16	25	130	30300	3.0
3.8	1.2	No	August 1996	80	80	17.5	17	31	183	51000	5.1
3.8	1.2	No	July 1996	19	80	16.0	17	21	66.3	16100	1.6
1.3	2.4	No	August 1996	80	80	16.7	16	31	126	43500	4.4
1.3	2.4	No	July 1995	0	74	27.2	21	12	76.9	17200	1.7
3.8	2.4	No	August 1995	165	80	29.1	11	54	154	106000	11
3.8	2.4	No	July 1995	0	74	27.9	56	5.5	31	6010	0.6
3.8	2.4	No	August 1995	165	80	27.9	13	39	99.8	72900	7.3
3.8	4.9	No	July 1996	19	80	28.0	21	25	39.8	15400	1.5
1.3	4.9	No	August 1996	80	80	26.2	20	32	103	43900	4.4
1.3	4.9	Yes	July 1995	0	87	27.3	76	6.1	14.4	2730	0.3
1.3	4.9	No	July 1995	0	83	27.3	51	3.0	15.8	768	0.08
1.3	4.9	Yes	August 1995	165	80	29.2	22	46	83.8	67900	6.8
1.3	4.9	No	August 1995	165	80	29.1	17	39	77.8	60200	6.0
1.3	4.9	No	July 1996	19	80	29.2	7	41	71.5	54600	5.5
3.8	4.9	No	August 1996	80	80	27.6	8	50	97.1	91200	8.9
3.8	4.9	Yes	July 1995	0	78	27.9	62	2.9	26.9	1860	0.2
3.8	4.9	No	July 1995	0	78	26.2	56	7.0	13.2	3510	0.4
3.8	4.9	Yes	August 1995	165	80	28.3	16	43	160	142000	14
3.8	4.9	No	August 1995	165	80	27.4	21	39	107	72500	7.2
3.8	4.9	No	July 1996	19	80	27.5	19	20	49.2	19800	2.0
3.8	4.9	No	July 1996	19	80	29.0	16	41	82	63800	6.4
3.8	4.9	No	August 1996	80	80	26.2	20	25	93.8	42700	4.3
3.8	4.9	No	August 1996	80	80	28.6	12	51	118	109000	11
7.6	4.9	No	July 1996	19	80	28.3	24	24	36.3	13100	1.3
7.6	4.9	No	August 1996	80	80	28.2	19	34	92.4	58800	6.1

[a]Natural rainfall recorded within seven days of simulated rainfall.

mecoprop (Table II), 0 to 7.8% for dicamba (Table III), 0 to 0.9% for chlorpyrifos (Table IV), 0.3 to 6.3% for NH_4-N (Table V), 0.3 to 4.1% for NO_3-N (Table VI), and 0.08 to 14% for PO_4-P (Table VII). Results of the 1995 simulated rainfall events were published in the Journal of Environmental Quality (25).

July 1995 Simulated Rainfall Event. No natural precipitation occurred within 12 d prior to or during the July 1995 rainfall simulation events. The antecedent soil moisture content of each rainfall simulation area ranged from 15 to 19% by weight. The rainfall simulator applied between 64 and 118 mm of water to each set-up in order to collect enough runoff for sampling. An average of 4 to 16% of the simulated rainfall water ran off during the events. Time to start of runoff and total runoff expressed on a volume basis for each treatment ranged from 21 to 76 min and 52 to 161 L, respectively. Variation in time to start of runoff and total runoff were most likely caused by differences in antecedent soil moisture or natural variation in soil properties among the plot areas.

There were no significant differences in chemical runoff losses or concentrations due to buffer mowing height or aerification, except for concentration of chlorpyrifos in runoff, which was increased in one comparison by the 1.3 cm mowing height and in another comparison by aerification (25). Numerical trends indicated that the 3.8 cm mowing height helped reduce chemical losses, compared with the 1.3 cm height, except when aerified. Also, aerification helped reduce runoff losses except when the buffer was mowed at the 3.8 cm mowing height. Aerification was performed in a direction parallel to the slope of the plots. As a result, it is possible that the aerification process created channels in the taller turf canopy, thus creating a less tortuous pathway for movement of surface runoff.

Treatments containing buffers were effective in reducing losses of all pesticides, NH_4-N, and PO_4-P in runoff water compared to no buffer. There were no significant differences in runoff concentrations or losses in comparisons between the 2.4 and 4.9 m buffers except for NO_3-N; however, trends showed greater runoff losses from the shorter buffer. Few significant treatment differences were observed for NO_3-N concentrations and losses most likely because very little of the urea was nitrified to NO_3-N within 24 h of application; therefore, most of the NO_3-N recovered in runoff probably pre-existed in soil. Moe et al. (5) noted similar findings when comparing N losses from urea and ammonium nitrate in surface runoff. Chaubey et al. (16) found that tall fescue buffers did not significantly reduce NO_3-N in runoff losses from a source area treated with swine manure; the buffers did significantly reduce runoff losses of NH_3-N and PO_4-P. Percent loss of pesticides and nutrients was ≤3.1% and 1.9%, respectively, based upon the total mass applied. In many instances, nutrient loss in runoff was greater from unfertilized plots compared to fertilized plots (data not shown).

August 1995 Simulated Rainfall Event. In contrast to July, 165 mm of natural precipitation was measured within 7 d of rainfall simulation events in August. No

Table II. Surface Runoff of Mecoprop from the 1995-96 Oklahoma State University Simulated Rainfall Experiments. Stillwater, OK.

Buffer Length	Buffer Height	Aerification	Date Applied	Natural Rainfall[a]	Simulated Rainfall	Soil Moisture	Time to Start of Runoff	Runoff Depth	Concentration	Mass	Percent of Total Applied
m	cm			mm	mm	(%)	min	mm	µg L^{-1}	µg	(%)
0	--	No	July 1995	0	83	17.0	56	7.2	164	16200	3.0
0	--	No	August 1995	165	80	18.4	15	63	97.8	56500	10
0	--	No	August 1995	165	80	17.7	17	56	88.1	45900	8.5
0	--	No	July 1995	19	80	17.7	16	25	60.5	14100	2.6
0	--	No	August 1996	80	80	17.5	17	31	79.5	22100	4.1
1.2	3.8	No	July 1996	19	80	16.0	17	21	28	6810	1.3
1.2	3.8	No	August 1996	80	80	16.7	16	31	54.8	18900	3.5
2.4	1.3	No	July 1995	0	74	27.2	21	12	44.9	10000	1.8
2.4	1.3	No	August 1995	165	80	29.1	11	54	89.4	61100	11
2.4	3.8	No	July 1995	0	74	27.9	56	5.5	15.3	3050	0.6
2.4	3.8	No	August 1995	165	80	27.9	13	39	59.7	44400	8.2
2.4	3.8	No	July 1996	19	80	28.0	21	25	13.5	5520	1.0
2.4	3.8	No	August 1996	80	80	26.2	20	32	44.2	18900	3.5
4.9	1.3	No	July 1995	0	87	27.3	76	6.1	6.3	1380	0.3
4.9	1.3	Yes	July 1995	0	83	27.3	51	3.0	14.4	874	0.2
4.9	1.3	No	August 1995	165	80	29.2	22	46	48.3	39400	7.3
4.9	1.3	Yes	August 1995	165	80	29.1	17	39	43.4	32900	6.1
4.9	1.3	No	July 1996	19	80	29.2	7	41	36.5	27500	5.1
4.9	1.3	No	August 1996	80	80	27.6	8	50	46	42400	7.8
4.9	3.8	No	July 1995	0	78	27.9	62	2.9	15.5	1050	0.2
4.9	3.8	Yes	July 1995	0	78	26.2	56	7.0	6.38	1730	0.3
4.9	3.8	No	August 1995	165	80	28.3	16	43	90.6	80500	15
4.9	3.8	Yes	August 1995	165	80	27.4	21	39	62.1	42000	7.8
4.9	3.8	No	July 1996	19	80	27.5	19	20	26.2	10400	1.9
4.9	3.8	No	July 1996	19	80	29.0	16	41	41.5	31900	5.9
4.9	3.8	No	August 1996	80	80	26.2	20	25	41.5	19000	3.5
4.9	3.8	No	August 1996	80	80	28.6	12	51	53.5	49800	9.2
4.9	7.6	No	July 1996	19	80	28.3	24	24	21.5	12000	2.2
4.9	7.6	No	August 1996	80	80	28.2	19	34	44.5	28700	5.3

[a]Natural rainfall recorded within seven days of simulated rainfall.

Table III. Surface Runoff of Dicamba from the 1995-96 Oklahoma State University Simulated Rainfall Experiments. Stillwater, OK.

Buffer Length	Buffer Height	Aerification	Date Applied	Natural Rainfall[a]	Simulated Rainfall	Soil Moisture	Time to Start of Runoff	Runoff Depth	Concentration	Mass	Percent of Total Applied
m	cm			mm	mm	(%)	min	mm	μg L^{-1}	μg	(%)
0	--	No	July 1995	0	83	17.0	56	7.2	16.5	1580	1.4
0	--	No	August 1995	165	80	18.4	15	63	10.8	6170	5.6
0	--	No	August 1995	165	80	17.7	17	56	9.82	4940	4.5
0	--	No	July 1995	19	80	17.7	16	25	7.5	1740	1.6
0	--	No	August 1996	80	80	17.5	17	31	13.8	3830	3.5
1.2	3.8	No	July 1996	19	80	16.0	17	21	3.5	857	0.8
1.2	3.8	No	August 1996	80	80	16.7	16	31	9.5	3290	3.0
2.4	1.3	No	July 1995	0	74	27.2	21	12	3.69	876	0.8
2.4	1.3	No	August 1995	165	80	29.1	11	54	9.7	6640	6.0
2.4	3.8	No	July 1995	0	74	27.9	56	5.5	1.34	341	0.3
2.4	3.8	No	August 1995	165	80	27.9	13	39	8.26	5480	5.0
2.4	3.8	No	July 1996	19	80	28.0	21	25	1	495	0.5
2.4	3.8	No	August 1996	80	80	26.2	20	32	7.75	3250	3.0
4.9	1.3	Yes	July 1995	0	87	27.3	76	6.1	1	3	0.003
4.9	1.3	No	July 1995	0	83	27.3	51	3.0	0	0	0.0
4.9	1.3	No	August 1995	165	80	29.2	22	46	5.58	4450	4.2
4.9	1.3	Yes	August 1995	165	80	29.1	17	39	5.67	4420	4.0
4.9	1.3	No	July 1996	19	80	29.2	7	41	6.25	4660	6.3
4.9	1.3	No	August 1996	80	80	27.6	8	50	7.5	6930	6.3
4.9	3.8	Yes	July 1995	0	78	27.9	62	2.9	4.79	156	0.1
4.9	3.8	Yes	July 1995	0	78	26.2	56	7.0	0	0	0.0
4.9	3.8	No	August 1995	165	80	28.3	16	43	9.65	8580	7.8
4.9	3.8	Yes	August 1995	165	80	27.4	21	39	7.32	4790	4.4
4.9	3.8	No	July 1996	19	80	27.5	19	20	4	1660	1.5
4.9	3.8	No	July 1996	19	80	29.0	16	41	6.5	5000	6.5
4.9	3.8	No	August 1996	80	80	26.2	20	25	7	3160	2.9
4.9	3.8	No	August 1996	80	80	28.6	12	51	8.75	8170	7.4
4.9	7.6	No	July 1996	19	80	28.3	24	24	3.25	1590	3.3
4.9	7.6	No	August 1996	80	80	28.2	19	34	8.25	5570	5.1

[a]Natural rainfall recorded within seven days of simulated rainfall.

Table IV. Surface Runoff of Chlorpyrifos from the 1995-96 Oklahoma State University Simulated Rainfall Experiments. Stillwater, OK.

Formulation	Buffer Length	Buffer Height	Aerification	Date Applied	Natural Rainfall[a]	Simulated Rainfall	Soil Moisture	Time to Start of Runoff	Runoff Depth	Concentration	Mass	Percent of Total Applied
	m	cm			mm	mm	(%)	min	mm	µg L^{-1}	µg	(%)
0.5 G	0	--	No	July 1995	0	83	17.0	56	7.2	34.8	2780	0.1
0.5 G	0	--	No	August 1995	165	80	18.4	15	63	37.2	18700	0.9
50 WP	0	--	No	August 1995	165	80	17.7	17	56	20.7	10300	0.5
50 WP	0	--	No	July 1995	19	80	17.7	16	25	0.278	60.5	0.003
50 WP	0	--	No	August 1996	80	80	17.5	17	31	1.4	370	0.02
50 WP	1.2	3.8	No	July 1996	19	80	16.0	17	21	0.225	51.3	0.003
50 WP	1.2	3.8	No	August 1996	80	80	16.7	16	31	0.27	95.9	0.005
0.5 G	2.4	1.3	No	July 1995	0	74	27.2	21	12	4.49	984	0.05
0.5 G	2.4	1.3	No	August 1995	165	80	29.1	11	54	7.48	4920	0.05
0.5 G	2.4	3.8	No	July 1995	0	74	27.9	56	5.5	0	0	0.0
0.5 G	2.4	3.8	No	August 1995	165	80	27.9	13	39	0	0	0.3
50 WP	2.4	3.8	No	July 1996	19	80	28.0	21	25	0.203	71.7	0.004
50 WP	2.4	3.8	No	August 1996	80	80	26.2	20	32	0.255	110	0.006
0.5 G	4.9	1.3	No	July 1995	0	87	27.3	76	6.1	5.73	963	0.05
0.5 G	4.9	1.3	Yes	July 1995	0	83	27.3	51	3.0	20.5	1070	0.001
0.5 G	4.9	1.3	No	August 1995	165	80	29.2	22	46	0	0	0.05
0.5 G	4.9	1.3	Yes	August 1995	165	80	29.1	17	39	0	0	0.04
0.5 G	4.9	1.3	No	August 1996	19	80	29.2	7	41	1.38	1030	0.0
0.5 G	4.9	1.3	No	July 1995	80	80	27.6	8	50	0.58	538	0.4
0.5 G	4.9	3.8	Yes	July 1995	0	78	27.9	62	2.9	4.64	27	0.0
0.5 G	4.9	3.8	No	July 1995	0	78	26.2	56	7.0	7.25	877	0.0
0.5 G	4.9	3.8	No	August 1995	165	80	28.3	16	43	9.11	7380	0.003
0.5 G	4.9	3.8	Yes	August 1995	165	80	27.4	21	39	0	0	0.005
50 WP	4.9	3.8	No	July 1996	19	80	27.5	19	20	0.208	64.8	0.05
0.5 G	4.9	3.8	No	July 1996	19	80	29.0	16	41	0.51	394	0.02
0.5 G	4.9	3.8	No	August 1996	80	80	26.2	20	25	0.208	94.7	0.03
0.5 G	4.9	3.8	No	August 1996	80	80	28.6	12	51	0.76	713	0.03
0.5 G	4.9	7.6	No	July 1996	19	80	28.3	24	24	0.777	541	0.04
0.5 G	4.9	7.6	No	August 1996	80	80	28.2	19	34	0.485	298	0.01

[a]Natural rainfall recorded within seven days of simulated rainfall.

Table VII. Surface Runoff of Phosphate-Phosphorus from the 1995-96 Oklahoma State University Simulated Rainfall Experiments. Stillwater, OK.

Buffer Length	Buffer Height	Aerification	Date Applied	Natural Rainfall[a]	Simulated Rainfall	Soil Moisture	Time to Start of Runoff	Runoff Depth	Concentration	Mass	Percent of Total Applied
m	cm			mm	mm	(%)	min	mm	mg L^{-1}	mg	(%)
0	--	No	July 1995	0	83	17.0	56	7.2	9.57	911	1.9
0	--	No	August 1995	165	80	18.4	15	63	6.52	3600	8.2
0	--	No	August 1995	165	80	17.7	17	56	8.14	4610	10
0	--	No	July 1995	19	80	17.7	16	25	4.5	1050	2.4
0	--	No	August 1996	80	80	17.5	17	31	10.3	2870	6.5
1.2	3.8	No	July 1996	19	80	16.0	17	21	2.79	673	1.5
1.2	3.8	No	August 1996	80	80	16.7	16	31	7.02	2390	5.4
2.4	1.3	No	July 1995	0	74	27.2	21	12	2.36	471	1.0
2.4	1.3	No	August 1995	165	80	29.1	11	54	6.06	4190	9.5
2.4	3.8	No	July 1995	0	74	27.9	56	5.5	1.94	276	0.5
2.4	3.8	No	August 1995	165	80	27.9	13	39	4.83	2890	6.6
2.4	3.8	No	July 1996	19	80	28.0	21	25	1.52	551	1.3
2.4	3.8	No	August 1996	80	80	26.2	20	32	6.2	2630	6.0
4.9	1.3	Yes	July 1995	0	87	27.3	76	6.1	1.02	136	0.2
4.9	1.3	No	July 1995	0	83	27.3	51	3.0	0.78	37	0.0
4.9	1.3	No	August 1995	165	80	29.2	22	46	3.35	2670	6.1
4.9	1.3	Yes	August 1995	165	80	29.1	17	39	4.01	2770	6.4
4.9	1.3	No	July 1996	19	80	29.2	7	41	3.63	2800	6.4
4.9	1.3	No	August 1996	80	80	27.6	8	50	5.72	5180	12
4.9	3.8	No	July 1995	0	78	27.9	62	2.9	1.35	71	0.05
4.9	3.8	Yes	July 1995	0	78	26.2	56	7.0	1.25	246	0.4
4.9	3.8	No	August 1995	165	80	28.3	16	43	3.87	3030	6.9
4.9	3.8	Yes	August 1995	165	80	27.4	21	39	4.47	3060	7.0
4.9	3.8	No	July 1996	19	80	27.5	19	20	1.39	576	1.3
4.9	3.8	No	July 1996	19	80	29.0	16	41	4.24	3440	7.8
4.9	3.8	No	August 1996	80	80	26.2	20	25	5.57	2550	5.8
4.9	3.8	No	August 1996	80	80	28.6	12	51	6.72	6280	14
4.9	7.6	No	July 1996	19	80	28.3	24	24	1.54	720	1.6
4.9	7.6	No	August 1996	80	80	28.2	19	34	5.23	3600	8.2

[a]Natural rainfall recorded within seven days of simulated rainfall.

natural precipitation occurred during the simulated rainfall events. The antecedent soil moisture content of each rainfall simulation area ranged from 25 to 30% by weight. All rainfall simulation events lasted 75 min. The rainfall simulator applied 80 mm of water to each set-up and an average of 49 to 80% of the simulated rainfall water ran off during the events. Time to start of runoff and total runoff expressed on a volume basis for each treatment ranged from 13 to 22 min and 498 to 819 L, respectively. Runoff occurred as much as six times earlier and total runoff was as much as 16 times greater in August compared to July. Higher antecedent soil moisture in August resulted in greater runoff concentrations and losses of pesticides and nutrients compared to July.

Unlike numerical trends observed in the July simulated rainfall events, significant differences in runoff concentrations and losses for comparisons of buffer mowing height indicated reduced runoff losses from buffers mowed at 1.3 cm compared to 3.8 cm. Linde et al. (8) attributed reduced surface water runoff from creeping bentgrass to its stoloniferous growth habit and greater turf density when compared to perennial ryegrass. It is possible that reduced runoff losses from bermudagrass turf mowed at 1.3 cm compared to 3.8 cm were related to increased turf density that is typically achieved by a lower height of cut.

Conversely, the taller height occasionally reduced chemical losses in runoff. This occurrence resulted most likely by the obstruction the flow of water off the plots. As such, results comparing buffer mowing height effects on runoff were inconclusive and require further experimentation. Fewer significant differences were observed for comparisons between buffer length and no buffer on pesticide and nutrient runoff in August compared to July. Thus, it appeared that the effectiveness of the buffer treatments was lessened by antecedent soil moisture conditions and the increased volume of surface runoff. The buffer treatments did cause significant reduction of chlorpyrifos in runoff water compared to the treatments containing no buffer. Chlorpyrifos is more water insoluble and strongly adsorbed compared to the herbicides; therefore, it is likely that the buffer treatments would contribute to reduced chemical loss even when the conditions favoring runoff were great.

No significant differences were observed for comparisons involving aerification. Reduced pesticide and nutrient runoff concentrations occurred from the wettable powder formulation of chlorpyrifos compared to the granular formulation.

July 1996 Simulated Rainfall Event. The site received 19 mm of natural precipitation within 7 d of the July 1996 rainfall simulation events. The antecedent soil moisture content of each rainfall simulation area ranged from 26 to 29% by weight. The rainfall simulator applied 80 mm of water to each set-up and an average of 25 to 51% of the simulated rainfall water ran off during the events. Time to start of runoff and total runoff for each treatment ranged from 7 to 24 min and 20 to 41 mm, respectively.

August 1996 Simulated Rainfall Event. The site received 80 mm of natural precipitation within 7 d of the August 1996 rainfall simulation events. The antecedent soil moisture content of each rainfall simulation area ranged from 28 to 29% by weight. The rainfall simulator applied 80 mm of water to each set-up and an average of 31 to 64% of the simulated rainfall water ran off during the events. Time to start of runoff and total runoff expressed in mm for each treatment ranged from 8 to 20 min and 25 to 51 mm, respectively. Higher antecedent soil moisture conditions prior to simulated rainfall resulted in greater surface runoff in August compared to July.

Buffer Length Experiment. No significant differences were found among the treatments in terms of time to runoff or runoff depth (Baird et al., Michigan State University, unpublished data). Significantly higher concentrations of 2,4-D, mecoprop, chlorpyrifos (August only), and PO_4-P were recovered in surface runoff from the non-buffered treatments compared to buffer treatments. The same trends were observed for runoff losses of the aforementioned chemicals. Overall, very few significant differences were found among the 1.2, 2.4, and 4.9 m buffer lengths.

Buffer Mowing Height Experiment. The 7.6 cm buffer mowing height was effective in reducing time to runoff, runoff depth, concentration, and losses of the chemicals compared to the 1.3 and 3.8 cm mowing heights. The mean times until runoff in July and August, respectively, were 7 and 8 min for the 1.3 cm buffer, 16 and 12 min for the 3.8 cm buffer, and 24 and 19 min for the 7.6 cm buffer. Mean runoff losses in July and August, respectively, were 24 and 34 mm for the 7.6 cm buffer compared to 41 and 50 mm for the 1.3 cm buffer and 41 and 51 mm for the 3.8 cm buffer. It was apparent during the simulated rainfall events that the 7.6 cm buffer canopy created a barrier to slow down or intercept overland flow from the 1.3 cm treated area.

Results from 1995 and 1996 showed greater surface runoff from the granular formulation of chlorpyrifos compared to the wettable powder. In a review of pesticide losses due to runoff, Wauchope (9) noted several studies where the wettable powder formulations of herbicides showed the greatest long-term losses in runoff. Most of the studies reviewed were on agricultural soils and compared wettable powders to other sprayable formulations not including granules. Although wettable powder formulations may be subject to runoff, granular formulations may be subject to greater runoff losses, especially when a heavy precipitation event closely follows application. Reduced runoff losses of chlorpyrifos from the wettable powder formulation was most likely a result of its ability to become adsorbed and incorporated into turf more rapidly than the granular formulation.

Similar to pesticide formulation, reduced concentrations and losses of NH_4-N and NO_3-N in runoff water occurred from sulfur-coated urea compared to urea. The reduction in N lost with the SCU formulation can be explained by the slow breakdown of the coating in water, thus reducing the soluble N source available to enter runoff water. Dunigan et al. (26) and Brown et al. (27) reported reduced runoff losses of N from slow- compared to quick-release N fertilizers.

Percent losses of pesticides and nutrients in all experiments were as great as 15% and 14%, respectively (Tables I-VII). Highest chemical losses in runoff occurred from either non-buffered treatments or when heavy natural rainfall preceded simulated rainfall events. Asmussen et al. (14) reported 2.5% and 10.3% runoff losses of 2,4-D under dry and wet soil moisture conditions, respectively. Similar effects of soil moisture on nutrient runoff were reported by Moe et al. (28) and Morton et al. (6).

Time Series Analysis. The multiplicative structure of the fitted concentration models suggested that approximate estimated treatment ratios and associated significance levels could be obtained from the antilogs of estimated contrasts in the log scale. Observed times until start of runoff ranged from 12.5 min to 24 min for the buffered plots and from 9 min to 23 min for non-buffered plots. Consequently, inferences about concentration ratios were deemed appropriate beginning (for convenience) at 15 min. Estimated ratios were obtained for 5-min intervals thereafter until the conclusion of the simulated rain event and were tested at the 5% significance level. Prediction curves for 2,4-D, mecoprop, and PO_4-P are shown in Figure 3. The concentrations of dicamba, chlorpyrifos, and nitrogen in the runoff samples were very low, only slightly above the detection limits of the analysis method, and therefore were not included in the time series analysis. Significant ratios for 2,4-D ranged from 2078.7 times higher for non-buffered plots at 15 min to 3.2 times larger at 40 min (Wilson et al., Oklahoma State Univeristy, unpublished data). Subsequent ratios ranged from 1.7 at 55 min to 3.1 at the conclusion of the rain event but were not significant at the 5% level. The buffer was found to reduce and delay the onset of 2,4-D concentration in runoff, with a peak concentration of 40.8 μg L^{-1} occurring approximately 51 min after the start of rainfall.

Similar results were found for mecoprop. Significant ratios were determined at each 5-min interval through 40 min, decreasing from 1190.3 times more concentration for the no-buffer treatment at 15 min to 3.4 times higher at 40 min. Subsequent ratios again suggested that runoff concentrations for non-buffered plots were higher than for buffered plots, although none were significant at the 5% level. As in the case of 2,4-D, peak concentration for buffered plots was delayed from the onset of runoff until approximately 49 min, at which time a predicted level of 10.4 μg L^{-1} occurred, still less than the lowest predicted concentration of 15.8 μg L^{-1} for non-buffered plots.

Observed buffer effects were more extensive for P. Unlike 2,4-D and mecoprop, significantly higher ratios were found throughout the first 65 min of the experiment for non-buffered plots, ranging from 59.3 times larger at 15 min to 1.9 times larger at 65 min. According to the fitted polynomial model, peaks in runoff

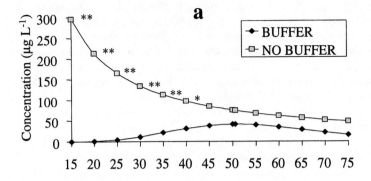

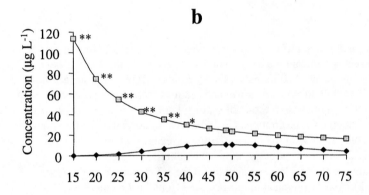

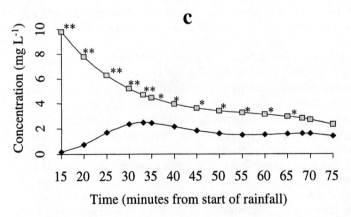

Time (minutes from start of rainfall)

Figure 3. Plots of predicted concentration versus time for: a) 2,4-D;
b) mecoprop; and c) phosphate-phosphorus. $**\alpha = 0.01$, $*\alpha = 0.05$.

concentration of 2.4 mg L^{-1} and 1.6 mg L^{-1} for the buffered plots were observed approximately 33 min and 68 min, respectively, after the onset of rainfall.

Analyses of differences in average predicted runoff losses (mass) for each of the contaminants showed no significant effect of the buffer, although the data suggested that the buffer strip was effective in reducing total contaminant runoff (data not shown). Experimental results suggest that the insignificance of each test can be attributed in part to the small number of set-ups used.

Cost considerations prohibit the examination of individual samples and promote less expensive volume-weighted composite assessments of buffer effect. Although the composite sampling method used in this research was appropriate for evaluation of the buffer treatments, individual sample analysis did provide a more complete picture on the effectiveness of the buffers in reducing surface runoff of chemicals.

Conclusions

These results suggest that turf is an effective filter of chemicals. Furthermore, use of turf buffers can significantly reduce chemical losses in runoff. Buffers help reduce surface runoff by: 1) diluting applied chemicals; 2) reducing surface flow velocity to reduce both the erosive power and sediment carrying capacity of runoff water; 3) providing a physical filtering of sediment or chemicals in solution; and 4) increasing potential for infiltration (17). Chemical dilution was an important factor in this study since addition of a buffer increased the plot area receiving simulated rainfall. The 3.8 cm buffer mowing height did not significantly contribute to reduced runoff losses compared to the 1.3 cm buffer. However, significant reductions in chemical runoff occurred from the 7.6 cm buffer. Aerification of the buffer did not significantly reduce surface runoff losses. Perhaps different aerification practices (e.g., hollow-tine), or application of these treatments on a larger scale watershed would cause a significant reduction in surface runoff of pesticides and nutrients.

The U. S. Environmental Protection Agency (EPA) has set Lifetime Health Advisory Levels in drinking water for dicamba, 2,4-D, mecoprop, chlorpyrifos, and nitrate in drinking water at 200 μg L^{-1}, 70 μg L^{-1}, 7 μg L^{-1}, 20 μg L^{-1}, and 10 mg L^{-1}, respectively (29). Typically, water features surrounding golf courses are not used for drinking purposes. However, by comparison in magnitude only, most all of the buffer treatments employed in the experiments reduced pesticide and nutrient concentrations in runoff to levels below EPA recommendations. The no buffer treatment and higher antecedent soil moisture conditions in August 1995 and 1996 increased pesticide and nutrient concentrations in runoff, especially 2,4-D and mecoprop, to levels exceeding EPA recommendations. Perhaps a better indication of the potential for contamination of surface waters on golf courses is the effect on aquatic organisms. For example, lethal concentrations (LC_{50}) found to kill 50% of bluegill, a common freshwater fish, are >2.4 μg L^{-1} for chlorpyrifos (30), >8 mg L^{-1} for 2,4-D (31), and >50 mg L^{-1} for dicamba (30). Most of the treatments containing buffers were effective in reducing pesticide concentrations below these critical values.

The concentration of nutrients in surface runoff from all treatments exceeded minimum levels that have been found to enhance eutrophication. However, in July 1995, surface runoff from the untreated control indicated that only a small amount of the nutrients lost in surface runoff, especially from treatments containing buffers, was from the applied nitrogen fertilizer. The correlation between the physico-chemical properties of pesticides and nutrients and their relative runoff potential was substantiated by this investigation. Greater runoff losses occurred from the herbicides and urea fertilizer which have higher water solubilities and lower adsorption compared to the insecticide and SCU fertilizer.

Based upon this investigation, the following management practices are recommended to reduce pesticide and nutrient runoff losses from turf: 1) establish a vegetation buffer between surface water resources and treated areas; 2) maintain bermudagrass buffers at a mowing height of at least 7.6 cm; 3) avoid application of pesticides and fertilizers when high soil moisture conditions exist; and 4) develop pest and nutrient management programs that utilize pesticide and fertilizer formulations with low runoff potential. Our results suggest that use of slow release N fertilizers, wettable powder instead of granular formulations of chlorpyrifos, and pesticides with lower water solubilities and stronger adsorption will reduce surface runoff potential.

Acknowledgments

This chapter reports research supported by a grant from the U.S. Golf Assoc., Far Hills, NJ. The authors wish to thank Dr. William Raun, Professor of Agronomy, Oklahoma State University, for insightful suggestions, and Steve Wilcoxen, Superintendent of Karsten Creek Golf Club, for use of a deep-tine aerator.

Literature Cited

1. Walker, W.J.; Branham, B. In *Golf Course Management and Construction: Environmental Issues*; Balogh, J.C.; Walker, W.J., Ed.; Lewis Publishers, Chelsea, MI, 1992, pp. 105-219.
2. Balogh, J.C.; Anderson, J.L. Environmental impacts of turfgrass pesticides. In *Golf Course Management and Construction: Environmental Issues*; Balogh, J.C.; Walker, W.J., Ed.; Lewis Publishers, Chelsea, MI, 1992, pp. 221-353.
3. Gross, C.M., Angle; J.S.; Welterlen, M.S. *J. Environ. Qual.* **1990**, 19(4), pp. 663-668.
4. Gross, C.M.; Angle, J.S.; Hill, R.L.;Welterlen, M.S. *J. Environ. Qual.* **1991**, 20(3), pp. 604-607.
5. Moe, P.G.; Mannering, J.V.; Johnson, C.B. *Soil Sci.* **1968**, 105(6), pp. 428-433.

6. Morton, T.G.; Gold, A.J.; Sullivan, W.M. *J. Environ. Qual.* **1998**, 17(1), pp.124-130.

7. Brown, K.W.; Duble, R.W.; Thomas, J.C. *Agron. J.* **1977**, 69(4), pp. 667-671.

8. Linde, D.T.; Watschke,T.L.; Jarrett, A.R.; Borger, J.A. *Agron. J.* **1995**, 87, pp. 176-182.

9. Wauchope, R.D. *J. Environ. Qual.* **1978**, 7(4), pp. 459-472.

10. Watschke, T.L.; Mumma, R.O. *A Final Report to the U.S. Dept. Int.*; U.S. Geol. Sur. ER 8904. Env. Res. Inst.; Penn State University, College Park, PA; 1989, 64 pp.

11. White, A.W.; Asmussen, L.E.; Hauser, E.W.; Turnbull, J.W. *J. Environ. Qual.* **1976**, 5, pp. 487-490.

12. Smith, A.E.; Bridges, D.C. *Crop Sci.* **1996**, 36, pp. 1439-1445.

13. Castelle, A.J.; Johnson, A.W.; Conolly, C. *J. Environ. Qual.* **1994**, 23(5), pp. 878-882.

14. Asmussen, L.E; White, A.W.; Hauser, E.W.; Sheridan, J.M. *J. Environ. Qual.* **1977**, 6(2), pp. 159-162.

15. Felsot, A.S.; Mitchell, J.K.; Kenimaer, A.L. *J. Environ. Qual.* **1990**, 19(3), pp. 539-545.

16. Chaubey, I., Edwards, D.R.; Daniel, T.C.; Moore, P.A. Jr.; Nichols, D.J. *Trans. ASAE* **1994**, 37(3), pp. 845-850.

17. Muscutt, A.D.; Harris, G.L.; Bailey, S.W.; Davies, D.B. *Agric. Ecosytems Environ.* **1993**, 45, pp. 59-77.

18. Dillaha, T.A.; Reneau, R.B.; Mostaghimi, S.; Lee, D. *Trans. ASAE* **1989**, 32(2), pp. 513-519.

19. Swanson, N.P. *Proceedings of the Rainfall Simulator Workshop*; Tucson, AZ; USDA-SEA Agricultural Reviews and Manuals ARM-W-10/July; 1979, pp. 166-169.

20. Di Corcia, A.; Marchetti, M. *Environ. Sci. Tech.* **1992**, 26, pp. 66-74.

21. Murphy, J.; Riley, J.P. *Anal. Chim. Acta.* **1962**, 27, pp. 31-36.

22. Lachat Instruments. *Quickchem method 12-107-04-1-B*; Lachat Instr., Milwaukee, WI; 1989.

23. Lachat Instruments. *Quickchem method 12-107-06-1-B*; Lachat Instr., Milwaukee, WI; 1990.

24. Littell, R.C.; Milliken, G.A.; Stroup, W.W.; Wolfinger, R.D. *SAS System for Mixed Models*; SAS Institute Inc., Cary, NC; 1996.

25. Cole, J.T.; Baird, J.H.; Basta, N.T.; Huhnke, R.L.; Storm, D.E.; Johnson, G.V.; Payton, M.E.; Smolen, M.D.; Martin, D.L.; Cole, J.C. *J. Environ. Qual.* **1997**, 26(6), pp. 1589-1598.

26. Dunigan, E.P.; Phelan, R.A.; Mondart, C.L. Jr. *J. Environ. Qual.* **1976**, 5(3), pp. 339-342.

27. Brown, K.W.; Thomas, J.C.; Duble, R.W. *Agron. J.* **1982**, 74(6), pp. 947-950.

28. Moe, P.G.; Mannering, J.V.; Johnson, C.B. *Soil Sci.* **1967**, 104(6), pp. 389-394.

29. Environmental Protection Agency. *Health Advisory Summaries*; U.S. Environmental Protection Agency, Office of Water, Washington, D.C.; 1989.

30. Johnson, W.W.; Finley, M.T. *Handbook of Acute Toxicity to Fish and Aquatic Invertebrates*; Resour. Publ. 137. Fish Wildl. Serv., U.S. Dept. Int., Washington, D.C.; 1980, 98 pp.

31. Hughes, J.S.; Davis, J.T. *Weeds* **1963**, 2(1), pp. 50-53.

Chapter 17

Hazard Evaluation and Management of Volatile and Dislodgeable Foliar Pesticide Residues following Application to Turfgrass

J. Marshall Clark[1], G. R. Roy[1], J. J. Doherty[1], A. S. Curtis[1], and R. J. Cooper[2]

[1]Massachusetts Pesticide Analysis Laboratory, Environmental Science Program, University of Massachusetts, Amherst, MA 01003
[2]Department of Crop Science, North Carolina State University, Raleigh, NC 27695

Volatilization can be a major route of pesticide loss following application to turfgrass. Consequently, a significant proportion of applied pesticides may be available for human exposure via volatile and dislodgeable foliar residues. Our research has established that there are volatile and dislodgeable residues available for golfer exposure following pesticide application to turfgrass and not all of these exposures can be deemed completely safe using the USEPA Hazard Quotient assessment. Of the 14 pesticides examined, 10 never resulted in an inhalation exposure situation that had a Hazard Quotient greater than 1.0. Five never resulted in a dermal exposure situation that resulted in a Hazard Quotient greater than 1.0 and after the first day following application, 9 had Hazard Quotients less that 1.0. Application of ethoprop, isazofos, diazinon and isofenphos, however, did result in Hazard Quotients greater than 1.0 over a period of 3 days post-application and hence the safety of exposure to these organophosphorous insecticides is less certain. We have evaluated the practical use of spray tank adjuvants, irrigation and the role of thatch accumulation on the dissipation of volatile and dislodgeable residues as means to mitigate the exposure potential of the organophosphorous insecticides that have high vapor pressures and inherent high toxicity. To date, the use of adjuvants and thatch management (aeration, dethatching) have not resulted in the attenuation of the exposure potential of these insecticides. However, the managed application of these insecticides in the combined presence of adjuvants and/or with post-application irrigation may hold some promise.

Volatilization may be defined as the loss of chemicals from surfaces in the vapor phase with the subsequent movement into the atmosphere (1). Post-application vaporization of pesticide residues was reported as early as 1946 (2) and numerous studies since then have established volatilization as a major avenue of pesticide loss following application (3). While pesticide volatilization following soil application has been fairly well documented, volatilization from plant canopies has been studied far less (4). A dense perennial ground cover, such as turfgrass, is quite different from a plowed field or field crop planting and is likely to provide a unique environment for volatilization. The high surface area associated with the dense turfgrass blades and the thatch layer, which retards the downward leaching of pesticides, is likely to result in a significant increase in volatile and dislodgeable pesticide residues (5, 6). Understanding the nature and magnitude of volatilization and dislodgeable foliar residues is important not only because of its impact on pesticide dissipation but also because of concerns regarding pesticide efficacy and human exposure via inhalation and dermal penetration. Unfortunately, research evaluating pesticide volatilization and dislodgeable foliar residues from turfgrass has been quite limited.

Volatilization of pesticide residues from plants under field conditions often exhibit marked diurnal fluctuations with maximum loss occurring at approximately solar noon (7-9). This diurnal variation is driven primarily by solar heating. During mid-afternoon, solar heating is at a maximum resulting in elevated surface temperatures and increased atmospheric turbulence. Prior to sunrise and after sunset, however, there is little insolation to elevate surface temperatures, thus resulting in minimal volatility during these periods (5).

Loss of pesticides by volatilization from foliage typically follows a diphasic decline with an initial rapid loss for approximately 1 week, followed by a period of much slower volatile loss (10, 9). The slower rate of volatile loss typically observed after the first week may be explained by two hypotheses (3). The first suggests that the remaining residues are less available because they lie deeper within the canopy and are trapped in irregular areas of leaves, stems and leaf-stem junctions. The second suggests that the latter residues are lost at a reduced rate because they are more strongly adsorbed or have penetrated the leaf surface. Both mechanisms contribute to volatile loss and the availability of dislodgeable foliar residues but the relative importance of each factor has not been determined (5).

While the benefits of pesticides in maintaining golf courses are clear, pesticide use also raises serious concerns regarding potential environmental pollution, human exposure risks and adverse impacts on wildlife. Substantial research has focused on assessing exposure of pesticide applicators (11) and harvesters reentering pesticide-treated crops (12). However, there is little research assessing golfer exposure to pesticides applied to golf course turfgrass.

During the golfing season, courses usually are open every day during the week, leaving little time between pesticide application and reentry into the treated area. Inhalation of volatile pesticides may be of toxicological concern given the high susceptibility of the human lung to airborne toxins, particularly those associated with aerosols. In addition, it has been shown frequently that dermal exposure of agricultural workers is related to the amount of pesticide present as dislodgeable foliar residues (13). The hands, arms and legs of golfers are often unprotected during play. The hands are likely the main route of dermal exposure since they are usually unprotected and are involved in a number of repetitive tasks that result in direct exposure to turf (e.g., picking up golf balls, repairing ball marks on greens, replacing divots in the fairway, cleaning club heads, etc). Thus, the potential for significant exposure to pesticides applied to golf courses certainly exists.

Because of the increasing environmental concern about pesticide use on golf courses and to the uncertainty of the extent of human exposure, we have conducted research for the past six years to develop means to assess the extent of pesticide volatilization and the availability of dislodgeable foliar residues following application to turfgrass. This chapter summarizes the exposure situations that we found, a hazard assessment of these exposures and preliminary findings on the effectiveness of selected processes to attenuate the exposure situations from pesticides of highest concern.

Small-Plot Techniques for Measuring Airborne and Dislodgeable Pesticide Residues following Application to Turfgrass

The volatile loss and the availability of dislodgeable foliar residues following application of pesticides to turfgrass has been examined using two small-plot techniques: The Field Chamber Method and the Theoretical Profile Shape Method.

Field Chamber Method. The initial experiments that first examined the volatile and foliar loss of pesticides following application to turfgrass were described in a series of papers co-authored by J.J. Jenkins and R.J. Cooper (14-16). Airborne and dislodgeable residues were determined following application of the dinitroaniline herbicide pendimethalin (N-(1-ethylpropyl)-2,6-dinitro-3,4-xylidine) to Kentucky bluegrass. In the first technique, airborne residues were measured using small field chambers consisting of 19 L Pyrex bottles without bottoms and fitted with Teflon cartridges containing XAD-4 polymeric polymer resin. Dislogeable foliar residues were determined from turfgrass plugs in which the grass blades were separated from soil and thatch and then extracted with methanol.

The focus of this initial research was to determine what relationship, if any, exists between dislodgeable foliar residues and pesticide volatility. If such a relationship exists, a mathematical model could be developed to estimate pesticide flux using only measurements of dislodgeable foliar residues, thereby eliminating the logistical and conceptual problems associated with estimating volatile loss using traditional micrometeorological or laboratory chamber methods. Also, the relationships between volatile loss and the physical parameters of temperature/solar radiation and windspeed could be investigated in a controlled and replicated fashion.

These studies found that it was possible to estimate total daily flux based on the diurnal pattern of volatility by making single flux measurements during the period of peak volatility (1300-1500 hrs). These authors concluded that there was a relationship between volatile loss and dislodgeable foliar residues as both flux and dislodgeable foliar residues exhibited similar biphasal patterns of decline over the course of the study. However, the measured values did not correlated well, suggesting that the relationship was more complicated than initially thought.

While the field chambers allowed the examination of what impacts different variables might play in the attenuation of volatile and dislodgeable residues, the modification of the chamber environment with regard to temperature and windspeed made it unclear whether the measured values of "chamber flux" actually reflected what was happening outside the chamber. It was concluded that the effects of increased temperature and decreased windspeed in the chambers were probably offsetting each other in this case, but that the chamber technique itself could not reasonably be expected to reflect environmental volatility in all cases. As such, the field chamber technique is not appropriate for studies that require an accurate determination of ambient pesticide air concentration (e.g. exposure studies) as opposed to an estimate of pesticide flux.

Theoretical Profile Shape Method. A second group of experiments was conducted to compare flux rates from the chambers with those obtained from an ambient method, the Theoretical Profile Shape (TPS) method, which is based on the trajectory-simulation (TS) model of Wilson et al., (17). Details of the TPS method and the TS model are given in Wilson et al., (17, 18) and Majewski et al., (19-21). Briefly, the TPS method employs the TS model, a 2-dimensional dispersion model, to estimate source strength $[F_z(0)]$ from a single measurement within the vertical profile of the horizontal flux at the center of a circular plot. Where:

$$F_z(0) = (uc)^{measured} / \Phi$$

$F_z(0)$ = source strength determined as the actual vertical flux rate (mg/m^2/hr)
$(uc)^{measured}$ = product of the measured windspeed and air concentration
Φ = normalized horizontal flux predicted by the TS model.

The measurement height (ZINST) is chosen based upon the plot radius, roughness length (z_0), and the Monin-Obukhov atmospheric stability length (L). In these experiments, the airborne residue and windspeed were measured at a height ZINST of 73 cm chosen according to a plot radius of 20 meters and a surface roughness length of 0.2 cm. Pendimethalin airborne residues were collected with a Staplex TF1A high volume air sampler onto XAD-4 polymeric resin. Dislodgeable foliar residues were determined as described for the field chamber experiments.

Previous work by Majewski and co-workers (19-21) has determined the TPS method to be comparable to other micrometeorological techniques for estimating evaporative flux of a number of pesticides from soil. In the case of pesticide application to turfgrass, the two small-plot techniques gave comparable results for pendimethalin flux. The important difference between the two methods is that while the TPS method has inherent error associated with the mathematical TS model used to calculate flux, its results are based on actual air concentrations, whereas the levels measured in the field chambers reflect an artificial environment and ultimately requires verification with an ambient technique. Additionally, the TPS method provides an accurate measure of pesticide air concentration, which can be used for purposes other than estimating flux, such as exposure estimates and hazard evaluations. The TPS method, likewise, uses plots small enough to allow for replication, which is a major limitation in the evaluation of information gathered using aerodynamic-based methods.

Fate of Volatile and Dislodgeable Residues following Pesticide Application to Turfgrass and Implications for Human Exposure

Estimation of Golfer Exposure and Hazard Assessment. We recently have addressed two significant concerns of the turfgrass industry (5, 6). First, we were interested in establishing the potential and the routes for golfer exposure to pesticides applied to turfgrass. Second, we wanted to prioritize those pesticides that may be of most concern following exposure.

Using a small circular plot design is conjunction with the TPS method, airborne pesticide concentrations were determined and used to estimate the inhalation and dermal exposure situations for golfers using the USEPA Hazard Quotient determination. An average daily inhaled dose of pesticide for a 70 kg adult playing a 4-hr round of golf was estimated by Eq. 1.

$$C \times R \times 4hr / 70 \text{ kg} = D_i \qquad \text{[Eq. 1]}$$

where C = measured air concentration of pesticide determined by high-volume air sampling ($\mu g \text{ m}^{-3}$), R = adult breathing rate during moderated activity (2.5 $m^3 \text{ h}^{-1}$, 22), and D_i = daily inhaled dose of pesticide ($\mu g \text{ kg}^{-1}$). The estimated inhaled dose (D_i) is divided by a chronic reference dose (Rfd, $\mu g \text{ kg}^{-1} \text{ d}^{-1}$, 23) resulting in the Inhalation Hazard Quotient (D_i / Rfd = IHQ).

Similarly, an average daily dermal dose was calculated using Eq. 2.

$$S \times P / 70 \text{ kg} \times 1000 \ \mu g \text{ mg}^{-1} = D_d \qquad \text{[Eq. 2]}$$

where S = calculated dermal exposure (mg). S is determined by multiplying the dislodgeable foliar residues determined from a cheesecloth wipe (24) by a dermal transfer coefficient of 5 x 10^3 cm h^{-1} (13), P = dermal permeability (0.1, USEPA default value, 25) and D_d = daily dermal dose of pesticide ($\mu g \text{ kg}^{-1}$). The estimated dermal dose (D_d) is divided by the chronic Rfd resulting in the Dermal Hazard Quotient (D_d / Rfd = DHQ).

Chronic Rfds were determined from daily doses shown to cause no observable effects on laboratory animals over their lifetime (NOELs) and then divided by safety factors of 10 to 10 000 depending on the completeness of the toxicological data set (23). Thus, Rfds are deemed to be safe doses that can be received over a lifetime without causing adverse effect. From this, HQs less than or equal to 1.0 indicate that the residues present are at concentrations below those that could cause adverse effects in humans. A HQ value greater than 1.0 does not necessarily infer the residues levels will cause adverse effects, but rather that the absence of adverse effects is less certain.

Hazard Assessment following Turfgrass Pesticide Applications. Using the above format, four pesticides were initially examined: Two insecticides, isazofos (O-5-chloro-1-isopropyl-H-1,2,4-triazol-3-yl O,O-diethyl phosphorothioate) and trichlorfon (dimethyl 2,2,2-trichloro-1-hydroxyethylphosphonate); a herbicide, MCPP (meoprop, (RS)-2-(4-chloro-o-tolyloxy) propionic acid); and a fungicide, triadimefon (1-(4-chlorophenoxy)-3,3-dimethyl-1(1H-1,2,4-triazol-1-yl) butanone). The hydrolytic degradative product of trichlorfon, DDVP (2,2-dichlorovinyl dimethyl phosphate), was also determined. The results of these experiment are summarized in Tables I and II.

As indicated in Tables III and IV, only the application of isazofos resulted in both IHQs and DHQs greater than 1.0 over the course of this study. Additionally, the hydrolytic degradative product of trichlorfon, DDVP, also resulted in a DHQ value greater than 1.0 (4.6, Table IV) but this only occurred in the combination with 1.3 cm of post-application irrigation. Using the USEPA Hazard Quotient criteria, it is apparent that exposure situations exist following application of pesticides to turfgrass, which cannot be deemed completely safe. Furthermore, the pesticides of most concern appear to be insecticides that have inherently high vapor pressure (high volatility) and relatively high inherent toxicity (low Rfds).

To validate these findings, a larger study was conducted that included 13 pesticides (plus DDVP) used extensively on turfgrass, which varied widely in terms of their vapor pressure, water solubility and inherent toxicity (Table V). Volatile and dislodgeable foliar residues were determined from small circular turfgrass plots and used to estimated IHQ and DHQ values exactly as described in the original studies (5, 6). For pesticide applications, the boom sprayer used in the initial studies was replaced with a Rogers Wind Foil[®] (Model GF 1500) equipped with a rubber skirt to retard spray drift. Methods for extraction and instrumental analysis of additional pesticides are given in Clark et al., (26). Post-application irrigation was reduced from 1.3 cm to 0.63 cm in this expanded study.

Of the 14 pesticides examined, only 3 (ethoprop, diazinon and isazofos) resulted in volatile residues at sufficiently high enough concentrations to produce IHQs greater than 1.0 over the entire time course of this study (Table VI). As in our initial study, all 3 pesticides were organophosphorous insecticides that have high vapor pressures (high volatility) and low Rfds (high inherent toxicity). As indicated in Table VII, 7 pesticides resulted in dislodgeable foliar residues at sufficiently high enough concentrations to produce DHQs greater than 1.0. Except for bendiocarb (a carbamate), all the rest were organophosphorous insecticides. Only ethoprop, isazofos, diazinon and isofenphos produced DHQs that exceeded 1.0 in the intervals beyond the first 24 hr period following application. Trichlorfon, chlorpyrifos and bendiocarb had DHQs only slightly above 1.0 and these fell below 1.0 after the first day following application. These results are consistent with our original findings and substantiate that there are exposure situations involving volatile and dislodgeable foliar residues following the application of selective pesticides (organophosphorous insecticides) to turfgrass that cannot be deemed completely safe using the USEPA HQ criteria. Additionally, increased hazard appears to be well correlated with pesticides that have high vapor pressures and low Rfd values and these characteristics

Table I. Summary of volatile and dislodgeable residues following triadimefon and MCPP application to turfgrass.

---

1. Less than 8% of applied triadimefon and less than 1% of applied MCPP were lost as measured volatile residues.

2. Nearly all of triadimefon's measured volatile residues were lost within the first two weeks following application.

3. Diurnal patterns of triadimefon volatility were observed. When surface temperature and solar radiation were greatest, volatile loss reached a maximum.

4. Dislodgeable residues of triadimefon and MCPP dissipated over time. By Day 5 post-application, dislodgeable residues of triadimefon and MCPP were less than 0.1% of applied compound.

5. Calculated HQs for triadimefon and MCPP from volatile and dislodgeable residues were below 1.0 for the entire 15-d experimental period.

6. Application of triadimefon and MCPP to turfgrass is not likely to result in a health hazard to golfers via volatile and dislodgeable residues. This reduced health hazard was due primarily to the low vapor pressures and relatively high Rfds associated with these two pesticides.

7. The vapor pressure, irrigation practices, water solubility and toxicity of a pesticide should be reviewed before application to golf courses.

Table II. Summary of volatile and dislodgeable residues following trichlorfon and isazofos application to turfgrass.

---

1. All insecticide applications resulted in less than 13% of applied compound being lost as measured volatile residues: trichlorfon/DDVP (12.7%), with irrigation (11.6%), isazofos (11.4%).

2. The large majority of detectable volatile residues were lost within the first week following application.

3. Diurnal patterns of volatility were observed. When surface temperature and solar radiation were greatest, volatile loss reached a maximum.

4. When irrigation followed application, volatile and dislodgeable residues increased on Days 2 and 3 compared to those residues on Day 1.

5. Irrigation enhanced the transformation of trichlorfon to DDVP, a more toxic insecticide, with maximum residues apparent at Days 2 and 3 post-application.

6. Isazofos volatile residues resulted in a calculated IHQ greater than 1 through Day 3 post-application.

7. DDVP and isazofos dislodgeable residues resulted in a DHQ greater than 1 through Days 2 and 3 following application.

Table III. Maximum air concentrations, doses, and inhalation hazard quotients (IHQ).

Insecticide	Maximum Air Concentration [a]	Dose	Rfd [b]	Hazard Quotient [c]
	(μg m^{-3})	(μg kg^{-1})	(μg kg^{-1}d^{-1})	(IHQ)
Trichlorfon				
without irrigation				
Day 3	1.5	0.2	125	0.002
with irrigation (1.3 cm)				
Day 2	1.5	0.2	125	0.002
DDVP				
without irrigation				
Day 3	0.9	0.1	0.5	0.2
Day 5	0.2	0.03	0.5	0.1
with irrigation (1.3 cm)				
Day 2	1.8	0.3	0.5	0.6
Day 3	0.9	0.1	0.5	0.2
Isazofos				
with irrigation (1.3 cm)				
Day 2	0.6	0.09	0.02	4.5
Day 3	0.2	0.03	0.02	1.5

[a] Maximum air concentration occuring after application and irrigation if applicable.
[b] Office of Pesticide Programs Reference Dose Tracking Report, January 1993.
[c] Hazard Quotient = Dose/Rfd.
(Reproduced with permission from ref. 5, copyright 1996, Crop Science).

Table IV. Maximum dermal exposures, doses, and dermal hazard quotients (DHQ).

Insecticide	Maximum Exposure [a]	Dose	Rfd [b]	Hazard Quotient [c]
	(mg)	($\mu g\ kg^{-1}$)	($\mu g\ kg^{-1}d^{-1}$)	(DHQ)
Trichlorfon				
without irrigation 3 h post-application	37	53	125	0.4
with irrigation (1.3 cm) Day 2	5.5	7.9	125	0.1
DDVP·				
without irrigation 3 h post-application	0.0	0.0	0.5	0.1
with irrigation (1.3. cm) Day 2	1.6	2.3	0.5	4.6
Isazofos				
with irrigation (1.3 cm) Day 2	0.2	0.34	0.02	17.1

[a] Estimated dermal exposure after application and irrigation if applicable using the model of Zweig et al., (1985).
[b] Office of Pesticide Programs Reference Dose Tracking Report, January 1993.
[c] Hazard Quotient = Dose/Rfd.
(Reproduced with permission from ref. 5, copyright 1996, Crop Science).

Table V. Physical and toxicological properties of turfgrass pesticides.

Pesticides	Molecular Weight [a]	Vapor Pressure [a] (mm Hg)	Water Solubility [a] (ppm)	NOEL [b] (mg/kg/day)	Uncertainty Factor [b]	OPP Rfd [b] (mg/kg/day)
Insecticides						
Organophosphorous						
DDVP	221.0	1.6×10^{-2}	8000	0.05	100	0.0005
Ethoprop	242.3	3.5×10^{-4}	750	0.015	1000	0.000015
Diazinon	304.4	9.0×10^{-5}	60	0.009	100	0.00009
Isazofos	313.7	5.6×10^{-5}	168	nd [c]	100	0.00002
Chlorpyrifos	350.6	2.0×10^{-5}	1.4	0.03	10	0.003
Trichlorofon	257.4	3.8×10^{-6}	120000	nd [c]	100	0.002
Isofenphos	345.4	3.3×10^{-6}	18	0.05	100	0.0005
Carbamates						
Bendiocarb	223.2	3.4×10^{-6}	280	0.5	100	0.005
Carbaryl	201.2	3.1×10^{-7}	120	1.43	100	0.014
Pyrethroids						
Cyfluthrin	434.3	2.0×10^{-9}	0.002	2.5	100	0.025
Contact Fungicide						
Chlorothalonil	265.9	5.7×10^{-7}	0.6	1.5	100	0.015
Systemic Fungicide						
Propiconizole	342.2	4.2×10^{-7}	100	1.25	100	0.0125
Thiophanate-Methyl	342.4	7.1×10^{-8}	< ppb	8.0	100	0.08
Iprodione	330.2	3.8×10^{-9}	13	6.1	100	0.061

[a] Molecular weights, vapor pressures, and water solubilities were taken from The Pesticide Manual, Tenth Edition (1994).
[b] Toxicological data and Rfd values were taken from USEPA the Office of Pesticide Programs Reference Dose Tracking Report, 4/14/95.
[c] No data available.

Table VI.   Inhalation hazard quotient (IHQs [a]) for turfgrass pesticides in the high (vapor pressures > 1.0×10^{-5} mm Hg), intermediate (vapor pressures between 1.0×10^{-5} mm Hg and 1.0×10^{-7} mm Hg), and low (vapor pressures < 1.0×10^{-7} mm Hg) vapor groups.

Pesticide [b]	Vapor Pressure (mm Hg)	OPP Rfd (mg/kg/day)	IHQs Day 1	IHQs Day 2	IHQs Day 3
High V.P.					
DDVP	1.6×10^{-2}	5.0×10^{-4}	0.06	0.04	0.02
Ethoprop	3.5×10^{-4}	1.5×10^{-6}	50	26	1.2
Diazinon	9.0×10^{-5}	9.0×10^{-5}	3.3	2.4	1.2
Isazofos	5.6×10^{-5}	2.0×10^{-5}	8.6	6.7	3.4
Chlorpyrifos	2.0×10^{-5}	3.0×10^{-3}	0.09	0.1	0.04
Intermediate V.P.					
Trichlorofon	3.8×10^{-6}	2.0×10^{-3}	0.02	0.004	0.004
Bendiocarb	3.4×10^{-6}	5.0×10^{-3}	0.02	0.002	0.002
Isofenphos	3.3×10^{-6}	5.0×10^{-4}	n/d	0.02	n/d
Chlorothalonil	5.7×10^{-7}	1.5×10^{-3}	0.001	0.001	0.0003
Propiconizole	4.2×10^{-7}	1.25×10^{-2}	n/d	n/d	n/d
Carbaryl	3.1×10^{-6}	1.4×10^{-2}	0.0005	0.0001	0.00004
Low V.P.					
Thiophanate-Methyl	7.1×10^{-8}	8.0×10^{-2}	n/d	n/d	n/d
Iprodione	3.8×10^{-9}	6.1×10^{-2}	n/d	n/d	n/d
Cyfluthrin	2.0×10^{-9}	2.5×10^{-2}	n/d	n/d	n/d

[a] The IHQs reported are the maximum daily IHQ measured, all of which occurred during the 11:00 a.m. to 3:00 p.m. sampling period.
[b] All pesticides were watered in following application using 0.63 cm post-application irrigation.
[c] n/d = not detected.

Table VII. Dermal hazard quotients (DHQs) for turfgrass pesticides listed with increasing Rfds from top to bottom through Day 3 post application.

Pesticide [a]	Rfd (mg/kg/day)	Day 1 (DHQs) 5 Hours	Day 1 (DHQs) 8 Hours	Day 2 (DHQs) 12:00 p.m.	Day 3 (DHQs) 12:00 p.m.
Ethoprop	0.000015	190	156	26	39
Isazofos	0.00002	135	112	18	24
Diazinon	0.00009	32	25	4.6	5.7
Isofenphos	0.0005	5.7	5.7	1.1	1.1
DDVP	0.0005	0.3	0.3	n/d [b]	n/d
Trichlorofon	0.002	0.8	1.1	0.9	0.5
Chlorpyrifos	0.003	2.3	1.8	0.3	0.4
Bendiocarb	0.005	0.7	1.1	0.7	0.09
Propiconizole	0.0125	0.3	0.02	0.05	0.02
Carbaryl	0.014	0.009	0.01	0.007	0.0002
Cyfluthrin	0.025	n/a [c]	n/a	n/a	n/a
Ipridione	0.061	0.03	0.03	0.04	0.03
Thiophanate-Methyl	0.08	n/a	n/a	n/a	n/a

[a] All pesticides were watered in immediately following application with 0.63 cm of post-application irrigation.
[b] n/d = not detected.
[c] n/a = not available.

may be useful in predicting hazard associated with other related pesticides not included in the present study.

It is also of interest that the volatile and dislodgeable foliar residues measured in the expanded study are generally greater than those measured in the initial study (compare IHQ and DHQ values in Tables III and IV versus VI and VII, respectively). This disparity is likely the result of two factors; the first is the use of a spray drift suppression skirt that resulted in more pesticide being applied to the turfgrass during applications and the second is the reduction in the amount of post-application irrigation used in the expanded study (1.3 to 0.63 cm). Careful examination of both features may provide the means to attenuate turfgrass residues and reduce the hazard associate with such exposures.

Effects of Post-Application Irrigation. Using the 1.3 cm level of post-application irrigation (Tables III and IV), two unwanted processes occurred. First, trichlorfon was converted into its more toxic, volatile and water-soluble metabolite, DDVP, at a higher level than with 0.63 cm of irrigation (Tables VI and VII). Second, the higher level of post-application irrigation seems to delay the appearance of the maximal level of hazard associated with volatile and dislodgeable foliar residues from the day of application (0.63 cm) to Days 2 and 3 following application (1.3 cm). In order to demonstrate this aspect independently of the conflicting factor of different application regimes, we applied isazofos with the boom sprayer used in the original study and compared the two different levels of post-application irrigation (Table VIII). To show the dramatic effect that post-application irrigation has on the initial levels of dislodgeable foliar residues, samples were collected 15 min following application but before irrigation. Both levels of irrigation greatly reduced the initial levels of dislodgeable foliar residues, (compare 15 min values to 3 h values), with the 1.3 cm level slightly more effective than the 0.63 cm level. After irrigation, however, the maximal hazard associated with isazofos occurred on Day 2 (DHQ = 17.1) in the presence of 1.3 cm of irrigation whereas the maximal hazard in the presence of 0.63 cm of irrigation occurred at 3 h post-application (DHQ = 12). These findings indicate that the judicial use of post-application irrigation in combination with managed spray volume and sprayer configurations may be an effective means to attenuate the hazards associated with exposures to volatile and dislodgeable foliar residues associated with pesticide-treated turfgrass.

Use of Adjuvants and Thatch Management to Reduce Volatile and Dislodgeable Pesticide Residues. In order to mitigate the exposure potential of the organophosphorous insecticides that have high vapor pressures and inherent high toxicity, we evaluated the practical use of spray tank adjuvants and the importance of thatch accumulation on the dissipation of volatile and dislodgeable foliar residues following application of these more problematic insecticides. Two adjuvants were examined in these preliminary studies: Aqua Gro-L$^®$, a non-ionic wetting agent/penetrant; and Exhalt 800$^®$, an encapsulating spreader/sticker. Because ethoprop consistently resulted in the highest IHQ and DHQ values, it was chosen as a problematic organophosphorous insecticide for study in the combined presence with adjuvants. It was applied at the established rate reported previously to our standard

Table VIII. Effect of post-application irrigation on dislodgeable foliar residues of isazofos.

Post Application Irrigation		Dislodgeable Residues ($\mu g/m^2$)	Dose (mg/kg)	Rfd (mg/kg/day)	DHQ (Dose/Rfd)
1.3 cm					
Day 1	15 min[a]	3,920	1.0×10^{-2}	0.00002	560.0
	3 h	20	5.7×10^{-5}	0.00002	2.86
	8 h	10	2.9×10^{-5}	0.00002	1.43
Day 2		120	3.4×10^{-4}	0.00002	17.14
Day 3		40	1.1×10^{-4}	0.00002	5.71
0.63 cm					
Day 1	15 min[a]	4,280	1.2×10^{-2}	0.00002	611.1
	3 h	84	2.4×10^{-4}	0.00002	12.04
	8 h	63	1.8×10^{-4}	0.00002	8.93
Day 2		61	1.7×10^{-4}	0.00002	8.65
Day 3		12	3.4×10^{-5}	0.00002	1.70

[a] Dislodgeable foliar residue samples were collected 15 min after application prior to irrigation.

small circular plots using a Rogers Wind Foil® sprayer. For those applications done using spray tank adjuvants, a 2 % v/v concentration of adjuvant was used. Volatile and dislodgeable foliar residues were collected, analyzed and IHQ and DHQ values calculated as before. As for previous experiments, pesticide applications were carried out at the University of Massachusetts Turfgrass Research Facility in South Deerfield, MA. A paired-plot experimental design was used in these comparisons with one group of plots being prepared on turf established in 1991, which was never aerated or dethatched (Mature Turf) and the other plots being prepared on turf established in 1995 (New Thatch). It was estimated by physical examination that the mature turf plots had a thatch layer that was approximately 3-times thicker than the more newly established turf plots. For Application #1, only ethoprop was applied to a mature turf plot whereas the new turf received ethoprop with the adjuvant. In Application # 2, the treatments were reversed with the new plots receiving only ethoprop and the mature plots receiving ethoprop plus adjuvant.

To evaluate the effects of thatch accumulation on the dissipation of volatile and dislodgeable foliar residues, two additional applications of ethoprop were made. In the first application, ethoprop was applied simultaneously to a mature turf plot and a new turf plot, both in the absence of any spray tank adjuvant. In the second, ethoprop was applied to a mature turf plot that has been dethatched by vericutting in two directions, and the results compared to results obtained from this same plot prior to dethatching.

Table IX presents the results of applying ethoprop to mature and more recently established turf plots in the presence and absence of the wetting agent, Aqua Gro®. In no instance did the addition of this adjuvant result in substantial reductions in the amount of volatile and dislodgeable foliar residues of ethoprop following its application as determined by IHQ and DHQ determinations. In fact, the addition of this adjuvant generally resulted in slight increases in the IHQ and DHQ estimations between treatments. Similar results were obtained using the spreader/sticker adjuvant, Exhalt 800® (data not shown). Under these experimental conditions, neither of the adjuvants tested appeared to reduce the exposure to volatile and dislodgeable foliar residues following the application of ethoprop or isofenphos (data not shown) as judged by IHQ and DHQ determinations, respectively.

Similarly, no substantial or consistent differences were found in levels of volatile or dislodgeable residues following the application of ethoprop or isofenphos to mature versus more recently established turf plots or to thatched versus dethatched turf plots (data not shown). These preliminary results indicated that thatch management may not be a meaningful approach to mitigating the exposure to volatile and dislodgeable residues following pesticide application to turfgrass.

Conclusions

The large majority of the turfgrass pesticides evaluated in this study were deemed safe using the USEPA Hazard Quotient criteria. Pesticides that were not deemed completely safe by these criteria were all organophosphorous insecticides with high

Table IX. Inhalation (IHQs) and dermal (DHQs) hazard quotients for ethoprop following application with and without the wetting agent (WA) Aqua-Gro^R to mature and newly established turfgrass.

| | ETHOPROP IHQs | | | |
| | Applications without WA | | Applications with WA | |
SAMPLE	Mature turf (#1) [a]	Mature turf (#1) [b]	New turf (#1)	New turf (#3)
Day 1				
9:00 – 11:00	99	34	116	51
11:00 – 15:00	72	27	123	32
15:00 – 19:00	32	10	63	13
Day 2				
9:00 – 11:00	6.0	5.3	41	17
11:00 – 15:00	5.1	4.7	26	11
15:00 – 19:00	c	2.4	c	4.9
Day 3				
9:00 – 11:00	0.6	3.1	0.4	3.2
11:00 – 15:00	0.2	5.2	1.0	3.4
15:00 – 19:00	0.5	1.9	0.7	1.9

| | ETHOPROP DHQs | | | |
| | Applications without WA | | Applications with WA | |
SAMPLE	Mature turf (#1)	Mature turf (#1)	New turf (#1)	New turf (#2)
Day 1				
2 hr post app.	57	25	69	49
5 hr post app.	35	56	35	56
Day 2				
12:00 p.m.	b/d	25	b/d	30
Day 3				
12:00 p.m.	b/d	b/d	b/d	b/d

[a] Application #1.
[b] Application #2.
[c] Sample period canceled.
[d] below detection limit.

vapor pressures and high inherent toxicity. Because effective organophosphorous and carbamate insecticide alternatives are available that do not share these problematic features, the use of ethoprop, isazofos and diazinon on turfgrass should be minimized and applied judiciously and only when a delayed reentry period is practical. Additionally, the use of wetting agents and spreader/sticker adjuvants and thatch management do not appear to be practical means to minimize the potential exposure to these pesticides once applied to turfgrass. Post-application irrigation and spray volume regimes, however, may merit additional consideration.

We have determined that selected pesticides, which possess high volatility and toxicity, may result in exposure situations that cannot be deemed completely safe as judged by the USEPA Hazard Quotient criteria. This assessment, however, must be viewed in terms of the assumptions that were used in making these estimations. In all instances, maximum pesticide concentrations were used for the entire 4 hour exposure period, maximum rates for pesticide applications were used, and dermal transfer coefficients and dermal penetration factors were taken from non-turfgrass situations that are likely to exceed those that would take place on a golf course. Because of this, we view such estimates as *worst case scenarios*. In order to more accurately predict the health implications of pesticide exposure on golfers, a relevant dosimetry/biomonitoring evaluation of golfers, playing golf on a golf course, needs to be carried out. With more accurate exposure estimates, it is our belief that the exposure levels reported here will be found to be in excess of the true exposure to pesticides on a golf course.

References

1. Spencer, W. F.; Cliath, M. M. Movement of pesticides from soil to the atmosphere. In *Long range transport of pesticides*; Kurtz, D. A. (ed.).; Lewis Publ.: Chelsea, MI, **1990**; pp. 1-16.

2. Staten, G. *Contamination of cotton fields by 2,4-D or hormone type weed sprays.* J. Am. Soc. Agron., **1946**, *38*:536-544.

3. Taylor, A. W. *Post-application volatilization of pesticides under field conditions.* J. Air Pollut. Control. Assoc., **1978**, *28*:922-927.

4. Cooper, R. J. Volatilization as an avenue for pesticide dissipation. In: Carrow, N. E. et al. (ed.); *Intl. Turfgrass Soc. Res.*; J., Intertec Publ. Corp.: Overland Park, KS, **1993**, *Vol. 7*; pp. 116-126.

5. Murphy, K. C.; Cooper, R. J.; Clark, J. M. *Volatile and dislodgeable residues following trichlorfon and isazofos application to turfgrass and implication for human exposure.* Crop Sci., **1996**, *36*:1446-1454.

6. Murphy, K. C.; Cooper, R. J.; Clark, J. M. *Volatile and dislodgeable residues following triadimefon and MCPP application to turfgrass and implication for human exposure.* Crop Sci., **1996**, *36*:1455-1461.

7. Cooper, R. J.; Jenkins, J. J.; Curtis, A. S. *Pendimethalin volatility following application to turfgrass.* J. Environ. Qual., **1990**, *19*:508-513.

8. Taylor, A. W.; Glotfelty, D. E.; Glass, B. L.; Freeman, H. P.; Edwards, W. M. *Volatilization of dieldrin and heptachlor from a maize field.* J. Agric. Food Chem., **1976**, *24*:625-631.

312

9. Taylor, A. W.; Glotfelty, D. E.; Turner, B. C.; Silver, R. E.; Freeman, H. P.; Weiss, A. *Volatilization of dieldrin and heptachlor residues from field vegetation.* J. Agric. Food Chem., **1977**, *25*:542-548.

10. Spencer, W. F.; Farmer, W. J.; Cliath, M. M. *Pesticide volatilization.* Residue Reviews, **1973**, *49*:1-47.

11. Fenske, R. *Nonuniform dermal deposition patterns during occupational exposure to pesticides.* Arch. Env. Contam. Toxicol., **1990**, *19*:332-337.

12. Knaak, J.; Iwata, Y. The safe levels concept and the rapid field method. In: Plimmer, J. (ed.) *Pesticide residue exposure*; ACS Sympositum Series 182; Am. Chem. Soc.: Washington, D.C., **1982**; pp. 23-39.

13. Zweig, G. Leffingwell, J. T.; Popendorf, W. *The relationship between dermal pesticide exposure by fruit harvesters and dislodgeable foliar residues.* Environ. Health, **1985**, *B20(1)*:27-59.

14. Jenkins, J. J.; Cooper, R. J.; Curtis, A. S. In: Kurtz, D. A. (ed.); *Long range transport of pesticides*; Lewis Publ.: Chelsea, MI, **1990**; pp. 29-46.

15. Jenkins, J. J.; Cooper, R. J.; Curtis, A. S. Bull. Environ. Contam. Tox. **1991**, *47*; pp. 594-601.

16. Jenkins, J. J.; Cooper, R. J.; Curtis, A. S. Two small-plot techniques for measuring airborne and dislodgeable residues of pendimethalin following application to turfgrass. In: Racke, K.D; Leslie, A. R. (eds.); *Pesticides in urban environments*; ACS Symposium Series 522; Am. Chem. Soc.: Washington, D.C., **1993**; pp. 28-242.

17. Wilson, J. D.; Thurtell, G. W.; Kidd, G. E.; Beauchamp, E.G. *Estimation of the rate of gaseous mass transfer from a surface source plot to the atmosphere.* Atmos. Environ., **1982**, *6:*1861-1867.

18. Wilson, J. D.; Catchpoole, V. R.; Denmead, O. T.; Thurtell, G. W. *Verification of a simple micrometeorological method for estimating the rate of gaseous mass transfer from the ground to the atmosphere.* Agric. Meteorol., **1983**, *29*:183-189.

19. Majewski, M. S.; Glotfelty, D. E.; Seiber, J. N. Atmos. Environ. **1989**, *23(5)*: 929-938.

20. Majewski, M. S.; Glotfelty, D. E.; Paw, K. T.; Seiber, J. N. Sci. Technol. **1990**, *24*: 1490-1497.

21. Majewski, M. S.; McChesney, M. M.; Seiber, J. N. Environ. Toxicol. Chem. **1991**, *10*: 301-311.

22. U.S. Environmental Protection Agency. *Exposure factors handbook.* Appendix 3. Office of Health and Environmental Assessment: Washington, D.C. **1989**.

23. U.S. Environmental Protection Agency. *Office of Pesticide Programs Reference Dose Tracking Report.* Washington, D.C. **1993**.

24. Thompson, D. G.; Stephenson, G. R.; Sears, M.K. *Persistence, distribution and dislodgeable residues of 2,4-D following its application to turfgrass.* Pest. Sci. **1984**. *15*:353-360.

25. U.S. Environmental Protection Agency. *Dermal exposure assessment: principles and applications.* Interim Report. EPA/600/8-91/011B. **1992**.

26. Clark, J. M. *Evaluation of management factors affecting volatile loss and dislodgeable foliar residues.* USGA, 1996 Turfgrass and Environmental Research Summary. USGA: Far Hills, NJ, 1996; pp. 60-63.

BIOTECHNOLOGY AND ALTERNATIVE PEST MANAGEMENT

Chapter 18

In Vitro Screening and Recovery of *Rhizoctonia solani* Resistant Creeping Bentgrass: Evaluation of Three In Vitro Bioassays

M. Tomaso-Peterson[1,3], A. Sri Vanguri[2], and J. V. Krans[2]

Departments of [1]Entomology and Plant Pathology and [2]Plant and Soil Sciences, Mississippi State University, Mississippi State, MS 39762

Over the past decade there has been increased environmental awareness of fungicide use on golf courses. A need has been identified within alternative pest management programs for rapid *in vitro* bioassays to screen and select turfgrasses with enhanced resistance to pathogenic fungi. The Host-Pathogen Interaction System (HPIS), fungal extracts, and direct fungal inoculation are *in vitro* bioassays evaluated for rapid screening and selection of *Rhizoctonia solani* resistance in creeping bentgrass. Penncross callus that survived co-cultures with *R. solani* regenerated reduced numbers of plantlets. A reduction in tissue necrosis was observed in all *in vitro* bioassays when *R. solani*-selected plantlets were screened with *R. solani* isolates compared to non-selected plantlets.

Rhizoctonia blight of creeping bentgrass (*Agrostis stoloniferous* var. *palustris* Huds.), incited by *Rhizoctonia solani* Kühn, has been a persistant turfgrass problem since the disease was first documented in the early 1920s (*6*). In cool season grasses such as creeping bentgrass, the disease is more prevalent during summer months. Initial symptoms in bentgrass turf appear as small rings (10-15 cm dia), bluish-gray in color. Necrotic lesions develop on leaf blades, and as they coalesce, blades become totally necrotic. As the disease progresses throughout the turf, brown patches up to 50 cm in diameter become evident. Rhizoctonia blight can be controlled by cultural practices including mowing height, fertility, and irrigation management, supplemented with an effective fungicide program.

'Penncross' creeping bentgrass set the standard as the premier putting green turf in the1950s. This is attributed to high density, dark color, aggressive growth habit, and uniformity in a putting green surface. However, Penncross is susceptible to *R. solani*. At least seven cultivars have been developed from Penncross using traditional breeding methods to improve turf density, color, and overall quality, while increasing tolerance to brown patch (*26*).

---

[3]Current address: Box 9655, Mississippi State, MS 39762.

Years of research and funding are spent developing new turfgrass cultivars. Traditional breeding methods include selection, re-selection, plot tests, seed production, and finally turf evaluations at selected sites before a cultivar is commercially released. A preliminary step that can increase efficiency in selecting improved plant types is *in vitro* screening. The foundation of *in vitro* screening is based on somaclonal variation. Somaclonal variation accounts for the phenotypic and genotypic variations that occur when plants are cultured *in vitro* and has been exploited as a tool to recover improved plant types (*12*). This selection tool is used to eliminate cells that cannot withstand the selection pressure. Many horticultural and agronomically important crops have improved disease resistance via *in vitro* selection (*1,9,19*). *In vitro* selection for enhanced disease resistance may employ direct (fungal inoculum) or indirect (culture filtrates, fungal extracts, or purified toxins) methods for selection (*7,13,15,18,22,25*). The objective of this research was to develop and evaluate direct and indirect *in vitro* bioassays for rapid screening of Rhizoctonia blight resistance in creeping bentgrass.

Materials and Methods

The host plant used throughout the bioassays was 'Penncross' creeping bentgrass. Penncross callus was obtained according to procedures described by Krans et al. (*11*). Isolates of *R. solani* were maintained on potato dextrose agar (PDA) (39 g L^{-1}, Sigma, St. Louis, MO) in the dark at 25°C.

Host-Pathogen Interaction System (HPIS). The HPIS is an *in vitro* cell selection bioassay developed primarily for increasing the efficiency of recovering callus that can survive the presence of a pathogenic fungus. The HPIS consists of a double-sided, two compartment, Lutri-plate modified for use with callus and a pathogenic organism (*28*). Compartment A was adapted to meet the requirements for pathogen growth. A gas exchange port was made in the lid of compartment A by cutting a hole using a heated #3 cork borer. A sterile, 13-mm glass fiber disk was sealed over the hole using a nontoxic glue. This port facilitated movement of gases out of the compartment once sealed and prevented the entrance of contaminants. Thirty-seven milliliters of sterile water agar (20 g agar L^{-1}, Difco, Detroit, MI) were poured into compartment A and allowed to solidify. Sterile bentgrass leaf blades were placed on the water agar surface, serving as a natural substrate for the fungus. This medium was inoculated with a 3-mm PDA plug of *R. solani* and incubated in the HPIS for 7 d in the dark at 25°C. The lid of compartment A was sealed with a caulking cord material to prevent the fungus from growing out of the compartment. Modifications for the host compartment B were accomplished by removing the false bottom of compartment A. A plastic grid was incorporated into the Lutri-plate for support of the medium once the false bottom was removed. A sterile, 86-mm membrane (0.2 μm pore) was sealed in place using nontoxic glue on the bottom side of the water agar-plastic grid. A sterile tissue culture medium containing 3 mg L^{-1} 2,4-dichlorophenoxyacetic acid (2,4-D) (*2*) was then poured into compartment B of the HPIS. Once the tissue culture medium was solidified, the membrane was sandwiched between the growth media, serving as a physical barrier, restricting the fungus to compartment A.

316

Complete assembly of the HPIS was carried out in a laminar flow hood to maintain sterility.

Callus and plantlet recovery. Penncross callus was co-cultured (period in which callus and fungus were simultaneously growing in the HPIS) with *R. solani* to determine if surviving callus could be recovered and maintain plantlet regeneration capabilities. Penncross stock callus was obtained according to callus induction and maintenance procedures described by Krans et al. (*11*). An 86-mm glass fiber support disc was placed on the top surface of the tissue culture medium contained in compartment B of the HPIS. Stock callus was plated onto discs at the rate of 0.3 g, with callus aggregates ranging between 0.25-0.5 mm (2). Compartment A of the HPIS, the pathogen side, contained either a highly virulent isolate (RVPI) or a weakly virulent isolate (R12) of *R. solani* as a treatment. Based on preliminary HPIS experiments, (data not shown), Penncross callus was co-cultured with RVPI or R12 for 12, 24, 36, or 48 h. Controls were callus populations plated into uninoculated HPIS. Following each co-culture period, callus populations were transferred from HPIS via glass fiber support discs to either determine callus viability or plantlet regeneration. A 2 x 2 factorial completely random experimental design was used, including twelve replicated plates per treatment; six replicates per treatment in callus viability determinations and six in plantlet regeneration.

Callus viability was determined following co-culture periods to quantify survival. This was accomplished by treating callus populations with 3 ml of a 2% 2,3,5 triphenyl tetrazolium chloride (TZ) vital stain. To measure callus viability, a transparent, circular grid (47-mm diam.) comprised of 50 random sites (2 mm^2) was designed to fit on the cover of a Petri dish (90-mm diam.). The grid-cover was placed over a Petri dish containing a co-cultured callus population. An initial count of callus tissue in random sites was determined and then treated with 3 ml TZ. Following a 12 h incubation in the dark, a second survey of the callus populations was performed to determine the number of stained callus in the random sites. Callus viability was calculated using the formula:

$$\% \text{ viable callus} = \frac{\text{no. stained callus sites}}{\text{initial callus count}} \times 100$$

Plantlet regeneration was evaluated on callus that had co-cultured 12, 24, 36, or 48 h with *R. solani* isolates in the HPIS. Following each co-culture period, callus populations were allowed to regenerate in plantlet regeneration medium that was similar to the tissue culture medium used in the HPIS but without 2,4-D. To induce plantlet regeneration, callus was maintained under continuous cool white flourescent light (41 μmol m^{-2}s^{-1}) at 25°C. The number of plantlets was recorded at the two-leaf stage. Callus viability and plantlet regeneration were analyzed using analysis of variance procedure of Statistical Analysis Systems (SAS), mean separation was carried out using Fisher's least significant difference (P$\leq$ 0.05) test (*23*).

Screening of *R. solani* selected and non-selected Penncross plantlets. The HPIS was used to recover isolate RVPI-selected and non-selected plantlets. To obtain RVPI-selected plantlets, Penncross stock callus was plated in the HPIS as previously described and co-cultured 24 h with isolate RVPI. Following co-culture, callus populations were transferred to regeneration medium and cultured for plantlet regeneration as previously described. Non-selected plants were derived from callus populations co-cultured in the HPIS without fungus. Selected and non-selected treatments were replicated 5 times.

All plantlets were transferred and maintained (subcultured every 5 wks) in Magenta boxes containing 25 ml regeneration medium under the same culture conditions as mentioned above. Selected and non-selected plantlets served as stock material in the evaluation of the three *in vitro* bioassays.

The HPIS was assembled as previously described with isolate RVPI growing in the pathogen compartment A. Six segments 20 to 30-mm long consisting of a single node, were harvested from selected and non-selected plantlets. Node sections were placed into the host compartment B of the HPIS and co-cultured 15 d under the same environmental conditions described for plantlet regeneration. Disease was rated at the end of the co-culture period using a visual scale of 0 to 5; where 0 = no necrosis, and 5 = total necrosis. Controls developed to establish a baseline for necrosis included nodes that were incubated in the HPIS with no fungus present. A completely random design included nodal tissue of RVPI-selected and non-selected plantlets replicated 30 times. Disease ratings were analyzed using analysis of variance procedure of SAS and comparisons were determined by an F-test (*23*).

Extract Bioassay. An extract bioassay consisted of screening RVPI-selected and non-selected stock plantlets, based on plant response to extracts obtained from liquid cultures inoculated with *R. solani* isolate RVPI.

Culture preparation. Treatments consisted of liquid media with leaf tissue, shredded wheat, or potato added as the carbon source. Isolate RVPI was the inoculum source, and liquid cultures were incubated 21 d at 25°C under continuous cool white fluorescent lights. The control contained sterile, distilled water. Liquid cultures consisted of Murashige and Skoog's basal medium (MS) (*17*) including 2.2 g salts, 30 g oven-dried, finely ground bentgrass leaves, 30 g finely ground shredded wheat, or 24 g potato dextrose broth (PDB) incorporated into one liter of distilled water. The pH was adjusted to 5.8, and the medium was autoclaved at 121°C at 100 psi for 15 min. After cooling, 1 ml of MS vitamins was added to media except that containing PDB. The media were inoculated with one mycelial plug each from a culture of isolate RVPI growing on PDA.

Extraction procedure. The extraction procedure, as reported by Mikel (*16*), was used to obtain the initial extracts. Liquid cultures were extracted by thoroughly blending the contents then neutralizing with sodium acetate or sodium hydroxide. Liquid cultures were extracted by initial mixing with 0.1% ethyl acetate (ratio of the liquid culture volumes to ethyl acetate, 1:3) in separatory funnels with agitation to facilitate extraction. A liquid and solid phase resulted, and the liquid phase was subjected to three additional extractions using equal volumes (1:1) of ethyl acetate. The final extract was dry-filtered over anhydrous sodium sulfate and absorbent cotton in a Buchner funnel placed in a vacuum-adapted flask. The precipitate was further dried in a rotation evaporator at 55°C. Precipitates were dissolved in 10 µl of acetone and distilled water containing a 0.1% surfactant to give a final volume of 1 mg extract ml $^{-1}$. Solutions were then sonicated 15-20 min in a Fisher Scientific FS 15 Sonicator to promote dissolving of extracts.

Bioassay. Extract evaluations were conducted on RVPI-selected or non-selected plantlets. A 24 well (20 x 30 mm) tissue culture plate contained a filter paper disc (20-mm-dia) that supported a single node harvested from each plantlet source. Two hundred microliter aliquots of each treatment extract was dispensed into wells containing nodal tissue. Saturated filter paper facilitated prolonged exposure of nodes to the extracts. The plates were covered with transparent lids to minimize evaporation of extracts. Selected and non-selected nodes were incubated with extract treatments in the light at 25°C for 6 days. Following incubation, disease was evaluated using a scale of 0 to 5 where 0 = no necrosis and 5 = total necrotic tissue. A completely random experimental design included 30 replicates per treatment for each node source. Disease ratings of RVPI-selected and non-selected nodes were analyzed using analysis of variance procedure of SAS and comparisons were determined by an F-test (*23*).

Direct Bioassay. RVPI-selected and non-selected plantlets were exposed to *R. solani* isolate RVPI by direct inoculation. Stem segments, 20 to 30 mm long, consisting of a single node were harvested from each plantlet source. Nodal tissue was placed onto tissue culture medium previously described in this chapter. Each node was inoculated with a plug of RVPI mycelium by placing the plug directly adjacent to the node. Inoculated tissues were incubated 15 d in continuous cool white flourescent light at 25°C. Controls consisted of a node inoculated with a PDA plug, with no fungus present. Following incubation, disease ratings were recorded based on a scale of 0 to 5 where 0 = no necrosis and 5 = total necrosis. A completely random experimental design was used including 30 replicates for each node source. Disease ratings were analyzed using analysis of variance procedure of SAS and comparisons of injury was determined by an F-test (*23*).

Results and Discussion

Callus and plantlet recovery. Recovery of Penncross callus inoculated with isolate RVPI was significantly reduced (P≤ 0.05) over all co-culture periods (46%) as compared with either R12 (86%) or the uninoculated control (87%). Callus viability was significantly reduced after each period when co-cultured with RVPI compared to R12 or uninoculated control treatments (Table I).

Callus viability in RVPI treatments decreased significantly, with a reduction of 75% observed across the 12 to 48 h period of co-culturing (Table I). As co-culture periods increased with RVPI, callus viability decreased in comparison to callus viability in R12 (10%) and uninoculated control (2%) treatments (Table I). RVPI may produce fungal exudates that exert a certain level of selection pressure on the callus. Callus viability in treatments with R12 that were co-cultured for 48 h was significantly reduced compared to the 12 h co-culture (Table I). Callus viability decreased over an extended exposure period to the weakly virulent isolate. An increased co-culture time may be required to

Table I. Percentage of viable bentgrass callus recovered from co-cultures with *R. solani* isolates in the Host-Pathogen Interaction System.

Treatment[†]	12 h[‡]	24 h	36 h	48 h
	---------- % viability ----------			
RVPI	81A[§] a[¶]	40A b	42A b	20A c
R12	95B a	74B c	87B ab	85B b
Control	92B a	80B b	87B ab	90B a

[†]RVPI = highly virulent and R12 = weakly virulent *R. solani* isolates; control = no fungus.

[‡]Co-culture periods.

[§]Callus viability within co-culture periods followed by the same capital letter do not differ according to Fisher's least significant difference (P≤ 0.05) test.

[¶]Callus viability within treatments followed by the same lower case letter do not differ according to Fisher's least significant difference (P≤ 0.05) test.

accumulate inhibitory levels of fungal exudates in the HPIS for weakly virulent isolates. Callus viability in uninoculated control treatments were similar across co-culture periods except at 24 h due to poor vigor of stock callus (Table I). Exposing callus at a level of undifferentiation to fungal exudates produced by a virulent pathogen has been shown to result in successful recovery of plants with enhanced disease resistance (*10,24*). The HPIS bioassay permits transfer of fungal exudates to a callus culture during concurrent growth (*5*). RVPI significantly reduced callus viability, and the number of regenerated plantlets was significantly reduced as compared to the uninoculated control (Figure 1).

Penncoss callus co-cultured with RVPI up to 36 h maintained the ability to regenerate plantlets at a low frequency. Results of similar *in vitro* selection studies have demonstrated a decreased plantlet regeneration frequency when the selection pressure threshold was high (*3*). There were no differences in callus vigor and plantlet regeneration among treatments at the 24 h co-culture period.

Screening of *R. solani* selected and non-selected Penncross plantlets. RVPI-selected and non-selected plantlets were screened with RVPI in the HPIS to determine whether desirable variants could be recovered with enhanced fungal tolerance. Following 15 d co-culture with RVPI, nodal sections of RVPI-selected and non-selected plantlets were rated for necrosis. Control nodes had no necrosis and were not included in the statistical analysis for comparison. RVPI-selected nodes resulted in significantly less necrosis compared to non-selected nodes (Table II).

Based on these results, it was concluded that plantlets derived from RVPI-selected callus were less sensitive to fungal exudates than non-selected plantlets. Additional research must be conducted to definitively demonstrate HPIS can be used as a rapid *in vitro* bioassay for recovering plants with enhanced resistance to *R. solani*. It has been demonstrated by other researchers that HPIS can be used as a rapid *in vitro* bioassay for determining deleterious effects of fungal exudates on seedlings of leguminous species (*5,20*) and detecting deleterious rhizobacteria for potential biological controls of leafy spurge (*Euphorbia esula* L.) (*27*).

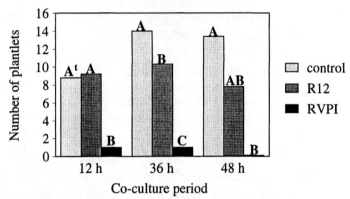

Figure 1. Number of plantlets of creeping bentgrass regenerated from callus populations co-cultured 12 to 48 h with *Rhizoctonia solani* isolate RVPI, R12, or uninoculated control in the Host-Pathogen Interaction System.

ᵗPlantlet means within a co-culture period followed by the same letter do not differ according to Fisher's least significant difference (P ≤ 0.05) test.

Table II. Tissue necrosis ratings of RVPI-selected and non-selected nodes of creeping bentgrass following a 15 day co-culture with *Rhizoctonia solani* isolate RVPI in the Host-Pathogen Interaction System

Nodal source[†]	Necrosis rating[‡]
RVPI-selected	2.8
Non-selected	4.4*

[†]RVPI-selected nodes harvested from plantlets regenerated from callus co-cultured 24 h with RVPI in the HPIS. Non-selected nodes harvested from plantlets regenerated from uninoculated callus co-cultured in the HPIS.
[‡]Rating based on a visual scale of 0 to 5; where 0 = no necrosis and 5 = total necrosis.
*Means differ significantly at P ≤ 0.05.

Extract Bioassay. Extracts of liquid cultures of *R. solani* were used to screen fungal tolerance in RVPI- selected and non-selected nodal tissues. Nodes were incubated with three RVPI-derived extracts for 6 d. RVPI-selected nodes displayed significantly less necrosis within each extract treatment compared to nodes from non-selected plantlets (Table III).

Table III. Tissue necrosis ratings of RVPI-selected and non-selected plantlet nodes of creeping bentgrass following a six day exposure to RVPI-derived extracts

Extract[†]	Nodal source[‡]	Necrosis rating[§]
Bentgrass leaf tissue	RVPI-selected	1.5
	Non-selected	2.9*
Shredded wheat	RVPI-selected	0.8
	Non-selected	2.2*
Potato dextrose	RVPI-selected	0.7
	Non-selected	3.4*

[†]Extracts derived from liquid cultures of *Rhizoctonia solani* containing ground bentgrass leaves, ground shredded wheat, or potato dextrose as the carbon source.
[‡]RVPI-selected nodes harvested from plantlets regenerated from callus co-cultured 24 h with RVPI in the HPIS. Non-selected nodes harvested from plantlets regenerated from uninoculated callus co-cultured in the HPIS.
[§]Rating based on a visual scale of 0 to 5; where 0 = no necrosis and 5 = total necrosis.
*Within extracts, non-selected nodes differ significantly at P ≤ 0.05 compared to RVPI-selected nodes.

Control nodes did not manifest necrosis and were not included in the statistical analysis of comparisons. Based on results of the extract bioassay, fungal extracts may be employed as a selection agent in an *in vitro* bioassay to successfully discriminate between resistance and susceptibility among plants. Several researchers (*4,13,14*) have successfully used toxin extracts or culture filtrates in bioassays to recover plants with improved resistance to pathogens.

322

Direct Bioassay. Direct inoculation of nodal tissues harvested from RVPI-selected and non-selected plantlets was conducted to determine whether this method was an effective rapid *in vitro* bioassay. Nodal tissue from RVPI-selected plantlets displayed significantly less necrosis compared to nodes of non-selected plantlets (Table IV).

Table IV. Tissue necrosis ratings of RVPI-selected and non-selected nodes following a 15 day direct inoculation/incubation with RVPI.

Nodal source[†]	Necrosis rating[‡]
RVPI-selected	2.7
Non-selected	4.3*

[†]RVPI-selected nodes harvested from plantlets regenerated from callus co-cultured 24 h with RVPI in the HPIS. Non-selected nodes harvested from plantlets regenerated from uninoculated callus co-cultured in the HPIS.
[‡]Rating based on a visual scale of 0 to 5; where 0 = no injury and 5 = total necrosis.
*Means differ significantly at $P \leq 0.05$.

Control nodes displayed no necrosis due to harvesting. The direct inoculation approach was effective in the differentiation of RVPI-selected and non-selected nodes. However, this method was considered the least desirable among the three *in vitro* bioassays. Fungal mycelium, growing saprophytically throughout the culture plates, obscured the nodes and made visual assessment difficult.

Summary

Environmental awareness over the past decade has increased concern over fungicide use in golf course environments. In 1991, the United States Golf Association initiated funding for research projects aimed at several environmental aspects in a golf course setting, including alternative pest management (*8*). A need has been identified within alternative pest management for rapid *in vitro* bioassays to select and screen turfgrasses with enhanced resistance to pathogenic fungi. Three *in vitro* bioassays discussed in this chapter were used to discriminated between pathogen-selected and non-selected plantlets. RVPI-selected plantlets were derived from callus exposed to isolate RVPI at the level of undifferentiation. A reduction in plant necrosis was observed in all *in vitro* bioassays when RVPI-selected nodes were screened with the pathogen. Based on these results, *in vitro* bioassays can be used to select and screen for enhanced resistance to *R. solani* in creeping bentgrass.

References

1.   Behnke, M. Selection of potato callus for resistance to culture filtrates of *Phytophthora infestans* and regeneration of resistant plants. Theor. Appl. Genet. **1979**, 55:69-71.

2.   Blanche, F.C., J.V. Krans, and G.E. Coats. Improvement in callus growth and plantlet formation in creeping bentgrass. Crop Sci. **1986**, 26:1245-1248.

3. Bolick, M., B. Foroughi-Wehr, F. Kohler, R. Schuchmann, and G. Wenzel. *In vitro* selection for disease resistance in potato and barley. *In* Nuclear techniques and *in vitro* culture for plant improvement. International Atomic Energy Agency. Vienna, Austria. 1986. pp. 275-285.

4. Bouharmont, J. Application of somaclonal variation and *in vitro* selection for plant improvement. Acta-Hortic. **1994**, 355:213-218.

5. Callahan, F.E. and D.E. Rowe. Use of a host-pathogen interaction system to test whether oxalic acid is the sole pathogenic determinant in the exudate of *Sclerotinia trifoliorum*. Phytopathology **1991**, 81:1546-1550.

6. Couch, H.B. Diseases of Turfgrasses. Krieger Publishing Company: Malabar, FL, 1995, 3 ed.; Chapter 1, pp. 3.

7. Daly, J.M.and H.W. Knoche. Advances in Plant Pathology. Academic Press: NY, NY, 1982: pp. 83.

8. Environmental Research Summary. United States Golf Association Golf House. 1991. Far Hills, NJ.

9. Gengenbach, B.G. and C.E. Green. Selection of T-cytooplasm maize callus cultures resistant to *Helminthosporium maydis* race T pathotoxin. Crop Sci. **1975**, 15:645-649.

10. Hammerschlag, F., D. Ritchie, D. Werner, G. Hashmil, L. Krusberg, R. Meyer, and R. Huettel. *In vitro* selection of disease resistance in fruit trees. Acta-Hortic. **1995**, 392:19-26.

11. Krans, J.V., V.T. Henning, and K.C. Torres. Callus induction, mantenance, and plantlet regeneration in creeping bentgrass. Crop Sci. **1982**, 22:1193-1197.

12. Larkin, P.J. and W.R. Scowcroft. Somaclonal variation and eyespot toxin tolerance in sugarcane. Plant Cell Tiss. Organ Cult.. **1983**, 2:111-121.

13. Ling, H.D., P. Vidyaseheran, E.S. Borromeo, F.J. Zapata, and T.W. Mew. *In vitro* screening of rice germplasm for resistance to brown spot disease using phytotoxin. Theor. Appl. Genet. **1985**, 71:133-135.

14. Malepszy, S. and A. El-kazza. *In vitro* culture of *Cucumis sativus* XI. Selection of resistance to *Fusarium oxysporum*. Acta-Hortic. **1990**, 280:455-458.

15. McCoy, R.J. and C.M. Kraft. Comparison of techniques and inoculum sources in evaluating pea (*Pisum sativum*) for resistance to stem rot caused by *Rhizoctonia solani*. Plant Dis. **1984**, 68:53-55.

16. Mikel, R.J. Microbial metabolism studies of praziquantel. M.S. Thesis. University of Mississippi, MS. 1987.

17. Murashige, T. and F. Skoog. A revised medium for rapid plant growth and bioassays with tobacco tissue cultures. Physiol Plant. **1962**, 15:473-497.

18. Nyange, N.E., B. Williamson, R.J. McNichol, G.D. Lyon, and C.A. Hackett. *In vitro* selection of *Coffea arabica* callus for resistance to partially purified phytotoxic culture filtrates from *Colletotrichum kahawae*. Ann. Appl. Biol. **1995**, 127:425-439.

19. Pauly, M.H., W.W. Shane, and B.G. Gengenbach. Selection for bacterial blight pathotoxin resistance in wheat tissue culture. Crop Sci. **1987**, 27:340-344.

20. Rowe, D.E. Oxalic acid effects in exudates of *Sclerotinia trifoliorum* and *S. sclerotiorum* and potential use in selection. Crop Sci. **1993**, 33:1146-1149.

22. Sacristan, M.D. and F. Hoffman. Direct infection of embryogenic tissue cultures of haploid *Brassica napus* with resting spores of *Plasmodiophora brassicae*. Theor. Appl. Genet. **1979**, 54(3):129-131.

23. SAS Institute Inc. SAS/STAT user's guide, release 6.03. Cary, NC. 1997.

24. Song, H.S., S.M. Lim, and J.M. Widholm. Selection and regeneration of soybeans resistant to the pathotoxic culture filtrates of *Septoria glycines*. Phytopathology. **1994**, 84:948-951.

25. Taylor, R.J. and G.A. Secor. Potato protoplast derived callus tissue challenged with *Erwinia carotovora* sub. sp. *carotovora*: Survival, growth, and identification of selected callus lines. Phytopathology. **1990**, 129:228-236.

26. The New Penn Pals. Tee-2-Green Corporation. Hubbard, OR. 1996.

27. Thouraya, S. and R.J. Kremer. Leafy spurge (*Euphorbia esula*) cell cultures for screening deleterious rhizobacteria. Weed Sci. **1994**, 42:310-315.

28. Tomaso-Peterson, M. and J.V. Krans. Evaluation of a new *in vitro* cell selection technique. Crop Sci. **1990**, 30:226-229.

Chapter 19

Development of Improved Turfgrass with Herbicide Resistance and Enhanced Disease Resistance through Transformation

F. C. Belanger, C. Laramore, S. Bonos, W. A. Meyer, and P. R. Day

Rutgers University, 59 Dudley Road, New Brunswick, NJ 08901

Genetic engineering of creeping bentgrass has the potential to provide alternative pest management approaches for golf courses. Bentgrass transformation is very efficient so transgenic plants can readily be produced for evaluation of promising genes. The effectiveness of genetically engineered herbicide resistance in creeping bentgrass has been demonstrated in multiple field tests. This trait is now ready to be incorporated into a commercial cultivar. This would allow the substitution of a nonselective herbicide with low environmental impact for herbicides with greater soil longevity and higher chance to contaminate groundwater. Transgenic bentgrass plants containing several potential disease resistance genes are currently being field tested. If any of these genes proves effective this would result in a reduction in the fungicide use required to maintain bentgrass.

There is great interest in improving the pest management options for creeping bentgrass through a combination of biotechnology and plant breeding. Two major categories of pests encountered on golf courses are weeds and diseases. There is significant potential for genetic engineering to have an impact on both of these problems. Our goal is to provide golf course managers with herbicide resistant cultivars for more effective weed control, and disease resistant cultivars which can be maintained in a more environmentally sound and cost-effective manner.

Bentgrass Transformation

We have a highly efficient transformation system for creeping bentgrass using particle bombardment (*1*). We use embryogenic callus as our target tissue. Our efficiency is such that we routinely obtain multiple independent transformed plants from each bombardment event. We currently have many independently transformed lines containing a herbicide resistance gene and potential disease resistance genes. These plants are being field tested.

The development of an improved cultivar of creeping bentgrass using genetic engineering will be more complex than the production of a transformed plant. Rather it will require the integration of the biotechnology program with the breeding program. Creeping bentgrass is a primarily self-incompatible cross-pollinating species. Cultivars are a population of genotypes produced by intercrossing several superior genotypes. Both plant fertility and stable expression of a transgene in the progeny will therefore be required for commercial application of genetic engineering in bentgrass. Following field evaluation of the original transgenic plants, those displaying the best turf qualities will be crossed with the most advanced bentgrass germplasm from the Rutgers breeding program. It will require at least two or more cycles of recurrent phenotypic selection and evaluation before commercial varieties can be developed.

Weed Control

An approach to weed control that has proved successful in a number of crop species is the engineering of resistance to nonselective herbicides, such as glyphosate (Roundup) or glufosinate (Finale). Resistance genes for both of these herbicides are known. Crop species, such as corn, soybean, and canola, engineered to express one or the other of these genes are commercially available. Fields of such engineered crops can thus be sprayed with the appropriate herbicide resulting in broad spectrum weed control and no damage to the crop. Both of these herbicides are very effective and considered to be environmentally benign. Both are rapidly degraded after application by soil bacteria and both are non-toxic to mammals (2-3). Because of the benefits of low environmental impact and broad spectrum effectiveness, genetic engineering of crops for resistance to these herbicides has received considerable attention.

This approach to weed control in bentgrass has tremendous potential. Bentgrass is grown as a single species and currently there is no effective method of controlling *Poa annua* which is a serious weed problem. It is thus an ideal candidate for the genetic engineering approach. The ability to spray a bentgrass green planted with herbicide-resistant transgenic plants with one of the nonselective herbicides would offer a convenient and effective method of weed control. This approach obviously requires the routine use of an herbicide. Because of the positive attributes of both glyphosate and bialaphos, there is little concern that their use will negatively affect the environment. An additional benefit of the use of herbicide resistant bentgrass for control of *Poa annua* would be the reduction in the amount of insecticide used to control the annual bluegrass weevil (*Listronotus maculicollis*).

The Rutgers turfgrass biotechnology program has produced herbicide-resistant creeping bentgrass by transformation with the *bar* gene, which confers resistance to glufosinate (4-5). These transformed plants were field tested and shown to be resistant at 1X and 3X the normal application rate (6). Additionally, the transformed plants were shown to be fertile and the resistance gene was transmitted to the progeny as would be expected for a nuclear encoded gene (6). Both the effectiveness of the herbicide resistance gene and the fertility of the original transformants was thus established. These results illustrate the potential of genetic engineering in providing a new approach to weed control in bentgrass.

For this approach to be commercially useful, transformed plants expressing either the glyphosate or glufosinate resistance genes would need to be incorporated into a breeding program for development of a cultivar with overall good turfgrass qualities, in addition to the herbicide resistance trait.

Disease Resistance

Another major pest problem encountered on golf courses is disease control. Bentgrass is highly susceptible to a large number of fungal diseases and generally requires considerable fungicide use (7). Golf courses throughout the country are under pressure to reduce their inputs of fungicides. A number of alternative approaches to disease control are being investigated, such as management strategies, biocontrol, and cultivar improvement through breeding. It is likely that such an integrated pest management (IPM) strategy to disease (8) will ultimately provide the best control. We are hopeful that genetic engineering may also contribute to an IPM approach.

We are specifically investigating the potential of genetic engineering in providing some disease control. The production of transgenic creeping bentgrass cultivars with enhanced disease resistance could help in reducing dependence on chemicals with potentially adverse environmental impacts. Unlike herbicide resistance, where two extremely effective resistance genes are known which function in virtually any plant species, disease resistance genes with such broad applicability are not yet known. There are, however, a number of genes which have shown promising results in some species. We are working with some of these to see if they may be useful in creeping bentgrass. We consider it wise to work with several genes at the same time since we do not know which, if any, of the genes will be most effective. Our transformation system is efficient, so we can readily obtain transgenic plants containing the genes of interest. Plants containing transgenes giving the best effect will be chosen for continuing in the breeding program. As new beneficial genes are identified in other species, we plan to incorporate them into our program. We have a number of independent transgenic lines expressing some potential disease resistance genes and we are currently at the stage of evaluating their effectiveness.

We do not expect any one gene to produce complete disease resistance or stress tolerance. Because the current use of fungicides is so high, however, genes which can confer measurable improvements in disease resistance will be extremely valuable in turfgrass maintenance.

Bacterio-opsin. Bacterio-opsin is a proton pump protein from the bacterium *Halobacterium halobium*. Mittler et al. (9) reported that expression of bacterio-opsin in tobacco protected the plants from viral and bacterial pathogens. Transgenic plants expressing bacterio-opsin were able to block the replication of tobacco mosaic virus and thus exhibited fewer symptoms of infection (9). Transgenic plants also prevented disease symptoms and growth of the bacterial pathogen *Pseudomonas syringae* pv *tabaci* (9). The mechanism of bacterio-opsin induced pathogen resistance may be through activation of the plant defenses since pathogenesis-related proteins are constitutively expressed in the bacterio-opsin expressing transgenic tobacco (9).

Expression of bacterio-opsin in potato resulted in dramatic resistance to the US1 isolate of the fungal pathogen which causes late blight disease, *Phytophthora infestans* (10). The plants were not, however, resistant to a more aggressive isolate (US8) of the pathogen (10). The degree of resistance conferred by a particular transgene is likely to vary depending on the strain or race of a pathogen. However, since bacterio-opsin expression can confer resistance to pathogens as diverse as fungi, viruses, and bacteria it is a good candidate gene to confer broad spectrum resistance to turfgrass diseases.

Pokeweed Antiviral Protein. The pokeweed antiviral protein (PAP) is a ribosome-inactivating protein from the plant *Phytolacca americana*. PAP expression in transgenic tobacco confers broad spectrum resistance to several plant viruses (*11*). Like bacterio-opsin, the mechanism of PAP-induced viral resistance may also be through activation of the plant defenses, since pathogenesis-related proteins are induced (*12*).

Expression of the wild type form of PAP in transgenic tobacco was toxic to the plants (*11*). A C-terminal deletion of the PAP coding sequence has reduced plant toxicity when expressed in transgenic plants, yet maintains the antiviral activity (*13*). Expression of the C-terminal deletion in tobacco also shows protection against the fungal pathogen *Rhizoctonia solani* (*12*). PAP, thus is another good candidate gene for inducing broad spectrum pathogen resistance in turf.

Glucose Oxidase. Glucose oxidase is an active oxygen species-generating enzyme from the fungus *Aspergillus niger* (*14*). It acts on the substrates glucose and oxygen yielding gluconic acid and hydrogen peroxide. Wu et al. (*15*) found that expression of glucose oxidase in potato resulted in resistance to the bacterial pathogen *Erwinia carotovora* subsp. *carotovora* and the fungal pathogen *Phytophthora infestans*.

The mechanism of action of glucose oxidase may be two-fold. Hydrogen peroxide, a product of glucose oxidase, is itself toxic to many microbial pathogens. In fact, glucose oxidase was found to be the active agent from the biocontrol fungus *Talaromyces flavus* (*16*). Hydrogen peroxide may thus directly inhibit invading organisms. Hydrogen peroxide also activates the plant defenses, inducing systemic acquired resistance (*17*) and expression of pathogenesis-related proteins (*18*). Glucose oxidase is thus another good candidate gene for inducing broad spectrum pathogen resistance in turfgrass

Field Test. In the summer of 1997 we established a field trial of some of our bentgrass transformants expressing the genes described above. In order to evaluate the plants under their normal use conditions they were maintained as mowed spaced plants. Our preliminary data obtained in the fall of 1997, based on natural dollar spot infection in the field, was promising. A number of transgenic lines showed significantly less disease than the controls.

We are currently pursuing more rigorous field evaluations based on inoculation with the pathogen. Any transgenic lines exhibiting enhanced disease resistance, relative to the controls, will be selected for the breeding program.

Summary

In summary, we feel that biotechnology, in combination with conventional breeding, has considerable potential for improving pest management options for creeping bentgrass. From experiences with other crops and from the field tests of transgenic bentgrass, genetically engineered herbicide resistance is likely to be a successful pest management strategy for weed control. This would allow the substitution of a nonselective herbicide with low environmental impact for herbicides with greater soil longevity and higher chance to contaminate groundwater.

Biotechnology may also contribute to disease control. If successful this would result in a reduction in the fungicide use required to maintain bentgrass. We currently have many plants transformed with candidate genes for disease resistance.

We will be evaluating the efficacy of the genes and incorporating the best plants into a breeding program for cultivar improvement.

Acknowledgments

This work was supported by the United States Golf Assoiation.

Literature Cited

1. Klein, T.M.; Fromm, M.; Weissinger, A.; Tomes, D.; Schaaf, S.; Sletten, M.; Sanford, J.C. *Proc. Natl. Acad. Sci. USA* **1988**, *85*, 4305-4309.
2. Franz, J.E. In*The Herbicide Glyphosate*; Grossman, E. and Atkinson, D., Eds.; Discovery, development and chemistry of glyphosate. Butterworths: London, **1985**. pp 3-17.
3. Kishore, G.M.; Shah, D.M. *Ann. Rev. Biochem.* **1988**, *57*, 627-663.
4. Hartman, C.; Lee, L.; Day, P.; Tumer, N. *Bio/technology*, **1994**, *12*, 919-923.
5. Lee, L.; Laramore, C.; Day, P.; Tumer, N. *Crop Sci.* **1996**, *36*, 401-406.
6. Lee, L.; Laramore, C.; Hartman, C.L.; Yang, L.; Funk, C.R.; Grande, J.; Murphy, J.A.; Johnston, S.A.; Majek, B.A.; Tumer, N.E.; Day, P.R. *International Turfgrass Society Research Journal* **1997**, 8:337-344.
7. Vargas, J.M. Jr. *Management of Turfgrass Diseases*, second edition. CRC Press: Boca Raton, FL, **1994**.
8. Schumann, G.L.; Vittum, P.J.; Elliott, M.L.; Cobb, P.P. *IPM Handbook for Golf Courses*. Ann Arbor Press, Inc.: Chelsea, MI, **1998**; pp 123-149.
9. Mittler, R.; Shulaev, V.; Lam, E. *Plant Cell* **1995**, *7*, 29-42.
10. Abad, M.S.; Hakimi, S.M.; Kaniewski, W.; Rommens, C.M.T.; Shulaev, V.; Lam, E.; Shah, D.M. *Molecular Plant Microbe Interactions* **1997**, *10*, 635-645.
11. Lodge, J.K.; Kaniewski, W.K.; Tumer, N.E. *Proc. Natl. Acad. Sci. USA* **1993**, *90*, 7089-7093.
12. Zoubenko, O.; Uckun, F.; Hur, Y.; Chet, I.; Tumer, N. *Nature Biotechnology* **1997**, *15*, 992-996.
13. Hur, Y.; Hwang, D-J.; Zoubenko, O.; Coetzer, C.; Uckun, F.M.; Tumer, N.E. *Proc. Natl. Acad. Sci. USA* **1995**, *92*, 8448-8452.
14. Frederick, K.R.; Tung, J.; Emerick, R.S.; Masiarz, F.R.; Chamberlain, S.H.; Vasavada, A.; Rosenberg, S.; Chakraborty, S.; Schopter, L.M.; Massey, V. *J. Biol. Chem.* **1990**, *265*, 3793-3802.
15. Wu, G.; Shortt, B.J.; Lawrence, E.B.; Levine, E.B.; Fitzsimmons, K.C.; Shah, D.M. *Plant Cell* **1995**, *7*, 1357-1368.
16. Kim, K.K.; Fravel, D.R.; Papavizas, G.C. *Phytopathology* **1988**, *78*, 488-492.
17. Chen, Z.; Silva, H.; Klessig, D.F. *Science* **1993**, *162*, 1883-1886.
18. Klessig, D.F.; Malamy, J. *Plant Mol. Biol.* **1994**, *26*, 1439-1458.

Chapter 20

The Use of Transgenic Plants to Confer Resistance to Brown Patch Caused by *Rhizoctonia solani* in *Agrostis palustris*

D. E. Green II[1], J. M. Vargas, Jr.[1], Benli Chai[2], N. M. Dykema[1], and M. Sticklen[2]

Departments of [1]Botany and Plant Pathology and [2]Crop and Soil Sciences and the Pesticide Research Center, Michigan State University, East Lansing, MI 48824

Turfgrass diseases are of major importance on most golf courses. More fungicides are applied to golf courses than all the other pesticides combined. It is important to find alternative methods to control turfgrass disease on golf courses. Improving disease resistance by insertion of resistance genes is one such alternative. An expression cassette containing a chitinase gene has been transferred into *Agrostis palustris*. Eleven transgenic lines of *A. palustris* were screened for their susceptibility to *R. solani* under controlled environmental conditions. Transgenic lines 711 and 9603 had significantly greater *Rhizoctonia* resistance compared to their parental cultivar, Penncross, providing a 20-80% reduction in disease severity.

Turfgrass diseases are serious problems in the management of turfgrass and can be the most important limiting factor in its successful establishment and growth (*1*). This is especially true on the golf course where high quality turf is required. Some of the most economically important turfgrass diseases are caused by fungal pathogens such as *Rhizoctonia solani* Kühn, *Sclerotinia homoeocarpa* F.T. Bennett, and *Pythium aphanidermatum* (Edson) Fitzp. the casual agents of brown patch, dollar spot, and Pythium blight, respectively (*1-3*). Preventative applications of chemical fungicides are the most reliable method of controlling these and other fungal pathogens. Although many efforts are being made by the government and chemical manufacturers to ensure the safest products possible, fungicides create certain risks to both humans and the environment. Alternative methods of disease control are needed to reduce the increasing expense, public concern, and risks associated with these chemicals.

Brown patch, or Rhizoctonia blight, is a serious disease of most turfgrass species grown in the United States (*2*). Recommended methods for control of brown patch include use of resistant cultivars, low nitrogen fertilization, limitation of long periods of leaf wetness (i.e. avoidance of shade, poor drainage, and late afternoon and evening irrigation), and applications of fungicides (*2-3*). Piper and Coe (*4*) recognized host resistance as one of the most promising methods to control brown patch in turf. Although numerous studies have been conducted to screen turfgrass species and cultivars for resistance to *R. solani*, only moderate levels of resistance have been identified (*5-7*). Unacceptable levels of brown patch frequently develop on even the most resistant turfgrass cultivars when maintained under conditions necessary to obtain the high quality turf required in golf course management.

Transformation of turfgrass species using techniques to insert genes from the DNA of one organism into another is a potential means of increasing resistance in turfgrass to fungal pathogens such as *R. solani*. Several studies have demonstrated the successful transformation of turfgrass species with genes conferring antibiotic and herbicide resistance or expression of the selectable marker β-glucuronidase (GUS)(*8-14*). Factors associated with resistance have been identified in many crops (*15-16*). One group of these resistance factors has been labeled pathogenesis response (PR) proteins because of their association with increased host resistance (*16*). Chitinases are one group of PR proteins found in plants and are believed to be involved in plant defense (*16-17*). Chitinase catalyzes the hydrolysis of chitin and has been shown to inhibit fungal growth *in vitro* causing lysis of the hyphal tips (*18-20*). Chitin, a linear β-1,4-linked N-acetylglucosamine polymer, is a major component of fungal cell walls but is not naturally found in plants (*21*). It is believed that induction of chitinase within the plant during biotic and abiotic stress, such as fungal infection, is one of a multitude of plant defense responses.

Numerous chitinases have been identified within plants, ranging from acidic extracellular to basic vacuolar forms (*17, 21-22*). Plants naturally express chitinase, generally encoding several different forms of these proteins within their genome. Although plants produce several forms of chitinase, the class I basic vacuolar chitinases have been shown to have the greatest activity against fungi (*20*). Because of their induction by abiotic as well as biotic stress factors, the role of chitinases in plant defense is not yet clear. One possible role in plant defense by this group of proteins may be direct inhibition of fungal pathogens by lysis of infecting hyphal tips (*17, 21, 23*). Toyoda et al. (*24*) found chitinase from *Streptomyces griseus* caused lysis of primary and secondary haustoria of the powdery mildew pathogen *Blumeria graminis* (DC.) E.O. Speer f.sp. *hordei* Marchal when micro-injected into infected barley (*Hordeum vulgare* L.). Another possible role in plant defense for plant chitinases may be their elicitation of other plant defense responses such as lignin deposition within the plant cell wall (*17, 21, 23*). Some of the extracellular forms of chitinases have been shown to release chitin oligomers by partial digestion of the fungal cell wall. These carbohydrate oligomers are believed to be active elicitors of plant defense responses (*17, 21, 23*).

Regardless of their exact function, constitutive expression of chitinase genes has the potential to increase resistance in turfgrass species. Transgenic plants containing constitutively expressed chitinase genes have been shown to increase resistance to fungal pathogens by as much as 50% in some crops (*25-33*). However, other studies found that constitutive expression of chitinase within transgenic plants did not improve resistance to fungal pathogens (*28,31, 34-35*). This variation in the effect of chitinase on host resistance is believed to result from a complex interaction between the host and the fungal pathogen. Variations in the form and site of insertion of the chitinase gene and interactions of the gene product with other host proteins have been suggested as possible host factors influencing the efficacy of the chitinase protein (*28, 31*). Composition of the fungal cell wall has also been suggested as a factor influencing efficacy of chitinase in transgenic plants (*28, 31*). The type, location, and proportion of various proteins and carbohydrates has been shown to vary within the cell wall of different fungal species (*36*). Therefore, other proteins or carbohydrates on the exterior of the fungal cell wall may protect the chitin elements from the plant chitinase (*28*).

 Rhizoctonia solani is one of the fungi shown to be susceptible to constitutively expressed chitinase. Studies comparing various untransformed plants with plants transformed with a chitinase gene have demonstrated increased resistance to this pathogen (*25-27, 29-32*). Recently, transgenic creeping bentgrass lines containing the pHS2 class I basic elm chitinase gene (*37*) were created from embryogenic calli of the moderately *R. solani* resistant cultivars Penncross and Putter (*38, 39*). The objectives of this study were to quantify the levels of increased *R. solani* resistance within transgenic creeping bentgrass clones containing a constitutively expressed chitinase gene.

Materials and Methods

Plant culture. Transgenic plants of *A. palustris* were regenerated from the embryogenic calli originated from single seedlings between 1995 and 1996 in E. Lansing, MI (*39*). Two plasmids were cotransferred into the plant genome using the Biolistic PDS-1000/He system (DuPont, Wilmington, DE). Plasmids used in transformation were pJS101, carrying the selectable marker for bialophos resistance (*bar* gene; provided by Dr. Ray Wu of Cornell University), and the plasmid vector pKYLX-71 containing the class I basic elm chitinase (pHS2) gene constitutively expressed by the CMV 35S promoter (*38*). Eleven transgenic lines containing the elm chitinase gene were selected from the transgenic plants to screen for *R. solani* resistance (*Table I*). Selection procedures involved screening the transgenic plants for *bar* gene expression at calli and plantlet stages on agar medium containing bialophos (L-2-amino-4-(hydroxymethyphosphinyl)butanoic acid-L-ananyl-L-alaninine), and at plant maturity with applications of glufosinate (L-phosphinylthricin)(*39*). Integration and expression of the elm chitinase gene within the host genome in the eleven transgenic lines was verified using Southern, Northern, and Western analyses (*39*).

 In addition to the eleven transgenic lines containing the chitinase gene, parental cultivars of *A. palustris*, Penncross and Putter, and a transgenic line of each parental cultivar containing only the *bar* gene, lines 9604 & 7201, were included as controls (*Table 1*). Twenty pots of each line or parental cultivar were vegetatively replicated by transplanting approximately 20 mature shoots into 10 cm diameter × 10 cm deep plastic pots containing a soil:sand:peat (5:3:2) media. Pots of bentgrass were raised in a greenhouse at 25-35 °C for 2 to 4 months to ensure maturity and foliar coverage of the soil surface. During this period the bentgrass was lightly watered daily, mowed weekly at 2.5 cm, and fertilized weekly at 12.2, 1.5, and 5.9 kg ha^{-1} of N, P, and K, respectively. The insecticides pyridabena (2-tert-butyl-5-(4-tert-butylbenzylthio)-4-chloropyridazin-

3(2H)-one), bifenthrin ((2 methyl[1,1'-biphenyl]-3-yl) methyl-3-(2-chloro-3,3,3-trifluoro-propenyl)-2,2-dimethylcyclopropane), avermectin (5-O-demethyl avermectin), and azadirachtin (Nortriterpenoid botanical insecticide from *Azadirachta indica*) were applied as needed to control fungus gnats (*Bradysia coprophila* [Lintner]) and mealy bugs (*Planococcus citri* (Risso)).

Inoculum production and inoculation. Inoculum consisted of an oat:wheat:corn (3:3:1) grain mix colonized by *R. solani* AG 1-Ia Rs13. This *R. solani* isolate from creeping bentgrass was kindly supplied by Dr. Wakar Uddin and Mike Soika at Penn State University. Inoculum production followed procedures described by Burpee and

Table I. Creeping bentgrass cultivars and transgenic lines screened for resistance to *R. solani*

Parental cultivar	Transgenic	Chitinase gene	Density[a]	Spread[b]
Penncross		-[c]	97 ab[d]	3
Penncross	9604	-	87 cde	4
Penncross	9606	+	97 ab	3
Penncross	9603	+	95 ab	4
Penncross	9601	+	70 h	1
Penncross	815-7	+	87 cde	2
Penncross	711	+	98 a	3
Putter		-	96 ab	3
Putter	7201	-	81 ef	3
Putter	7204	+	74 gh	2
Putter	7205	+	79 fg	2
Putter	7208	+	84 def	3
Putter	815-1	+	92 abc	4
Putter	9106	+	80 efg	1
Putter	910-12	+	91 bcd	5

[a] Shoot density based on visual estimation of percent coverage of soil surface.
[b] Lateral stoloniferous growth estimated using a 0-4 visual scale, where
0 is no spread and 4 is rapid lateral spread.
[c] Designates presence (+) or absence (-) of the chitinase gene.
[d] Letter designates significant differences between treatment means according to Tukey's mean separation technique.

Goulty (*40*). Briefly, this entailed inoculating the double autoclaved grain mix with three 2-cm^2 agar pieces of *R. solani* Rs13 previously grown on acidified potato dextrose agar for 48 h. Inoculated cultures were then incubated at 26 °C for 10-14 days to allow colonization of the grain mix.

Controlled environment experiments were conducted 15 May through 27 July, 1998. Prior to inoculation, pots of creeping bentgrass were mowed at the 2.5 cm height and foliar coverage visually estimated as the percent soil coverage per pot. The center of each pot was then inoculated with approximately 3 g of grain colonized by *R. solani*. After inoculation, plants were placed in plastic trays (8 tray^{-1}) containing 4 cm of distilled water, then lightly misted with distilled water, and sealed in plastic bags to help maintain 100% relative humidity. To reduce variability in leaf wetness due to foliar contact with the plastic bag, 30 cm wide × 60 cm long × 45 cm tall wooden frames were used to maintain a constant distance between the interior surface of the plastic bag and turf canopy. The turf canopy was lightly misted with distilled water at 48 h intervals to maintain 100% relative humidity. Plastic bags containing trays of inoculated bentgrass were situated inside a growth chamber set to maintain a diurnal cycle of 14 h light at 28 °C and 10 h dark at 25 °C. Twenty replicates of each line or parental cultivar were screened in a randomized complete block design with 5 blocks over time.

Data collection and statistical analysis. Disease severity was visually estimated at 3, 5, and 7 days post-inoculation using the Horsfall-Barrett scale (*41*). Prior to statistical analysis, the visual severity estimates were transformed to midpoint percent values to maximize the advantages of the Horsfall-Barrett scale (*42*). Values of disease severity were used to calculate areas under disease progress curves (AUDPC) from the formula:

$$\Sigma_i^{n-1}[(y_i + y_{i+1})/2][t_{i+1} - t_i]$$

where n is the number of assessment times, i = 1,2,3,...,n - 1, y_i is the amount of disease (necrosis) at the ith rating, and t_i is the time of the ith rating (*42*). The General linear model procedure (Statistical Analysis Systems Institute, Inc., Cary, NC) and Dunnett's two-tailed t-test were used to detect differences in *Rhizoctonia* susceptibility between parental cultivars or *bar*-only lines (lacking the pHS2 gene) and transgenic lines carrying the chitinase gene.

Results

Plant growth of transgenic creeping bentgrass varied tremendously, as demonstrated by differences in shoot density at 8 weeks after propagation (*Table I*). Only three lines, 711, 815-1, and 9606, had growth rates equivalent to the parental cultivars Penncross and Putter (*data not shown*). Other transgenic lines varied from slightly to greatly reduced rates of growth, as observed with shoot density in lines 7204, 7205, and 9601 (*Table I*). Lines exhibiting the slowest rate of growth were characterized by severe deformation of the young shoots, similar to growth patterns due to hormonal imbalances. Although the shoots were able to grow out of this deformation, these lines continued to exhibit a reduced growth rate. Turfgrass color and quality also varied tremendously within the eleven transgenic lines screened in this study. Lines 711, 815-1, 9603, 9606, and 9604 were equivalent in turfgrass quality compared to their parental cultivars. Coarse texture or low shoot density resulted in poor turfgrass quality in the other transgenic lines.

Severity of brown patch was influenced early in the epidemic by differences in shoot density among the transgenic lines estimated visually as percent cover of the soil surface (*Table II*). However, no significant interaction was observed between the creeping bentgrass lines/cultivars and shoot density at any sampling period. At 3 days post-inoculation, each 10% increase in estimated soil cover resulted in a 0.4% increase in disease severity across all creeping bentgrass lines and cultivars. Shoot density also significantly ($P < 0.04$) influenced brown patch severity summarized over the entire epidemic with AUDPC values, where each 10% increase in soil cover resulted in a 0.1 unit increase in the AUDPC value. Shoot density did not significantly influence the severity of brown patch after 3 days post-inoculation (*Table II*).

Comparisons between *bar*-only transgenic lines and parental cultivars revealed no significant differences in the severity of brown patch except at 3 days post-inoculation where line 7201 had significantly ($P < 0.05$) less brown patch compared to its parental cultivar Putter (*Table III*). No other significant differences in brown patch severity were observed between the parental cultivars and *bar*-only lines (*Tables III and IV*). At 3 days post-inoculation, comparison of Putter creeping bentgrass with transgenic lines derived from Putter revealed that all six transgenic lines containing the pHS2 gene had increased levels of *R. solani* resistance (*Table III*). Only the transgenic

Table II. Probability values from analysis of covariance of shoot density and transgenic creeping bentgrass lines on brown patch severity

Source	df	Disease severity (%)[a]			AUDPC[c]
		3 days[b]	5 days	7 days	
Block	3	<0.01	<0.01	<0.01	<0.01
Line[d]	14	<0.01	<0.01	<0.01	<0.01
Density[e]	1	0.03	0.09	0.13	0.04
Line × Density	14	0.09	0.62	0.10	0.53

[a] Based on mid-point transformations of visual severity estimates using the Horsfall-Barrett Scale.
[b] Days post-inoculation.
[c] Area under disease progress curve.
[d] Cultivar or line of creeping bentgrass.
[e] Shoot density based on visual estimation of percent coverage of soil surface.

Table III. Comparison of mean brown patch severity of Putter creeping bentgrass and transgenic creeping bentgrass derived from Putter containing an elm chitinase gene

Cultivar or transgenic line	Disease severity (%)[a]			AUDPC[c]
	3 days[b]	5 days	7 days	
Putter	42.6 a[d]	85.4 a	84.3 a	1.63 a
7201	20.5 bc	57.9 a	71.8 a	1.11 ab
7204	16.6 bc	74.9 a	85.5 a	1.31 ab
7205	22.4 bc	72.1 a	83.4 a	1.32 ab
7208	21.9 bc	56.5 a	79.2 a	1.14 ab
815-1	25.4 b	58.6 a	77.2 a	1.18 ab
9106	11.2 bc	54.3 a	75.1 a	1.01 b
910-12	25.3 b	65.3 a	74.5 a	1.24 ab

[a] Based on mid-point transformations of visual severity estimates using the Horsfall-Barrett Scale.
[b] Days post-inoculation.
[c] Area under disease progress curve.
[d] Letter designates significant differences between treatment means according to Dunnett's two-tailed t-test.

line 9106 had significantly ($P \leq 0.05$) improved levels of resistance when compared to Putter based on AUDPC values (*Table III*). However, at all sampling dates the Putter derived transgenic lines containing the pHS2 chitinase gene were not significantly different in their resistance to *R. solani* as compared to the Putter derived *bar*-only line 7201. Brown patch severity was not significantly altered on transgenic lines containing the chitinase gene compared to their parental cultivar, Putter, at any other sampling period (*Table III*).

At all three sampling periods and based on AUDPC values, significant ($P \leq 0.01$) differences were detected in the susceptibility to *R. solani* (Rs13) between the transgenic creeping bentgrass line 711 and both its parental cultivar, Penncross, and the Penncross derived *bar*-only line 9604 (*Table IV*). The transgenic creeping bentgrass line 9603 was also found to have significantly ($P \leq 0.05$) greater levels of resistance to *R. solani* compared to its parental cultivar, Penncross, based on disease severity at all three sampling dates and AUDPC values (*Table IV*). However, *Rhizoctonia* resistance in transgenic line 9603 was only significantly ($P < 0.05$) improved at 7 days post-inoculation compared to the *bar*-only line 9604. Other transgenic lines derived from the creeping bentgrass cultivar Penncross did not have significantly improved resistance to *R. solani* (*Table IV*).

Table IV. Comparison of mean brown patch severity of Penncross creeping bentgrass and transgenic creeping bentgrass derived from Penncross containing an elm chitinase gene

Cultivar or transgenic line	Disease severity (%)[a]			
	3 days[b]	5 days	7 days	AUDPC[c]
Penncross	38.9 a[d]	75.9 a	78.1 a	1.47 a
9604	24.0 ab	62.0 ab	69.1 a	1.17 ab
9606	25.2 ab	63.0 ab	69.3 a	1.19 ab
9603	13.0 b	41.0 b	41.9 b	0.80 b
9601	21.1 ab	55.6 ab	62.5 a	0.98 ab
815-7	36.9 a	69.4 a	74.2 a	1.37 a
711	6.8 c	29.3 c	27.0 b	0.49 c

[a] Based on mid-point transformations of visual severity estimates using the Horsfall-Barrett Scale.
[b] Days post-inoculation.
[c] Area under disease progress curve.
[d] Letter designates significant differences between treatment means according to Dunnett's two-tailed t-tests.

Discussion

Two of eleven transgenic lines carrying a constitutively expressed class I elm chitinase gene provided improved levels of *Rhizoctonia* resistance in creeping bentgrass. Disease resistance was improved in transgenic lines 711 and 9603 as compared to Penncross and Putter creeping bentgrass. Infection of mature turfgrass species by *R. solani* generally initiates on the foliage and then proceeds into the crowns as the brown patch epidemic progresses (*2-3*). This pattern of infection by *R. solani* was observed on all transgenic lines containing the chitinase gene except lines 711 and 9603. The foliage of transgenic lines 711 and 9603 were initially blighted by *R. solani*, although disease progress was significantly delayed. When compared to the parental cultivars, this delay in the brown patch epidemic resulted in approximately a 3 and 1 fold improvement in resistance to *R. solani* over the entire epidemic in lines 711 and 9603, respectively. Additionally, *R. solani* was not observed to infect the crowns of line 711, even under severe disease pressure provided in this study where 60-95% of the crowns were blighted on the parental cultivars at 7 days post-inoculation. Improved levels of *Rhizoctonia* resistance in line 9603, however, were not adequate to prevent crown infection at 7 days post-inoculation.

The delay in disease progress in transgenic lines 711 and 9603 suggests that their improved resistance was from increased levels of some form of plant defense, such as constitutive expression of the chitinase gene. Research in M. Stricklen's lab has demonstrated the presence and expression of the phS2 gene in the creeping bentgrass lines examined in this study (*39*). However, the question still remains as to whether the improved levels of *Rhizoctonia* resistance are due to increased production of chitinase within these lines, although this seems possible.

Precisely why other transgenic lines containing the chitinase gene did not consistently provide improved resistance to *R. solani* is not fully understood. Transgenic lines derived from Putter containing the phS2 gene (7204, 7205, 7208, 815-1, 9106, & 910-12) did show improved *Rhizoctonia* resistance at 3 days post-inoculation compared to their parental cultivar. However, these transgenic lines showed similar levels of susceptibility to *R. solani* compared to the *bar*-only line 7201. It is possible that constitutive expression of this elm chitinase gene in the transgenic lines of Putter did provide some resistance to *R. solani* early in the epidemic. This does not explain the higher levels of resistance in the *bar*-only line 7201.

Penncross and Putter creeping bentgrass are heterogenic cultivars. This heterogenicity results in a diverse array of biotypes with different phenotypic expression within each cultivar. This was exemplified in our study with the variety of color, texture, and growth patterns observed between transgenic lines derived from individual genotypes (single seeds) within the heterogenic population of each parental cultivar. It is possible that the improved resistance to *R. solani* may be a result of selecting individual genotypes with greater levels of *Rhizoctonia* resistance from the populations of each cultivar.

Transformation procedures used to produce the transgenic creeping bentgrass lines insert the gene cassette at random locations within the plant genome (*33*). Random insertion can disrupt or change the level of expression of other natural host genes near the insertion site (*43-44*). Plant transformation techniques have also been found to lead to suppression or inaccurate expression of the inserted gene (*43, 45*). Normal gene regulatory processes within the plant, which are designed to protect the plant from mutation and foreign DNA such as plant viruses, can remove or suppress the inserted gene (*45*). Frequently, multiple copies of the gene cassette are inserted within a plant genome during transformation. Insertion of multiple copies or host recognition of specific sequences within the inserted gene cassette can lead to host regulation and reduction or silencing of both inserted and host genes with similar DNA sequences (*45-47*). Damage to the DNA sequence of the gene cassette, which can occur during the transformation procedures, can be improperly repaired leading to gene silencing or expression of proteins with modified tertiary and quaternary protein structure (*44*). Modifications in protein structure can lead to inactivity or reduced activity of protein function. Although southern blot analysis confirmed the presence and expression of the elm chitinase gene in each of the transgenic lines, modifications in protein structure, host gene regulation, or somaclonal variations are possible explanations for the inefficiency of the chitinase gene to improve *Rhizoctonia* resistance in some of our transgenic lines.

Conclusions

Although production of transgenic plants does not always result in the desired expression of the inserted gene, as seen in this study, careful screening of the transformed plants can result in viable transgenic turfgrass lines with improved disease resistance. In this study, insertion of an elm chitinase gene within the creeping bentgrass cultivar Penncross produced two transgenic lines with improved *Rhizoctonia* resistance. Improved resistance to *R. solani* in transgenic line 711 appears to be greater than the moderate levels of resistance to this pathogen currently available in creeping bentgrass cultivars, although this needs to be verified in future research. Furthermore, chitinase levels within these transgenic lines need to be quantified and their impact on fungal pathogens defined to verify the role of constitutive expression of chitinase in host resistance.

Several other aspects must also be carefully considered before these transgenic creeping bentgrass lines can be released. Concerns have been raised as to whether transgenic crops are any safer than chemical pesticides. These concerns include risk to human health, development of "super pests", and environmental risks *(48)*. Human health concerns center on production of foreign cellular compounds within transgenic plants with high mammalian toxicity. Two theories have been proposed for potential development of "super pests". The first theory being natural selection within the pest population, such as fungal pathogens, against the cellular compound(s) encoded by the inserted gene(s)*(48)*. A second possible theory in the development of "super pests" is the production of weeds with resistance to currently available pesticides. This can occur through out-crossing of the transgenic plant with other plant species or from the transgenic plant itself becoming a pest in another crop *(48)*. Creeping bentgrass, however, has been shown to have relatively few plant species with which it can out-cross, and is not likely to become a weed species in other agronomic crops *(9)*. The final concern which must be addressed is the risk to natural ecosystems where the transgenic turf will be used *(48)*. Disruption of natural ecosystems can occur through either competition or potential toxic effects on other organisms within an ecosystem.

Transgenic plants do offer several advantages as pesticide alternatives. Compounds produced within the plant remove the risk of injury to the applicator and are less likely to move long distances through drift, leaching, or run-off. Selection of natural plant defense genes, such as chitinase, also avoids problems with mammalian toxicity. Therefore, transformation of turfgrass by the insertion of plant resistance genes is likely to be a viable alternative or supplement to pesticides for control of turf diseases in the near future.

Acknowledgments

Funding for this research was provided by United States Golf Association and Michigan State Agriculture Extension Service.

340

Literature Cited

1. Vargas, J.M. *Management of Turfgrass Diseases*; Lewis Publishers; Ann Arbor, MI, 1994; 2nd edition.
2. Smiley, R.W., Dernoeden, P.H., and Clarke, B.B. *Compendium of Turfgrass Diseases.* The American Phytopathological Society; St. Paul, MN; 1992.
3. Smith, J.D., Jackson, N., and Woolhouse, A.R. *Fungal Diseases of Amenity Turf Grasses;* E&FN Spon.; London, England; 1989.
4. Piper, C.V. and Coe, H.S. *Phytopathology* 1919, *9*, 89-95.
5. Burpee, L.L. *Plant Dis.* 1992, *96*, 1065-1068.
6. Morris, K. *National Tall Fescue Test-1992, Final Report 1993-1995;* USDA-ARS; Beltsville, MD; NTEP No. 96-13.
7. Yuen, G.Y., Giesler, L.J., and Horst, G.L. *Crop Prot.* 1994, *13*, 439-441.
8. Asano, Y. and Ugaki, M. *Plant Cell Reports* 1994, *13*, 243-246.
9. Day, P.R. and Lee, L. *Northeastern Weed Sci. Soc.* 1997, *51*, 173-176.
10. Ha, S.B., Wu, F.S., and Thorne, T.K. *Plant Cell Reports* 1992, *11*, 601-604.
11. Lee, L., Laramore, C.L., Day, P.R., and Tumer, N.E. *Crop Sci.* 1996, *36*, 401-406.
12. Sugiura, K., Inokuma, C., Imaizumi, N., and Cho, C. *J. Turfgrass Mgt.* 1997, *2*, 43-53.
13. Wang, Z., Takamizo, T., Iglesias, V.A., Osusky, M., Nagel, J., Potrykus, I., and Spangenberg, G. *Bio/technology* 1992, *10*, 691-696.
14. Zhong, H., Bolyard, M.G., Srinivasan, C., and Sticklen, M.B. *Plant Cell Reports* 1993, *13*, 1-6.
15. Parlevliet, J.E. *Ann. Rev. Phytopath* 1979, *17*, 203-222.
16. Van Loon, L.C. *Plant Mol.* 1985, *4*, 111-116.
17. Punja, Z.K. and Zhang, Y. *J. Nematology* 1993, *25*, 526-540.
18. Benhamou, N., Broglie, K., Broglie, R., and Chet, I. *Can. J. Mirobiol.* 1993, *39*, 318-328.
19. Schlumbaum, A., Mauch, F., Vogeli, U., and Boller, T. *Nature* 1986, *324*, 365-367.
20. Sela-Buurlage, M.B., Ponstein, A.S., Bres-Vloemans, A.B., Melchers, L.S., van den Elzen, P.J.M., and Cornelissen, B.J.C. *Plant Physiol.* 1993, *101*, 857-863.
21. Flach, J., Pilet, P.-E., and Jolles, P. *Experientia* 1992, *48*, 701-716.
22. Graham, L.S. and Sticklen, M.B. *Can. J. Bot.* 1994, *72*, 1057-1075.
23. Collinge, D.B., Kragh, K.M., Mikkelsen, J.D., Nielsen, K.K., Rasmussen, U., and Vad, K. *The Plant J.* 1993, *3*, 31-40.
24. Toyoda, H., Matsuda, Y., Yamaga, T., Ikeda, S., Morita, M. Tamai, T., and Ouchi, S. *Plant Cell Reports* 1991, *217*, 217-220.
25. Benhamou, N, Broglie, K., Chet, I., and Broglie, R. *The Plant J.* 1993, *4*, 295-305.
26. Broglie, K., Chet, I., Holliday, M., Cressman, R., Biddle, P., Knowlton, S., Mauvais, C.J., and Broglie, R. *Science* 1991, *254*,1194-1197.
27. Dunsmuir, P., Howie, W., Newbigin, E., Joe, L., Penzes, E., and Suslow, T. In *Advances in Molecular Genetics of Plant-Microbe Interactions*; Nester, E.W. and Verma, D.P.S., Eds.; Kluwer Academic Press: Netherland, 1993; pp. 567-571.
28. Grison, R., Grezes-Besset, B., Schneider, M., Luncante, N., Olsen, L., Leguay, J.-J., and Toppan, A. *Nature Biotech.* 1996, *14*, 643-646.
29. Howie, W., Joe, L., Newbigin, E., Suslow, T., and Dunsmuir, P. *Transgenic Res.* 1994, *3*, 90-98.
30. Jach, G., Gornhardt, B., Mundy, J., Logemann, J., Pinsdorf, E., Leah, R., Schell, J., and Maas, C. *The Plant J.* 1995, *8*, 97-109.
31. Punja, Z.K. and Raharjo, S.H.T. *Plant Dis.* 1996, *80*, 999-1005.
32. Vierheilig, H., Alt, M., Neuhaus, J.-M., Boller, T., and Wiemken, A. *MPMI* 1993, *6*, 261-264.
33. Zhu, Q., Maher, E.A., Masoud, S., Dixon, R.A., and Lamb, C.J. *Bio/technology* 1994, *12*, 807-812.
34. Neilsen, K.K., Mikkelsen, J.D., Kragh, K.M., and Bojsen, K. *MPMI* 1993, *6*, 495-506.
35. Neuhaus, J.-M., Ahl-Goy, P., Hinz, U., Flores, S., and Meins, F. *Plant Mol. Biol.* 1991, *16*, 141-151.

36. Bartnicki-Garcia, S. *Annu. Rev. Microbiol.* **1968**, *22*, 87-108.
37. Hajela, R.K., Graham, L., and Stickland, M.B. In: *Dutch Elm Disease Research: Cellular and Molecular Approaches*; Springer-Verlag; New York, NY, 1993; pp. 193-207.
38. Warkentin, D., Chai, B., Liu, A.-C., Hajela, R.K., Zhong, H., and Sticklen, M.B. In: *Turfgrass Biotechnology: Cellular and Molecular Genetic Approaches to Turfgrass Improvement*; Ann Arbor Press; Chelsea, MI, 1998; pp. 153-161.
39. Chai, B. *Transformation Studies of Creeping Bentgrass (Agrostis palustris Huds) Using a Dutch Elm Disease Chitinase Gene for Disease Resistance;* Michigan State University: East Lansing, MI, 1999.
40. Burpee, L.L. and Goulty L.G. *Phytopathology* **1984**, *74*, 692-694.
41. Horsfall, J.G. and Cowling, E.B. In *Plant Disease: An Advanced Treatise*; Horsfall, J.G. and Cowling, E.B.; Eds.; Academic Press: New York, NY, 1978, Vol. II; pp. 120-136.
42. Campbell, C.L., and Madden, L.V. *Introduction to Plant Disease Epidemiology;* John Wiley & Sons; New York, NY, 1990.
43. Jones, J.D.G, Dunsmuir, P., and Bedbrook, J. *EMBO J.* **1985**, *4*, 2411-2418.
44. Eckes, P., Schell, J., and Willmitzer, L. *Molec. Gen. Genet.* **1985**, *199*, 216-221.
45. Finnegan, J. and McElroy, D. *Bio/Technology* **1994**, *12*, 883-888.
46. Assaad, F.f., Tucker, K.L., and Signer, E.R. *Plant Mol. Biol.* **1993**, *22*, 1067-1085.
47. Linn, F., Heidmann, I., Saedler, H., and Meyer, P. *Mol. Gen. Genet.* **1990**, *222*, 329-336.
48. Sticklen. M.B. In: *Pesticides for the Next Decades: The Challenges Ahead;* VPI Publ.; Blacksburg, VA, 1991; pp. 522-566.

Chapter 21

Microbial Strategies for the Biological Control of Turfgrass Diseases

Eric B. Nelson and Cheryl M. Craft[1]

Department of Plant Pathology, Cornell University, Ithaca, NY 14853

Common approaches for implementing biological control strategies involve the use of microbial inoculants and organic amendments. The goal with these strategies is to increase populations and activities of disease-suppressive microbes associated with turfgrass plants. Microbial inoculants are used to temporarily increase soil or plant populations of disease-suppressive microbes, which require repeated applications to maintain populations at levels necessary for disease control. The use of organic amendments attempts to produce a long-term change in the soil environment whereby the activities of indigenous disease-suppressive microbes are favored. The application of some composted organic amendments not only introduces a varied and diverse microbial community into soils, but also provides a substrate for microbes already present in turfgrass soils, enhancing their population and activities. The mechanisms by which these different microbial approaches affect disease severity in turfgrasses are discussed along with a discussion of the present status and future of biological control strategies for disease control in turf.

For over 80 years, traditional turfgrass management programs have relied heavily on synthetic chemical fungicide applications for disease control. It has only been since the late 1980's that a more visible trend toward non-chemical strategies has become apparent. Not only are turfgrass managers seeking alternatives to chemical fungicides, but an increasing number of research laboratories around the world are now focusing efforts on biological methods of disease control in response to these increasing demands from turfgrass managers. Although we have much to learn about implementing biological disease control strategies successfully, these approaches are scientifically sound and, in a growing number of cases, are an accepted and viable option for controlling diseases.

[1]Current address: Bionique Testing Laboratories, RR#1, Box, Bloomingdale Road, Saranac Lake, NY 12983.

Golf course superintendents view biological control as a desirable alternative to fungicide treatments for a number of different reasons. One of the more important is that biological control can be a rational means of extending and augmenting the efficacy of fungicides and, at the same time, reducing the environmental load of pesticides. Most importantly, however, biological control is increasingly being viewed as an effective sustainable solution for maintaining turfgrass health.

Approaches to Biological Control

The most common approaches for implementing biological control strategies have involved either a bioaugmentative approach involving the use of microbial inoculants or a biostimulatory approach involving the use of organic amendments to encourage the activities of native pathogen-suppressive microorganisms (*1,2*). The goal with both of these strategies is to increase the populations and activity of disease-suppressive microbes associated with turfgrass plants. However, each approach differs fundamentally in mechanisms and sustainability.

With bioaugmentative approaches, microbial inoculants are used to temporarily and dramatically increase soil or plant populations of specific disease suppressive microbes. Generally the microorganisms used as inoculants have specific pathogen targets and operate under relatively narrow modes of action. As with fungicide applications, the use of inoculants requires repeated applications to maintain populations of introduced microorganisms at levels necessary for disease control. In contrast, however, the application of inoculants must be handled quite differently from fungicides since precautions must be taken to maintain maximum viability of the inoculant.

With biostimulatory approaches, organic amendments are used to produce a long-term and hopefully permanent qualitative change in soil microbial communities whereby the activities of indigenous disease-suppressive microbes are stimulated. For example, the application of some composted amendments not only introduces a diverse consortium or microorganisms into soils, but also provides a substrate for those microbes already present in turfgrass soils, enhancing their population and activities. Periodic applications of these amendments are generally required to sustain populations and activities.

Bioaugmentation with Microbial Inoculants

Microbial inoculants have been studied for a number of years in turfgrass management for thatch reduction (*3*), fertilizer enhancement (*4*), insect control (*5*), weed control (*6*), and disease control (*7*). The most frequently studied microorganisms for biological disease control in turfgrasses have been species of the bacterial genera *Pseudomonas, Bacillus, Enterobacter,* and *Streptomyces* as well as species of the fungal genera *Trichoderma* and *Gliocladium*. However, few studies have defined ecological relationships, mechanisms of control, or formulation and application technologies, each of which are important to the success of introduced inoculants. Despite these deficiencies, research results combined with the experience of turfgrass managers point to the potential for introducing microorganisms to manage diseases and improve turfgrass health.

A number of field studies have demonstrated the potential for bacterial and fungal inoculants to control foliar diseases such as dollar spot and brown patch. For example, in studies on bentgrass/*Poa annua* putting greens, top dressings fortified with *Enterobacter cloacae* significantly reduced dollar spot disease development (*8*). Monthly applications of this bacterial preparation introduced approximately 10^7-10^9 CFU/g dry wt of thatch and provided up to 63% disease control. This level of control was as effective as iprodione or propiconazole in reducing dollar spot severity. The dollar spot control provided by this bacterial inoculant was evident for 2 mo. after application and was more effective as a preventive treatment than a curative treatment. In other studies, weekly applications of *Fusarium heterosporum* to Penncross putting greens reduced dollar spot incidence nearly 80% (*9*).

Another well-studied microbial inoculant for foliar diseases of turfgrasses is *Trichoderma harzianum* strain 1295-22 (*10-12*). This fungus is the active ingredient in the recently registered granular product, BioTrek 22G. This fungus is effective for control of dollar spot, brown patch, and *Pythium* diseases on creeping bentgrass. In field trials over a 4-yr period, monthly applications of granular or peat-based formulations of strain 1295-22 reduced early-season dollar spot severity by as much as 71% and delayed disease development up to 30 days.

One of the more intriguing properties of *T. harzianum* is the ability to persist in the rhizosphere of creeping bentgrass, where monthly applications have been effective in maintaining populations at ca. 10^6 cfu/g of thatch/soil. In some experiments, populations increased with each successive application (*12*). *T. harzianum* was also shown to persist at population levels between 10^5-10^6 cfu/g, levels that were adequate to achieve a low level of biological control. However, if populations fall below 10^5 cfu/g, biological control efficacy is lost (*11*).

Many other agents have been tested in both laboratory and field studies for control of brown patch, Pythium blight, southern blight, leaf spots, and Typhula blight with varying degrees of success (*7*). Part of the variability in performance may be due do to factors influencing population level, activity, and survival. For example, environmental factors such as temperature, moisture, and UV light change dramatically in a foliar canopy and can be detrimental to microbial growth and activity. Furthermore, management practices such as grooming, irrigation, fertilization, and pest control may also influence microbial growth and activity. Therefore, in order to maintain consistent levels of biocontrol, considerable efforts are needed to develop suitable formulations and application methods that protect microbial inoculants from environmental extremes and are compatible with management practices.

Among the more challenging problems for biological control is the management of root diseases such as summer patch, take-all patch, necrotic ringspot, spring dead spot, and Pythium root rot. These are particularly destructive diseases and among the most difficult to control. Furthermore, their infection courts are not typically accessible to chemical or biological agents, making control erratic and unpredictable.

The impact of soil organisms on the severity of take-all patch is quite clear. Sarniguet and Lucas (*13,14*) reported that the central regions of take-all patches recolonized by healthy grasses contained substantially higher soil populations of fluorescent *Pseudomonas* spp. than the surrounding diseased areas or disease-free turf. Of all the *Pseudomonas* spp. isolated, between 44-82% were antagonistic to *Gaeumannomyces graminis* var. *avenae* as compared with only 12-24% antagonistic strains in the adjacent disease free area. Treatment of bentgrasses with *Pseudomonas* spp. and *Phialophora* spp. has been effective in reducing the severity of take-all (*15-18*).

Biological control of summer patch on Kentucky bluegrass has been observed with bacterial inoculants (*19-21*). Strains of *Serratia marcescens* and *Stenotrophomonas maltophilia* provided greater than 50% suppression of summer patch symptoms in greenhouse trials, but were not tested further in the field (*20*). Significant levels of control were observed only when populations were greater than 10^8 cells/ml. Similarly, strains of *Bacillus subtilis* and *Enterobacter cloacae* were effective in suppressing summer patch symptoms in greenhouse experiments and reduced summer patch symptoms from 34 to 53% in field trials (*19*).

Pythium root rot is another important disease for which chemical and other cultural controls have been inconsistent. Recent studies have demonstrated that weekly applications of *T. harzianum* (strain 1295-22) significantly reduced foliar symptoms of Pythium root rot in both greenhouse and field experiments and enhanced turf quality (*11*). In other studies, strains of actinomycetes from *Pythium*-suppressive composts were effective in reducing Pythium root rot on field-collected turf tested in the greenhouse (*22*).

Importance of Maintaining Populations of Introduced Inoculants.

One of the more difficult obstacles to the successful bioaugmentation of turfgrass soils has been inconsistent performance of microbial inoculants in the field. There are numerous examples where biological control is as effective as fungicide applications in laboratory tests, but inconsistent and unpredictable in field tests. One of the more important factors contributing to reduced efficacy in the field is the failure to maintain adequate populations of microbes in turfgrass soils. It is apparent from numerous laboratory experiments that populations of biocontrol organisms must be maintained at high levels (usually $>10^6$ CFU/g soil). If populations drop below this level, control efficacy is jeopardized (*11,12*). To overcome this, one or more of several things must happen: 1) applications must be made more frequently, 2) management practices and/or environmental conditions must favor activity of the inoculant over that of pathogenic microorganisms, or 3) the inoculant must be formulated and applied in a way that favors its activity and survival.

Because of the need to maintain high populations of microbial inoculants, some have questioned the sustainability and environmental impacts of frequent applications of high levels of microorganisms. The frequent application of inoculants may provide the only effective approach for suppressing foliar diseases and maintaining populations on foliage since this habitat is generally more unfavorable for microbial persistence than a soil environment. Selection of inoculants adapted specifically to the turfgrass environment may facilitate their activity and the maintenance of stable populations.

Mechanisms of Pathogen and Disease Suppression.

Inconsistent performance of microbial inoculants may also be related to the expression of specific mechanisms by which these microorganisms suppress pathogens and diseases. Knowledge of the manner in which pathogens and diseases are suppressed by microorganisms and how the expression of these mechanisms is regulated is critical to the development of strategies for ensuring consistent and efficacious performance. Nearly all of the studies on biological control mechanisms have come from studies with crop plants other than turfgrasses. However, many of the same processes may also play important roles in turfgrass ecosystems.

Of the traits common to soil microbes, there are at least four that have been consistently linked with biological control processes. These include 1) antibiotic biosynthesis, 2) resource competition, and 3) hyperparasitism. Some microorganisms may express various combinations of these mechanisms for successful biological control.

Antibiotic Biosynthesis. Antibiotic biosynthesis, particularly in bacterial systems, has been the most commonly-studied trait related to biological control. Antibiotic compounds produced by species of *Pseudomonas, Bacillus,* and *Streptomyces* are known to play key roles in the biological control of *Pythium, Rhizoctonia,* and *Gaeumannomyces* diseases and, more recently, have been shown to play roles in the suppression of *Sclerotinia homoeocarpa* and *Drechslera poae* in bentgrass and Kentucky bluegrass, respectively (*23*).

A number of ecologically important antibiotics have been described in biocontrol systems, including pyrrolnitrin, pyoluteorin, 2,4-diacetylphloroglucinol, phenazine-1-carboxylic acid, and oomycin A (*24*). New antibiotics are constantly being described that could potentially play important roles in the biological control of plant diseases. The majority of studies on antibiotic biosynthesis and its relation to biological control have been limited primarily to fluorescent *Pseudomonas* species; the majority of research encompassing only 5 to 10 strains of *Pseudomonas fluorescens*, a few strains of *P. aureofaciens*, and a couple of strains of *P. putida*. Compelling evidence now exists for the link between production of antibiotics in the rhizosphere and the role of these antibiotics in suppressing root (*25,26*) and foliar diseases (*23*).

Resource Competition. Microbial competition often occurs when two or more microorganisms compete for the same nutrient resource. Despite the fact that competition has been proposed as a mechanism of biological control, no definitive evidence to support this hypothesis currently exists. Although intuitively, competition should be an important mechanism of biological control, the complexities of the soil environment make the proof of competition mechanisms difficult to obtain. The best current examples of competition in biological control are competition for iron and seed exudate fatty acids.

Siderophore-Mediated Iron Competition. Siderophores are low molecular weight iron chelates that are produced by many soil microbes under iron-limiting conditions. They chelate ferric iron and serve as a major vehicle for iron transport into microbial cells. Nearly all organisms produce siderophores, but those produced by species of *Pseudomonas* and enteric bacteria generally have higher affinities for iron than do other fungal siderophores. In a number of specific pathosystems, the biological control of soilborne pathogens has been attributed to the production of siderophores (*27-32*), whereas in other cases no role for siderophores in biological control processes can be found (*33-36*). Although siderophore competition has generally been considered a direct form of biological control, it is possible that some siderophores may be acting indirectly by enhancing natural plant defense mechanisms under iron-limiting conditions (*31,37*). More work is needed to resolve the functional relationships between siderophore production and biological control.

Fatty Acid Competition. Seed and root exudates play an important role in the initiation of soilborne plant diseases by serving as stimulants of fungal propagules (*38,39*). Without the release of stimulatory molecules in these exudates, pathogenic relationships between plants and soil pathogens do not occur. Recently, there has been interest in understanding the interaction of biocontrol organisms with stimulatory components of seed and root exudates. Since soilborne fungal pathogens are highly dependent on exudate molecules to initiate plant infections, microbial interference with the production and activity of exudate stimulants could be an effective mechanism of biological control among seed-applied spermosphere bacteria (*40*). Recent studies have shown that strains of *Enterobacter cloacae* and other seed-associated bacteria that are effective in suppressing Pythium seed rot and damping-off of various plant species can also reduce the stimulatory activity of plant seed exudates to sporangia of the seed-rotting fungus, *Pythium ultimum (41)*. Many of the seed-associated bacterial strains tested could reduce the stimulatory activity within 24 hours to levels supporting germination of less than 20% of *Pythium* sporangia. In particular, strains of *E. cloacae* can rapidly metabolize linoleic acid, a major Pythium-stimulatory molecule found in seed exudates (*42*), making it unavailable for the stimulation of Pythium pathogenesis (*24,43*). It is possible that metabolism of seed released fatty acids during the establishment of some bentgrass cultivars may enhance resistance to Pythium damping-off (Kageyama and Nelson, unpublished).

Hyperparasitism. Hyperparasitism is a complex process by which microorganisms parasitize other microorganisms. This process is generally mediated by the attachment and production of cell wall degrading enzymes, such as chitinases, glucanases, and proteinases, by the biocontrol strain that destroys the host (i.e., turfgrass pathogen) cell. Species of the biocontrol fungus *Trichoderma* are well known for this property (*44*). However, to date, there is no proof that this process plays a significant role in the biological control of diseases by *Trichoderma* and other mycoparasitic fungi.

Bacteria may also be parasitic to pathogenic fungi. In these cases biocontrol efficacy is mediated by the production of chitinase enzymes. *Serratia marcescens* is known not only for its chitinase production, but also for its activity as a biological control agent for soilborne diseases, including summer patch on Kentucky bluegrass caused by *Magnaporthe poae* (*20,21*). Much of the success of *S. marcescens* as a biological control agent is believed to result from its production of chitinolytic enzymes inhibitory to chitin-containing soilborne fungi. Additionally, species of *Enterobacter* (=*Pantoea*) also produce chitinolytic enzymes that have been directly associated with biological control activity (*45*). A number of reports have been published in which chitinase genes from *S. marcescens* have been cloned and used for the development of transgenic bacteria and plants, each with enhanced suppression of soilborne diseases (*46-55*).

It has long been the goal of many researchers in the field of biological control to develop more effective biocontrol strains or those that perform more consistently. The expression of chitinase genes in transgenic bacteria and fungi is currently the best example of the development of genetically engineered biological control agents for the suppression of soilborne diseases. This can serve as a suitable model for illustrating the possibilities of utilizing transgenic microorganisms for the control of turfgrass diseases.

Biostimulation with Compost Amendments

Unlike microbial inoculants, the goal with compost amendments is to enhance the native populations of soil organisms at the expense of plant pathogenic fungi so that disease severity can be reduced. Composts have been used for centuries in soil management and their addition to soil is considered an effective and sustainable means of improving productivity and overall plant health. Results of studies conducted over the past 10-15 years have shown the potential for compost amendments to reduce the severity and incidence of a wide variety of turfgrass diseases when applied either as a topdressing or as a root zone amendment (*56*).

For example, numerous studies have shown that monthly topdressing applications of composts at rates as low as 10 lbs/1000 ft^2 are effective in suppressing diseases such as dollar spot, brown patch, Pythium root rot, Pythium blight, necrotic ringspot, red thread, and Typhula blight. Levels of disease control can vary from 0-94%, depending on the target disease, the type of compost and the manner and degree to which the material is composted (56).

Root-zone amendments with composts have the potential to induce much higher and longer-lasting disease suppression of root-infecting pathogens, than do topdressing amendments. Studies have shown that amending sand-based greens with municipal biosolids compost, brewery sludge compost, or an uncomposted reed-sedge peat induces a high level of suppression of Pythium root rot. These amendments have provided complete control for 6 mo. after incorporation and retained disease suppressive properties for up to 4 years (1,57).

Mechanisms of Disease Suppression with Compost Amendments. Despite the fact that a number of composts have been shown to suppress turfgrass diseases, some types and batches of compost are not suppressive. A key to understanding disease suppressiveness lies in the microbiology of the composts themselves and in the soils to which they are applied. Immature (1-3 mos) composts still undergoing thermophilic decomposition and low in microbial activity are not suppressive to *Pythium* and other pathogens (58,59). However, when allowed to age for a suitable period of time (2-3 years), many composts become highly suppressive. These suppressive composts contain relatively high populations of heterotrophic bacteria, actinomycetes, and fungi compared with non-suppressive composts. However, these quantitative differences alone are not sufficient to explain the suppressive properties of composts in the field.

Studies have shown that control of *Pythium graminicola* with compost amendments is dependent on the microbial properties of the amendment (59), and the soil microbial responses following application of the amendment (1). Although microbial populations and activity levels can be correlated with *Pythium*-suppressiveness in laboratory studies (59), these properties may not be entirely predictive of the expected level of disease suppression in the field. This is particularly true for poultry composts in which suppression of Pythium root rot may result not from the microbial activity of the compost but from the direct enhancement of resident soil microorganisms (1). In general, a greater frequency of *Pythium*-suppressive bacteria can be recovered from compost-induced suppressive soils than in amended or non-amended non-suppressive soils (E.B. Nelson and C.M. Craft, *unpublished*). Furthermore, the level of *Pythium* suppression by individual bacterial strains from these suppressive soils is also greater than from strains recovered from non-suppressive soils. While it is not yet known whether these suppressive bacteria arise from compost microbial communities, soil microbial communities, or combinations of populations from each of these habitats, it is likely that unique guilds of microorganisms are associated with such suppressive environments. Characterization of these suppressive and non-suppressive communities is currently under investigation.

It should also be pointed out that whereas the biostimulatory properties of compost amendments play a central role in the suppression of *Pythium*-incited diseases, it is not yet clear whether these same relationships are true for other turfgrass pathogens and diseases or whether the same populations of microorganisms are responsible for the suppressiveness observed for other diseases. It is likely that the same mechanisms could apply equally to other turfgrass pathogens. However, different mechanisms of disease suppression with compost amendments have been described for different fungal plant pathogens (*60*).

Biological Control Products for the Turfgrass Industry

Today, the turfgrass manager has a number of bioaugmetative and biostimulatory products available for use. The numbers and types of organic amendments and microbial inoculants being marketed for disease control in turfgrasses are overwhelming. In many cases, it is difficult to know which of these products should be taken seriously. Biological control products currently available to the turfgrass industry can be grouped into four different classes: 1) EPA-registered inoculants, 2) unregistered inoculants for which documented levels of disease reduction are claimed, 3) unregistered inoculants for which disease reduction claims have not been validated, and 4) products that are known to suppress diseases but are marketed for other purposes. These latter products are commonly marketed as natural organic fertilizers and compost amendments, whereas many of the microbial inoculants fall into the first three categories.

The first class of products is registered in the same manner as chemical fungicides. Currently, there are only two microbial-based products registered on turfgrasses as biological fungicides worldwide. Binab T is a preparation of *Trichoderma harzianum* (ATCC 20476) and *Trichoderma polysporum* (ATCC 20475) available in Sweden and the UK, and BioTrek 22G is a preparation of *T. harzianum* (1295-22) available in the United States. A number of other products that are registered for other crops will likely see registrations on turfgrasses in the future. Additionally, there are other microorganisms being developed specifically for turfgrass applications that are currently in the registration process.

Through registration as a biological fungicide, claims may be made about the control of specific diseases. However, products not registered with the EPA but with labels that claim control of specific diseases cannot be sold or used legally for turfgrass applications. Nonetheless, a number of these types of products are currently available to turfgrass managers. Some of these products are registered on other crops but currently lack a turfgrass registration whereas others are not registered on any crop.

The second class of products is one of the more difficult groups to assess. These unregistered products are marketed at least in part for disease control. Although specific diseases that are controlled are not normally listed on the label, label-wording frequently infers that the use of the product will nonspecifically reduced the incidence or severity of turfgrass diseases. By making such claims, regardless of how vague they may be, they require EPA registration to be used legally (J. Anderson, EPA, *personal communication*). Currently, there are dozens of these types of products available to golf course superintendents, with many new ones appearing every year. Some have been thoroughly validated by research and are likely to be effective inoculants, but are not widely known throughout the industry.

The third class of products includes a large group of biologically-based materials sold for a variety of turfgrass ailments, including disease control. In many cases there is little or no logic to the selection of the "active" microbial strains contained in the product and no logical development of appropriate application strategies for the product. Although it is difficult to know how much testing has gone into the development of these products, it is doubtful that many have ever been scientifically tested for turf applications. These products rely primarily on marketing shrewdness and testimonials to support sales. It is this group of products that poses the greatest risk to the future health of biological control as a management strategy in turfgrasses since failures with these types of products can instill skepticism of the entire concept among turfgrass managers. The fourth group of products includes a variety of natural organic fertilizers, root enhancers, soil inoculants, and organic amendments. Many of these materials have been used in the turfgrass industry for a number of years. Most are not marketed for disease control but may have some disease control efficacy. In some cases, these products may be well-tested whereas others have no documented efficacy. Although products such as this may have a high degree of quality control as far as fertility contents, little or no quality control is maintained over disease-suppressive properties.

The Future of Biological Control in Golf Course Turf

Turfgrass management has clearly entered an age where microbiological solutions are being sought for biological problems. It is becoming increasingly apparent that maintaining active microbial communities in turfgrass soils is a vital part of overall turfgrass health. Studies on biological control clearly show the potential to affect disease control through both of the microbial-based technologies described. Currently, there are more questions than answers about how to optimize these technologies. Nonetheless, interest in and commercialization of biological control products continues to grow.

Golf course turf represents one of the most intensively managed plant-soil ecosystems. Many of the demands placed on golf course superintendents have forced them to manage turf in a manner that is detrimental to plant health. For example, the trends toward agronomically unrealistic cutting heights on rootzone mixes low in microbial activity, the ever-increasing amount of traffic on putting greens, and the low nutrient inputs to maintain high green speeds, have placed unprecedented stresses on turfgrass plants, making them highly susceptible to diseases. Along with these demands has come increased fungicide use, which, in turn, has spawned additional negative impacts on golf turf that may also require corrective treatment. As we become increasingly concerned with environmental contamination and a gradual decline of soil and plant health, a return to more biological-based approaches to turfgrass management will provide additional tools for maintaining a more sustainable and healthier turfgrass ecosystem.

Literature Cited

1)Nelson, E. B. *Turfgrass TRENDS* **1996**, *5*, 1-15.
2)Nelson, E. B.; Burpee, L. L.; Lawton, M. B. *Biological control of turfgrass diseases*; Leslie, A., Ed.; Lewis Publishers, Inc.: Chelsea, Michigan, 1994, pp 409-427.
3)Mancino, C. F.; Barakat, M.; Maricic, A. *Hortscience* **1993**, *28*, 189-191.
4)Peacock, C. H.; Daniel, P. F. *Hortscience* **1992**, *27*, 883-884.
5)Villani, M. G. *Turfgrass TRENDS* **1995**, *4*, 1-6.
6)Zhou, T.; Neal, J. C. *Weed Technology* **1995**, *9*, 173-177.
7)Nelson, E. B. *International Turfgrass Society Research Journal* **1997**, *8 (part 1)*, 791-811.
8)Nelson, E. B.; Craft, C. M. *Plant Disease* **1991**, *75*, 510-514.
9)Goodman, D. M.; Burpee, L. L. *Phytopathology* **1991**, *81*, 1438-1446.
10)Harman, G. E.; Lo, C. T. *Turfgrass TRENDS* **1996**, *5*, 8-14.
11)Lo, C.-T.; Nelson, E. B.; Harman, G. E. *Plant Disease* **1996**, *80*, 736-741.

12)Lo, C.-T.; Nelson, E. B.; Harman, G. E. *Plant Disease* **1997**, *81*, 1132-1138.

13)Sarniguet, A.; Lucas, P. *Phytopathology* **1991**, *81*, 1202.

14)Sarniguet, A.; Lucas, P. *Plant and Soil* **1992**, *145*, 11-15.

15)Wong, P. T. W.; Siviour, T. R. *Annals of Applied Biology* **1979**, *92*, 191-197.

16)Wong, P. T. W.; Baker, R. *Phytopathology* **1981**, *71*, 1008.

17)Wong, P. T. W.; Baker, R. *Control of wheat take-all and Ophiobolus patch of turfgrass by fluorescent pseudomonads*; Parker, C. A., Rovira, A. D., Moore, K. J., Wong, P. T. W. and Kollmorgen, J. F., Ed.; The American Phytopathological Society: St Paul, 1985, pp 151-153.

18)Baldwin, N. A.; Capper, A. L.; Yarham, D. J. *Evaluation of biological agents for the control of take-all patch (Gaeumannomyces graminis) of fine turf*; Beemster, A. B. R. e. a., Ed.; Elsevier Science Publishers: Amsterdam, 1991; Vol. 23, pp 231-235.

19)Thompson, D. C.; Clarke, B. B.; Kobayashi, D. Y. *Plant Disease* **1996**, *80*, 856-862.

20)Kobayashi, D. Y.; Guglielmoni, M.; Clarke, B. B. *Soil Biology & Biochemistry* **1995**, *27*, 1479-1487.

21)Kobayashi, D. Y.; El-Barrad, N. E.-H. *Current Microbiology* **1996**, *32*, 106-110.

22)Stockwell, C. T.; Nelson, E. B.; Craft, C. M. *Phytopathology* **1994**, *84*, 1113.

23)Rodriguez, F.; Pfender, W. F. *Phytopathology* **1998**, *87*, 614-621.

24)Nelson, E. B. *Microbial mechanisms of biological disease control*; Sticklen, M. B. and Kenna, M., Ed.; Ann Arbor Press: Chelsea, MI, 1997, pp 55-92.

25)Thomashow, L. S. *Applied and Environmental Microbiology* **1990**, *56*, 908-912.

26)Bonsall, R. F.; Weller, D. M.; Thomashow, L. S. *Applied and Environmental Microbiology* **1997**, *63*, 951-955.

27)Kloepper, J. W.; Leong, J. *Current Microbiology* **1980**, *4*, 317-320.

28)Becker, J. O.; Cook, R. J. *Phytopatholgy* **1988**, *78*, 778-782.

29)Elad, Y.; Barak, R. *Journal of Bacteriology* **1983**, *154*, 1431-1435.

30)Loper, J. E. *Phytopathology* **1988**, *78*, 166-172.

31)Buysens, S.; Heungens, K.; Poppe, J.; Höfte, M. *Applied and Environmental Microbiology* **1996**, *62*, 865-871.

32)Lemanceau, P.; Bakker, P. A. H. M.; Dekogel, W. J.; Alabouvette, C.; Schippers, B. *Applied and Environmental Microbiology* **1992**, *58*, 2978-2982.

33)Ahl, P.; Voisard, C.; Defago, G. *Journal of Phytopathology* **1986**, *116*, 121-134.

34)Kraus, J.; Loper, J. E. *Phytopathology* **1992**, *82*, 264-271.

35)Paulitz, T. C.; Loper, J. E. *Phytopathology* **1991**, *81*, 930-935.

36)Trutmann, P.; Nelson, E. B. *Phytopathology* **1992**, *82*, 1120.

37)Leeman, M.; den Ouden, F. M.; van Pelt, J. A.; Dirkx, F. P. M.; Steijl, H.; Bakker, P. A. H. M.; Schippers, B. *Phytopathology* **1996**, *86*, 149-155.

38)Curl, E. A.; Truelove, B. *The Rhizosphere*; Springer-Verlag: New York, 1986; Vol. 15.

39)Nelson, E. B. *Plant and Soil* **1990**, *129*, 61-73.

40)Maloney, A. P.; Nelson, E. B.; van Dijk, K. *Genetic complementation of a biocontrol-negative mutant of Enterobacter cloacae reveals a potential role of pathogen stimulant inactivation in the biological control of Pythium seed rots*; Ryder, M. H., Stephens, P. M. and Bowen, G. D., Ed.; Graphic Services: Adelaide, 1994, pp 135-137.

41)van Dijk, K.; Nelson, E. B. *Soil Biology and Biochemistry* **1998**, *30*, 183-192.

42)Ruttledge, T. R.; Nelson, E. B. *Phytochemistry* **1997**, *46*, 77-82.
43)van Dijk, K. *Seed exudate stimulant inactivation by Enterobacter cloacae and its involvement in the biological control of Pythium ultimum*; M.S. Thesis; Cornell University, 1995, 96 pp .
44)Papavizas, G. C. *Annual Review of Phytopathology* **1985**, *23*, 23-54.
45)Chernin, L.; Ismailov, Z.; Haran, S.; Chet, I. *Applied and Environmental Microbiology* **1995**, *61*, 1720-1726.
46)Haran, S.; Schickler, H.; Pe'er, S.; Logeman, S.; Oppenheim, A.; Chet, I. *Biological Control* **1993**, *3*, 101-108.
47)Fuchs, R. L.; McPherson, S. A.; Drahos, D. J. *Applied and Environmental Microbiology* **1986**, *51*, 504-509.
48)Jones, J. D. G.; Grady, K. L.; Suslow, T. V.; Bedbrook, J. R. *EMBO Journal* **1986**, *5*, 467-473.
49)Shapira, R.; Ordentlich, A.; Chet, I.; Oppenheim, A. B. *Phytopathology* **1989**, *79*, 1246-1249.
50)Sundheim, L.; Poplawsky, A. R.; Ellingboe, A. H. *Physiological and Molecular Plant Pathology* **1988**, *33*, 483-491.
51)Broglie, K.; Chet, I.; Holliday, M.; Cressman, R.; Biddle, P.; Knowlton, S.; Mauvais, C. J.; Broglie, R. *Science* **1991**, *254*, 1194-1197.
52)Cornelissen, B. J. C.; Melchers, L. S. *Plant Physiology* **1993**, *101*, 709-712.
53)Howie, W.; Joe, L.; Newbigin, E.; Suslow, T.; Dunsmuir, P. *Transgenic Research* **1994**, *3*, 90-98.
54)Zhu, Q.; Maher, E. A.; Masoud, S.; Dixon, R. A.; Lamb, C. J. *Bio - Technology* **1994**, *12*, 807-812.
55)Koby, S.; Schickler, H.; Chet, I.; Oppenheim, A. B. *Gene* **1994**, *147*, 81-83.
56)Nelson, E. B. *Biological and Cultural Tests for the Control of Plant Diseases* **1998**, *13*, 1-8.
57)Thurn, M. C. *Organic source effects on disease suppression and physical stability of putting green root zone mixes*; M.S. Thesis; Cornell University, 1993, 60 pp.
58)Kuter, G. A. *Mycologia* **1984**, *76:*, 936-940.
59)Craft, C. M.; Nelson, E. B. *Applied and Environmental Microbiology* **1996**, *62*, 1550-1557.
60)Hoitink, H. A. J.; Boehm, M. J.; Hadar, Y. *Mechanisms of suppression of soilborne plant pathogens in compost-amended substrates*; Hoitink, H. A. J. and Keener, H. M., Ed.; Renaissance Publications: Worthington, OH, 1993, pp 601-621.

Chapter 22

Potential for Use of *Stenotrophomonas maltophilia* and a Related Bacterial Species for the Control of Soilborne Turfgrass Diseases

D. Y. Kobayashi, J. D. Palumbo, and M. A. Holtman

Department of Plant Pathology, Cook College, Rutgers University, New Brunswick, NJ 08901

Strains of *Stenotrophomonas maltophilia* and a related bacterium, proposed as *Lytobacter mycophilus* gen. nov., sp. nov., are capable of controlling summer patch, a turfgrass disease caused by *Magnaporthe poae*. Strains of these bacteria produce an abundance of extracellular enzymes that have the potential to degrade fungal cell walls, and are capable of colonizing the turfgrass rhizosphere. Growth chamber studies indicate repeated application of *S. maltophilia* improves disease suppression compared to standard applications. *S. maltophilia* populations are reestablished above 10^7 colony forming units/g rhizosphere sample following each repeated application, suggesting these higher populations are critical for disease suppression. Field studies indicate that *S. maltophilia* populations can be established in turfgrass at these levels, providing support that disease control in the field can be achieved.

Growing concerns for environmental and health safety issues have led to the anticipation of stricter regulations for the use of chemical pesticides to control diseases. As a result, investigations into the development of alternatives to fungicides for disease control have increased. Biological control provides one potential alternative; however, its development has been faced with several challenges. One of the most significant challenges is overcoming inconsistencies in disease control performance (*1*), a problem that is accentuated for turfgrass diseases due to the lack of tolerance for disease by the industry. Maintaining high aesthetic quality of turfgrass is often further complicated by intense management practices that cause severe environmental and physical stress to the plant.

The rigorous requirements for control of turfgrass diseases can contribute to frequently reported failures of biocontrols. Furthermore, the nature of biocontrol hinders its effective use within traditional paradigms for pesticide usage, where methods for application and evaluation of performance are often similar to those used for evaluating chemical products. However, the expectation of biocontrols to perform in the same manner and at levels similar to chemicals is unlikely. Therefore, it is necessary to develop alternative models to those used for chemicals for the use of biocontrol of turfgrass diseases. Alternative approaches can be envisioned to incorporate all levels of the development of biocontrol, including developing new methods for their application, to incorporating its use with other management practices to reduce disease.

Diseases of Turfgrass Caused by Root-Infecting Fungi

Summer patch, caused by *Magnaporthe poae*, is an important disease that affects cool season turfgrass species such as Kentucky bluegrass (*Poa pratensis*) and fine fescues (*Festuca* spp.) (*2*). The disease occurs during periods of high soil temperatures and sustained high water potential in the soil, that are combined with other environmental conditions leading to root stress, such as soil compaction and low mowing heights (*3*). Summer patch is similar to other patch diseases caused by root-infecting pathogens, including take-all patch, necrotic ring spot and spring dead spot (*2*). The life cycles of the pathogens causing these diseases are presumed to be similar. Ectotrophic colonization of the host roots by the pathogen is followed by infection of roots and subsequent colonization of the vascular tissue. Under the appropriate conditions, infection leads to foliar symptom development, and eventual death of the plant. Pathogen spread between plants likely occurs through ectotrophic growth on roots. For summer patch, ectotrophic growth can begin during the early spring, well before disease symptoms appear during summer months. Ectotrophic growth, as well as infection, can continue throughout the disease season into late fall. The pathogen is presumed to overwinter either ectotrophically on roots, or within the roots of infected plants.

The endophytic and ectotrophic stages of the pathogen represent two different stages of the *M. poae* life cycle, and thus represent two different stages targeted for disease control. Current methods for control of summer patch and related diseases rely heavily on the use of systemic fungicides. Contact fungicides have proven less effective (*3*), since they do not function as curative controls, presumably due to the inability to affect the pathogen after infection has occurred. In addition to chemical controls, some cultural and management practices have been demonstrated to alleviate symptoms (*3,4*), which indicate promise for the development of integrated disease management strategies in the future.

Patch Disease Severity Evaluations in Relation to Biocontrol Efficacy

Some success has been reported for biocontrol efficacy of patch diseases (e.g. 5,6), although control levels are not comparable to the most effective fungicides. It is likely that many biocontrols function similarly to contact fungicides, in which control results from antagonism occurring during the ectotrophic growth phase of the pathogen life cycle. Disease control at this level may not be as easily detected using standard rating methods used for evaluating chemical products.

A variety of methods are used to measure the severity of summer patch and other turfgrass patch diseases. Two methods involve visual ratings that either estimate the percentage of foliar necrosis within the affected area or determine the size of the area affected by disease. Disease severity, however, is more accurately measured as the product of both measurements, i.e., the percent necrosis *within* the affected patch area (% necrosis/measurement[2]). Significant differences in disease severity among treatments that are not detectable on a given observation date can sometimes be detected by comparing disease severity over time, as measured by calculating the area under the disease progress curve (AUDPC). For example, in field studies evaluating summer patch disease control using treatments of bacterial biocontrol agents, significant differences in patch severity were either not detected on any observation date, or detected on only one of three observation dates. However, significant differences were detected for these same treatments when total disease as measured by AUDPC was compared (5,6). The observed differences in total disease using more sensitive methods of disease evaluation such as AUDPC may indeed reflect effects of treatments that represent significant reduction in pathogen populations and/or infections that are not as easily detected on single observation dates. Observations such as these may prove significantly useful in devising effective disease management strategies that integrate the use of biocontrols with other practices, while at the same time reducing chemical input.

Isolation and Taxonomic Characterization of Bacterial Strains for Biocontrol of Summer Patch Disease

A modified method that used mycelia of the pathogen as bait, originally designed by Scher and Baker (7), was combined with an enrichment culture procedure to isolate bacteria with parasitic traits to the summer patch pathogen, *M. poae* (8,9). The objective of this strategy was to isolate potential biocontrol bacteria which function by directly attacking the fungal pathogen. Using this approach, several bacterial strains were isolated from sources originating from golf course turf that demonstrated good biocontrol activity against summer patch under controlled environmental conditions (8,9, Kobayashi, D.Y. and El-Barrad, N., Rutgers University, unpublished data).

Stenotrophomonas (formerly *Xanthomonas*) *maltophilia* has been identified as a potential biocontrol agent for several agronomic crops. This bacterium is also known to possess several traits important in biocontrol. For example, *S. maltophilia* is competent within the rhizosphere of a variety of plant species (*10,11,12*), and is known to produce several extracellular enzymes, including chitinase, protease and lipase, that have the potential to degrade fungal cell wall components (*8,13,14*). Furthermore, this bacterium has been previously shown to form parasitic relationships with fungi (*15*).

One bacterial strain, 34S1, demonstrated to have good biocontrol activity for summer patch disease, was identified as *Stenotrophomonas maltophilia* based on a variety of tests, including comparisons to the Biolog nutritional utilization database (Microlog, Hayward, CA), and the fatty acid methyl ester (FAME) profile database (Microbial ID, Newark, DE) (*8*). Taxonomic characterization of a second strain, N4-7, revealed species similarity matches to strain 34S1 based on the results of Biolog. The near identical nutritional utilization profile suggests that the two strains occupy the same nutritional niches. However, distinct differences were determined between the two strains based on FAME profiles, 16S rDNA sequence, and serological comparisons. Physiological and morphological differences are also detected between the two strains. Strain N4-7 is nonmotile and lacks flagella, whereas strain 34S1 is motile and possesses polar flagella. Furthermore, strain N4-7 grows optimally between 25-30 C, with essentially no detectable growth at 37 C, while strain 34S1 grows well at 37 C. Based on these significant differences, strains N4-7 and 34S1 warrant taxonomic separation at the genus level (*16*, Holtman, M.A., Goyal, A., Zylstra, G. and Kobayashi, D.Y., Rutgers University, manuscript submitted), for which we propose the name *Lytobacter mycophilus* gen. nov., sp. nov., for strain N4-7.

Strain N4-7 shares a variety of traits with strain 34S1 that are thought to contribute to biocontrol activity. Both strains are capable of colonizing the turfgrass rhizosphere at populations greater than 10^6 colony forming units (cfu)/g (fresh weight) rhizosphere sample (*8,9*). In addition, both strains produce a variety of extracellular enzyme activities, including chitinase, protease and lipase. Strain N4-7 also produces significant β-1,3-glucanase activity, as detected by hydrolysis of the β-1,3-glucan substrates laminarin and zymosan. The substrates that are degraded by these enzymes constitute major components of the cell wall of *M. poae* as well as other fungal pathogens.

The extensive work on biocontrol activity of *S. maltophilia* 34S1, along with other isolates of the same species, suggests that environmental isolates of this species offer potential as biocontrol agents. However, their use beyond research purposes as biocontrol strains is subject to regulatory scrutiny due to their association with clinical strains that bear the same species name. Since significant similarities are observed between *S. maltophilia* 34S1 and strain N4-7, including traits involved in biocontrol activity as well as field efficacy data (Kobayashi, D.Y., El-Barrad, N. and MacDonald, G., Rutgers University, unpublished data), the taxonomic distinction of strain N4-7 reflects the importance and necessity to more clearly define potentially useful environmental isolates at taxonomic levels for use as bacterial biocontrol agents, thereby separating them from bacterial species classified as potentially hazardous clinical isolates.

Repeated Applications of Biocontrol Strains Improve Summer Patch Disease Control Efficacy in Controlled Environment Studies

Disease control efficacy studies in growth chambers indicate that strains N4-7 and 34S1 are both capable of significantly reducing summer patch disease by greater that 70% compared to disease in untreated control plants on single observation dates (*8,9*). Examination of disease progress curves indicates that treatment of pathogen-inoculated Kentucky bluegrass plants with strain 34S1 significantly delays the onset of summer patch, but does not change the rate of disease progression (*8*). In these studies, increasing bacterial concentrations further delays the onset of disease. Therefore, disease suppression results from a delay in disease onset, or a shift in time of disease progress curves. Similar observations are observed with strain N4-7 (Kobayashi, D.Y., El-Barrad, N. and MacDonald, G., Rutgers University, unpublished data), indicating that strains N4-7 and 34S1 both have similar effects on the suppression of summer patch.

The observations that onset of summer patch disease in pathogen-inoculated plants is delayed by treatment with biocontrol bacteria, and that increased concentrations of bacteria further delay disease onset, suggest that repeated application of biocontrol bacteria may continuously delay onset of disease. To test this hypothesis, strain 34S1 was repeatedly applied to *Magnaporthe poae*-inoculated Kentucky bluegrass var. Baron grown in 9" containers in controlled environmental conditions similar to the method described (*8,9*). Briefly, 25 ml of strain 34S1, at a concentration of 10^8 cfu/ml, was applied to each container on a bimonthly schedule beginning two weeks after seeding and continuing to the end of the experiment. This treatment was compared to a standard treatment of strain 34S1 applied two and three weeks after seeding. Plants were moved to the growth chamber, set at disease conducive conditions of 28 C and 70% humidity, four weeks after seeding and were scored regularly for disease symptom development by rating the percent foliar necrosis in each container as previously described (*8,9*). Figure 1a shows the results of mean rating values of 10 replicates for each treatment four weeks after plants were moved to disease conducive conditions. Summer patch disease suppression was significantly improved by repeated applications of strain 34S1 when applied on a bimonthly schedule over the eight week experimental period, compared to a standard experimental treatment of two single applications at weeks 2 and 3 of the experiment. The experiment was repeated a second time with similar results (data not shown).

Examination of bacterial populations in the rhizosphere of turfgrass, according the method described previously (*8,9*) indicated that repeated applications boosted strain 34S1 populations to above 10^7 cfu/g rhizosphere sample upon each application. In contrast, populations of strain 34S1 on plants treated by standard applications

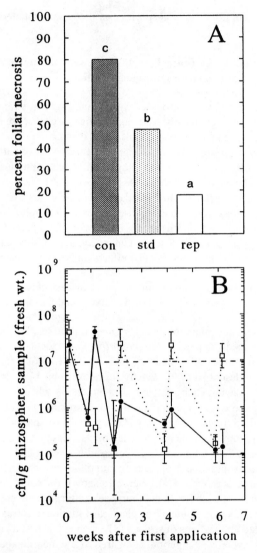

Figure 1. Effect of repeated and standard applications of
Stenotrophomonas maltophilia 34S1 to *Magnaporthe poae*-inoculated
Kentucky bluegrass var. Baron. A) Foliar disease symptom severity. Con
= pathogen-inoculated, untreated disease control plants; std = standard
treatment application of strain 34S1; rep = repeated application of strain
34S1. Different letters above bars indicate significant differences
according to Duncan's multiple range test (P=0.05). B) Rhizosphere
populations of strain 34S1 sampled on a weekly basis for standard
application (solid line with close circles) compared to bimonthly (repeated)
applications (dotted line with open squares). Error bars represent one
standard deviation from the mean.

decreased to 10^5 cfu/g rhizosphere sample after initial applications. These observations suggest that establishing the bacterium at populations of 10^7 cfu/g rhizosphere sample is necessary and critical to achieve significant disease control.

Field Trials for Summer Patch Disease Control

During the summer of 1995, strain 34S1 was applied at a concentration of 10^8 cfu/ml at a rate of 1 L/m² on a weekly and bimonthly basis in field trials of 3-year-old stand of Kentucky bluegrass var. Baron inoculated with *M. poae* in New Brunswick, NJ, similar to the procedure described (5,6). The plots were maintained at conditions similar to those for landscape turf, with a mowing height of 1.5 inches. Summer patch disease was not significantly reduced by any treatment within this field trial. However, enumeration of strain 34S1 populations in the rhizosphere indicated that the bacterium did not reach critical levels at any point during the season, even 24 h after application (Figure 2a).

During the summer of 1996, field application of strain 34S1 was repeated at a separate site from 1995 on turfgrass more intensely managed under golf course greens conditions (mowing height of 1/16") in New Brunswick, NJ. In this study, treatments of weekly and bimonthly bacterial applications were conducted on a mixed stand of pathogen-inoculated bentgrass and annual bluegrass. No summer patch disease was observed on any pathogen-inoculated plots used for the experiment. However, mean population levels of strain 34S1 in the turfgrass rhizosphere generally reached higher levels in 1996 than those observed in 1995 (Figure 2b). On a few occasions, populations reached above 10^7 cfu/g rhizosphere sample. However, values still decreased to below 10^5 cfu/g sample on a regular basis. Nonetheless, overall population sizes from weekly applications appeared to be greater than those from bimonthly applications.

In an effort to increase rhizosphere populations, field studies were repeated in 1997, in which a single treatment of strain 34S1 was applied on a weekly basis at a higher concentration of 5×10^8 cfu/ml. The experiment was conducted on pathogen-inoculated plots of annual bluegrass maintained at identical conditions to plots used in 1996. Under these conditions, populations repeatedly achieved levels above 10^7 cfu/g rhizosphere sample, and on only a few occasions dropped below 10^5 cfu/g sample (Figure 2c). These observations indicate that under the appropriate conditions, rhizosphere population levels deemed critical for biocontrol in growth chamber studies can be established in the field.

Future Directions for Biocontrol on Turfgrass

The expectation of biocontrols to perform at consistent levels equivalent to current chemical disease control methods, using similar application methods, are major impediments for the development of biological controls. As a consequence, few

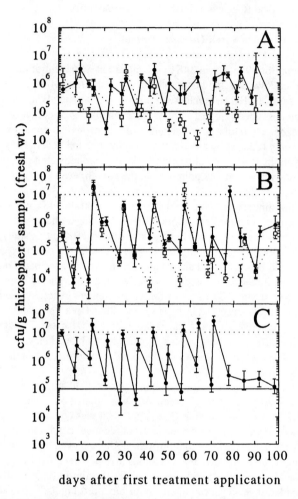

Figure 2. Turfgrass rhizosphere populations of *Stenotrophomonas maltophilia* 34S1 applied to field plots. A) 1995 field study comparing weekly applications (solid line with closed circles) and bimonthly applications (dotted line with open squares). B) 1996 field study comparing weekly applications (solid line with closed circles) and bimonthly applications (dotted line with open squares). C) 1997 field study of weekly applications. Error bars represent one standard deviation from the mean.

commercial biocontrol products are currently available for use on turfgrass. Infrequent application rates and poor establishment of biocontrol agent populations most likely contribute to failures that are often observed in the field. It is likely that the successful contribution of biocontrol for control of turfgrass diseases will require management practices that are adapted to its use and maximize its potential. Methods to apply biocontrols are already being developed and focus on integration with current management practices, such as delivering the agent through methods similar to those developed for fertigation and chemigation. There is ample evidence that management practices, such as aerifying soil-compacted areas, raising mowing heights and using different nitrogen fertilization sources, can reduce diseases such as summer patch. The development of disease management programs that incorporate these practices, along with the use of biologicals, can very likely lead to disease control at levels acceptable to the industry.

In efforts to develop disease management programs that successfully incorporate biocontrol, strong efforts will be needed to educate both users and the public concerning its benefits and safety. This includes clear taxonomic distinction of new beneficial biocontrol strains, as well as acceptance for the use of biocontrols under specific conditions. A clear understanding of how biocontrols can function as components of integrated disease management approaches is important to their success. Under these conditions, it is likely that biologicals can be utilized for turfgrass diseases, not necessarily to replace chemical pesticides, but to reduce their input, while still maintaining high efficacies for disease control expected by the industry.

Acknowledgments

Portions of this work was funded in part by United States Golf Association, Rutgers University Center for Turfgrass Sciences, The New Jersey Turfgrass Association, and The New Jersey Agricultural Experiment Station. We thank N. El-Barrad and G. MacDonald for excellent technical assistance with disease assays, and B. Dickson and J. Clarke for excellence assistance with maintenance of field plots. We also are grateful to B. Clarke for helpful suggestions throughout these studies.

Literature Cited

1. Weller, D. *Ann. Rev. Phytopathol.* **1988**, *26*, 379-407.
2. *Turfgrass Patch Diseases caused by ectotrophic root-infecting fungi*; Clarke, B.B.; Gould, A.B., Eds.; APS Press: St. Paul, MN, 1993.
3. Landschoot, P.J.; Clarke, B.B.; Jackson, N.; *Golf Course Manag.* **1989**, *57*, 38-41.
4. Thompson, D.C.; Clarke, B.B.; Heckman, J.R.; *Plant Dis.* **1995**, *79*, 51-56.
5. Thompson, D.C.; Clarke, B.; Kobayashi, D.Y.; *Plant Dis.* **1996**, *80*, 856-862.
6. Thompson, D.C.; Kobayashi, D.Y.; Clarke, B.B.; *Soil Biol. Biochem.* **1997**, *30*, 257-263.

362

7.   Scher, F.; Baker, R.; *Phytopathology* **1980**, *70*, 412-417.
8.   Kobayashi, D.Y.; Guglielmoni, M.; Clarke, B.; *Soil Biol. Biochem.* **1995**, *27*, 1479-1487.
9.   Kobayashi, D.Y.; El-Barrad, N.; *Cur. Microbiol.* **1996**, *32*, 106-110.
10. Debette, J.; Blondeau, R.; *Can. J. Microbiol.* **1980**, *26*, 460-463.
11. Lambert, B.; Leyns, F.; Van Rooyen, L.; Gosselé, F.; Papon, Y.; Swings, J.; *Appl. Environ. Microbiol.* **1987**, *53*, 1866-1871.
12. Lambert, B.; Meire, P.; Joos, H.; Lens, P.; Swings, J.; *Appl. Environ. Microbiol.* **1990**, *56*, 3375-3381.
13. Nord, C.-E.; Sjöberg, L.; Wadström, T.; Wretlind, B.; *Med. Microbiol. Immunol.* **1975**, *161*, 79-87.
14. O'Brien, M.; Davis, G.H.G.; *J. Clin. Microbiol.* **1982**, *16*, 417-421.
15. Willoughby, L.G.; *Trans. Br. Mycol. Soc.* **1983**, *80*, 91-97.15.12.

Chapter 23

Allelopathy versus *Neotyphodium (Acremonium)* Endophytes versus Competition Effects on Crabgrass Suppression by 12 Perennial Ryegrass Cultivars

John W. King[1], Donghoon Lee[1], Michael D. Richardson[1], Terry L. Lavy[2], Brigs Skulman[2], Melody L. Marlatt[2], and Charles P. West[2]

Departments of [1]Horticulture and [2]Crop, Soil, and Environmental Sciences, University of Arkansas, Fayetteville, AR 72701

The determination of allelopathic affects from 12 perennial ryegrass, Lolium perenne L., cultivars was attempted. Field trials of overseeding crabgrass, Digitaria sanquinalis (L.) Scop., into ryegrass stands did not show differences among the 12 cultivars. Crabgrass overseeding trials with 99 other ryegrasses showed a range of crabgrass percent cover and inhibitions in bioassays. Laboratory investigations centered on the effects of extracts of leaf-stem tissue on growth of duckweed, Lemna minor L., fronds. Extracts were prepared by macerating frozen tissue samples in water, centrifuging and filtering. Full (10 g/30 ml), one-half and one-quarter strength extract concentrations generally resulted in decreasing inhibition of frond growth, but sometimes stimulations occurred. Various bioassays of crabgrass seed germination showed some inhibitions. Overall results as to inhibition of duckweed and crabgrass by specific cultivars were inconsistent.

This paragraph serves as an expanded abstract. Twelve perennial ryegrass cultivars were selected for moderate to high stand density and zero to 95% endophyte infection and planted in field plots in late October 1993. A sero-immunoassay measured percent of Acremonium (renamed Neotyphodium genus) endophyte infection in plants sampled from the field plots. Percent of infection for several cultivars differed greatly from that expected based on testing seed. We did not find a correlation between endophyte infection level and inhibition of duckweed growth or crabgrass seed germination. None of the 12 cultivars affected the percent cover of crabgrass when crabgrass was overseeded into one half of each cultivar plot. No differences in percent crabgrass cover occurred where the 12 cultivars were overseeded into bermudagrass turf in the fall and overseeded to crabgrass in early spring. When a portion of each plot was overseeded to crabgrass, percent crabgrass cover differed significantly among 99 ryegrass cultivars included in the 1994 National Turfgrass Evaluation Program Perennial Ryegrass Test. But field studies do not differentiate between competition and possible allelopathic effects. Our primary method of screening for allelopathy was the duckweed bioassay. Frozen leaf-stem tissue samples were macerated in water, centrifuged, and filtered. Aliquots of extracts were placed into cell-plate cells with nutrient media and three-frond duckweed plants were added and frond numbers were counted one week later. Full (10 g/30 ml), half, and quarter strength extract concentrations from each cultivar generally resulted in decreasing inhibition of frond growth in the various duckweed bioassays, but sometimes stimulations occurred. Results of duckweed bioassays were also inconsistent over times of sampling. In surface soil from under the 12 ryegrasses-crabgrass seed germination bioassays, no differences in germination occurred. When air dried powdered leaf-stem tissue was mixed with soil in petri dishes, tissue from six of the 12 ryegrasses inhibited crabgrass seed germination strongly. Inhibition level increased with increasing rates of tissue incorporated. From among the 99 NTEP ryegrasses overseeded with crabgrass, APM and Top Hat allowed only 8% crabgrass cover while DVS NA 9402 and Linn were worst with more than 41% crabgrass cover. Linn and DVS NA 9402 extracts caused more inhibition of duckweed frond growth than APM and Top Hat. In an agar/extract-crabgrass seed bioassay Linn and Top Hat inhibited crabgrass seed germination more than APM and DVS NA 9402. Ryegrass stand density of Linn and DVS NA 9402 was low. Top Hat and APM stand density was high and medium, respectively. Over the totality of the many bioassays, the results as to inhibition of duckweed and crabgrass by specific cultivars were inconsistent.

This paragraph begins the introduction of these investigations. Allelopathy holds the potential to reduce the need for crabgrass control herbicides in culturing good quality turfgrass areas. This idea has driven considerable research to identify cultivars of tall fescue and perennial ryegrass which have useful levels of allelopathy against crabgrass.

Kentucky 31 tall fescue, Festuca arundinacea Schreb., has been reported to possess allelopathy against crabgrass (1). This report was the key to our receiving a grant from the University of Arkansas Alternative Pest Control Center to evaluate allelopathy in 12 tall fescue cultivars. Our results as well as results from many other allelopathy studies are summarized in a special report of the Alternative Pest Control Center (2). A few allelopathic responses to Kentucky 31 occurred in several bioassays, but not in others. Other cultivars only occasionally produced inhibition responses in duckweed or crabgrass seed germination bioassays. Thus inconsistencies of results over time made it difficult to interpret our tall fescue data.

A literature search revealed no publication on either allelopathy or endophyte levels in perennial ryegrass in relation to weed control. However, Acremonium (renamed Neotyphodium) endophyte infection of perennial ryegrasses caused allelopathy against white clover (3). Endophyte infection has been added to many turfgrasses in recent years to improve insect, disease and drought tolerance (4). This information, plus our on-going laboratory duckweed bioassay (5) procedure for evaluating allelopathic responses in duckweed growth from water extracts of plant tissues and endophyte testing (6) procedure, resulted in a grant from the United States Golf Association Green Section to study allelopathy of perennial ryegrasses against duckweed growth and crabgrass seed germination and growth.

Twelve perennial ryegrass cultivars were chosen for detailed study on the basis of their range of endophyte content and density of turfgrass stand. These somewhat arbitrary selections were based on National Turfgrass Evaluation Program and other performance reports, consultation with seed company representatives and seed company tests for endophytes infection levels in the seed of cultivars considered for inclusion. Trials involving crabgrass growth within field plots, growth responses of duckweed after treatment with extracts from leaf-shoot tissue, and crabgrass seed germination bioassays were conducted as well as sero-immunoassay tests of levels of endophyte infection.

Overseeding Crabgrass into Field Plots of the 12 Perennial Ryegrass Cultivars

Two types of crabgrass overseeding tests were conducted. One was the overseeding of crabgrass into plots of the 12 cultivars in early spring and evaluating ryegrass stand density and percent crabgrass cover in the plots during the spring and summer. The second was the overseeding of common bermudagrass "fairway" turf in the fall with the 12 cultivars, then overseeding crabgrass in the spring and evaluating density and crabgrass cover. It has been demonstrated repeatedly in practical turfgrass culture that the denser the turf the fewer the weeds. Overseeded bermudagrass turfs have greater stand density in the spring-time However, it is unclear whether allelochemicals in leaf-shoot debris in the turf and on the soil also contribute to weed control. Competition and any allelopathic effects are not separated in ordinary field studies; thus the need for the duckweed and crabgrass seed bioassays.

Culture of Field Plots and Overseeding Crabgrass. The standard cultural program was similar for both field tests. Fertilizer applications of 1 lb N/1000 ft^2 were made in early December, mid February, early September, and late October. Generally the fertilizer used was a 24-8-16 or similar grade with about 40% sulfur coated urea slow release nitrogen. Broadleaf winter weeds were controlled by an application of 2,4-D, meccoprop, and dicamba (Trimec) in mid March. The process for planting ryegrass was to vertical groove to a 0.25 inch depth, lightly rake off any debris, spread seed by hand and cross rake to cover seed. Ryegrasses were seeded at 5 lb/1000 ft^2 in the plots seeded in late October of 1993 and at 10 lb/1000 ft^2 when overseeded into bermudagrass in late October of 1994 and 1995. After planting either ryegrasses or crabgrass, irrigation water was applied lightly and frequently as needed to insure good germination and establishment. Thereafter watering was deep and infrequent as needed to prevent drought stress. All plots were 5 x 5 ft. Crabgrass was overseeded into the east or west half of each plot in late March of 1994, 1995 and 1996. The process was to spike several times and spread seed with a drop spreader at 1.1 lb/1000 ft^2. The side of the plot seeded to crabgrass was alternated each year to allow turf recovery to good uniformity and density before the next overseeding. The opposite half of the plot was treated with benefin (Balan 2.5 G) preemergence crabgrass herbicide in late March and given a supplemental application in early June. The crabgrass plots were given three MSMA applications at weekly intervals in late July and early August each year to control crabgrass. This fostered maximum ryegrass stand density recovery during the fall and winter seasons. Mowing was done regularly with a mulching mower. In the ryegrass test, mowing was at 0.75 inch height from mid March to late July - the crabgrass growing season. In the ryegrass overseeded into bermudagrass test, the 0.75 inch mowing was from early October to late July. During the remainder of the year mowing height was raised to 2.5 inches to allow recovery from summer stresses. A randomized complete block design was used in both tests with six replications in the ryegrass overseeded to crabgrass test overseeded to crabgrass test and four replications for the ryegrass overseeded into bermudagrass test.

Results of Overseeding Crabgrass into the 12 Ryegrass Cultivars. Our objective was to determine whether any of the 12 ryegrass cultivars affected percent crabgrass cover development. Visual estimates of ryegrass density and percent crabgrass cover in the overseeded plots were made in mid-May and late-June to mid-July of 1994, 1995 and 1996, respectively. Since this data showed no significant differences in 1994, density counts of shoots of ryegrass and crabgrass per square decimeter (a 4" diameter plug/cup cutter) were made in 1995 and 1996. Since tree root competition damaged the plots in replication one, statistical analysis was performed on data from the five remaining replications. Data were analyzed according to RCB ANOVA and means were separated by multiple t tests at the 5% level. As shown in Table I, no significant differences in crabgrass percent cover ratings or density counts occurred in 1994 or 1995 (or in 1996 - data not shown). Some differences in density ratings and counts occurred in 1995, but not in 1994 (or 96). Thus we conclude that, among the 12 ryegrasses chosen for study, any differences in inherent stand density, endophyte infection or possible allelochemical content were insufficient to affect crabgrass stand development.

Results of Overseeding Crabgrass into the 12 Ryegrass Cultivars Overseeded into Common Bermudagrass "Fairway" Turf. No significant differences in crabgrass percent cover ratings or density counts occurred in 1995 or 1996 (data not presented). We conclude as above that these 12 ryegrasses did not affect crabgrass development.

Preparation of Water Extracts from the 12 Ryegrass Cultivars for Duckweed and Other Bioassays

Since field plot tests do not distinguish between competition and allelopathic effects, it is necessary to conduct bioassays that isolate allelopathic effects. The duckweed bioassay system measures the effects of an aliquot of water extract from leaf-shoot tissue of individual ryegrass cultivars on the multiplication/growth rate of duckweed fronds. The system is capable of measuring inhibition at levels of allelochemicals ranging from 50 to 1000 μmol.

Duckweed are quite sensitive to their chemical environment. They multiply or reproduce by a process called budding. Duckweed are three-frond plants which grow by producing "buds" which enlarge into fronds and three new fronds split off as another plant. Other plants can be used in bioassays by measuring growth and/or determining fresh or ash weight, but counting fronds is a much simpler laboratory procedure and has good sensitivity.

Methods for Preparing Water Extracts. Leaf-shoot samples were cut about 1 inch above the ground from the field plots of each ryegrass cultivar. Plastic bags of these samples were stored in a freezer until processing through the extract preparation procedure for duckweed or other bioassays. Then, 20 g of frozen shoot tissue (combined from equal parts from each field rep sample) were chopped with scissors into

Table 1.   Effect of 12 Perennial Ryegrass Cultivars on Growth of Crabgrass Overseeded into Half of Each Ryegrass Plot.

Ryegrass Cultivar	1994		1995		1996	
	Ryegrass Density Rating* May 10	Crabgrass % Cover Rating** June 28	Ryegrass Density Rating* May 15	Crabgrass % Cover Rating** July 21	Ryegrass Density Counts*** May 14	Crabgrass Density Counts*** July 20
Loretta	5.8	22	5.8 cd	54	231 bcd	151
Gator	6.0	22	6.0 bcd	58	249 a-d	149
Derby	5.2	22	5.6 d	64	210 cd	177
Derby Supreme	5.4	23	5.8 cd	62	208 d	173
Envy	5.6	24	6.6 b	60	259 ab	155
Omega II	4.8	23	6.0 bcd	50	274 ab	170
Manhattan II (E)	5.6	24	6.2 bcd	48	253 abc	174
Saturn	5.0	19	6.2 bcd	62	244 bcd	136
SR 4200	5.8	17	6.4 bc	54	262 ab	165
Brightstar	6.2	18	7.4 a	66	292 a	157
Assure	5.4	19	6.0 bcd	68	260 ab	151
Yorktown III	6.2	22	6.0 bcd	56	250 a-d	135
Pr > F	0.14	0.37	0.0028	0.59	0.0191	0.40
LSD	NA	NA	0.75	NA	44	NA
C.V.	14	24	10	26	14	20

*   Visual estimate of stand density on 1 to 9 scale; 9 = very dense.
**   Visual estimate of percent crabgrass cover in ryegrass plot.
*** Number of stems per square decimeter.

1 cm or smaller pieces. The tissue was then homogenized in a small blender with 60 ml deionized water. The slurry was filtered through cheesecloth and filter paper. Nutrients were added, pH adjusted to 4.4, and total volume adjusted back up to 60 ml with deionized water. The extract was divided into two 30 ml centrifuge tubes and centrifuged at 3800 rpm for 15 minutes. The supernatant was filtered through 1 um and 0.45 um syringe filters and was then considered full strength extract concentration. By diluting with blank assay media (containing nutrients) half and quarter strength extracts were prepared without altering nutrition levels.

Methods of the Duckweed Bioassay System. Three-frond duckweed plants were placed into cells in a cell plate with nutrient media and a small aliquot of extract. One week later fronds were counted and compared to frond number in control cells. Upon statistical

analysis, this indicates whether allelochemicals in the extract inhibited, had no effect, or stimulated frond growth.

To insure a sterile environment for assays, E medium, water, transfer dish, cell plates and pipettes were autoclaved prior to use. Transfers of growth media, extract aliquots, and duckweed were done aseptically under a laminar hood equipped with ultraviolet light. (To further minimize the growth in the cell plates of any contaminating micro-organisms, sucrose and tartaric acid is withheld from the otherwise standard E medium for growing duckweed) A corollary lab process maintained a duckweed stock.

Cell plates containing 24 cells were partially filled with 1.5 ml of E growth medium. Six cells were "controls" and receive 5 ul of extra medium. Of the remaining 18 cells, six received 5 ul of full, six received one-half, and six received one-quarter strength extract concentration from a particular ryegrass cultivar. The location of each treatment within the 24-cell cell plate was randomized. A three-frond duckweed plant was placed in each cell. Care was used to select uniform plants. The cell plates were placed in a growth chamber at 26 C under constant light. After one week, the fronds in each cell were counted. The lab scheduling worked most efficiently when two cell plates were processed together. This allowed the frond count from 12 control cells to be averaged and used to calculate the percent of control for each extract rate for each cultivar. Each cell plate processed had six replications per treatment and percent control of duckweed was calculated over the average of these six replications. Thus each cell plate of treatments was considered as one "run" and used as one replication in the statistical analysis of the duckweed frond count data over seasons and years.

Results of the Duckweed Bioassays. Results of duckweed bioassays are shown in Table II. Leaf-shoot tissue samples were collected in mid-April and September 1994 and mid to late June of 1995 and 1996. The variability and inconsistencies of results among cultivars between April and September of 1994 were a serious concern. Therefore, for June 1995 samples six "runs" (see above) were conducted and four "runs" in 1996. The results are presented as percent control of duckweed. Since there were no replications of "runs" for April and September 1994 bioassays, the statistical mean separation analysis for April and September 1994 data as well as June 1995 and 1996 data is based on the pooled error term from the pooled analysis for the data from the June 1995 and 1996 bioassays.

The numerical values presented in Table II are percent of control (Number of duckweed fronds in treated cells divided by average number in control cells x 100.) The three extract concentration levels were used as a means for evaluating the relative strength of allelopathy of a cultivar. For instance, if all three extract concentrations inhibit duckweed strongly, then that cultivar has stronger allelochemicals than another cultivar that produces inhibition only by its full strength concentration.

As an example, let's examine Table II entries for Loretta for June 1995. The 31 ** for the full concentration indicates highly significant inhibition of duckweed growth compared to the 100 % value of the control. The 68 * indicates significant inhibition for the half strength concentration. The 90 does not show significant inhibition. The a, b, b indicate that full concentration gave results significantly different from the one-half and one-quarter strength concentrations.

Comparison of the Allelopathy of the 12 Ryegrasses. The most obvious result is the inconsistency of results over different sampling times (Table II). For most cultivars extracts from April 1994 samples produced stimulation of duckweed. Perhaps the relatively immature ryegrasses (planted in late October 1993) had not yet developed their allelochemicals fully. Thus the low strength allelochemicals may have acted as hormonal stimulants to duckweed growth. Ryegrass samples gathered in September 1994 and June of 1995 and 1996 generally produced inhibition regardless of extract concentration, but extracts from four cultivars produced stimulation. Based on the conjecture of young plants producing less allelochemicals, Derby and Omega II may have been the slowest to mature.

For the June 1995 and 1996 as well as to a lessor degree the September 1994 samples, the general tendency was for amount of inhibition of duckweed to decrease as extract concentration strengths decreased through full, one-half, to one-quarter. For most cultivars a significant difference occurred between full and one-half and one-quarter strength extracts. For the June 1996 samples for Gator and Derby, the three concentration levels produced responses bracketing strong inhibition to stimulation which suggests great sensitivity of duckweed to allelochemicals in these samples.

Extracts from all 12 cultivars inhibited duckweed growth at some sampling date. This pattern was most consistent for June 1995 and 1996 sampling dates. Loretta and Envy were the most consistent over four sampling dates in inhibiting duckweed growth. But as discussed above, none of the 12 cultivars affected crabgrass stand in field conditions.

Crabgrass Seed Germination On Surface Soil From Under The 12 Ryegrass Cultivar

To determine if the ryegrasses transmitted allelochemicals to the soil they were growing in we devised a soil-crabgrass seed germination bioassay. The surface 3\8th inch of soil from each ryegrass plug sample was placed in a petri dish, seeded to crabgrass, watered and percent germination counted.

Table II.   Effects of Extracts of Leaf-Shoot Tissue of 12 Ryegrass Cultivars on Growth of Duckweed Fronds.

Cultivar	Extract Conc.	April 94	Sept 94	June 95	June 96
			Percent of Control		
Loretta	Full	0 ** a	10 ** a	31 ** a	1 ** a
Loretta	Half	0 ** a	55 a	68 * b	22 ** b
Loretta	Quarter	107 b	51 a	90 b	53 ** b
Gator	Full	149 a	7 ** a	23 ** a	5 ** a
Gator	Half	191** a	46 * a	77 * b	106 c
Gator	Quarter	159 * a	63 a	82 b	148 ** b
Derby	Full	165 * a	140 a	6 ** a	24** a
Derby	Half	225 ** a	134 a	74 * b	143 ** b
Derby	Quarter	191 ** a	79 a	100 b	154 ** b
Derby Supreme	Full	36 * a	82 a	18 ** a	123 a
Derby Supreme	Half	137 b	73 a	84 b	120 a
Derby Supreme	Quarter	147 b	66 a	98 b	106 a
Envy	Full	0 ** a	24 ** a	22 ** a	10 ** a
Envy	Half	57 b	44 * a	90 b	29 ** b
Envy	Quarter	126 a	59 a	100 b	58 ** b
Omega II	Full	146 a	91 a	15 ** a	5 ** a
Omega II	Half	182 ** a	133 a	75 * b	24 ** b
Omega II	Quarter	168 ** a	114 a	96 b	50 ** b
Manhattan II (E)	Full	92 a	2 ** a	29 ** a	5 ** a
Manhattan II (E)	Half	182 ** b	39 ** a	105 b	58 ** b
Manhattan II (E)	Quarter	152 ** b	48 ** a	93 b	76 ** b
Saturn	Full	168 ** a	0 ** a	27 ** a	1 ** a
Saturn	Half	230 ** a	46 * a	84 b	7 ** a
Saturn	Quarter	203 ** a	45 * a	91 b	28 ** a

Percent less than 100 indicate inhibition and more than 100 stimulation.
* or ** indicate significant difference from the control (100%) at 5% or 1% for cultivar.
a, b or c indicate significant difference (5%) among full, one-half, or one-quarter extract concentration for cultivar.

Table II. *Continued.* **Effects of Extracts of Leaf-Shoot Tissue of 12 Ryegrass Cultivars on Growth of Duckweed Fronds.**

Cultivar	Extract Conc.	Percent of Control			
		April 94	*Sept 94*	*June 95*	*June 96*
SR 4200	Full	95 a	15 ** a	49 ** a	2 ** a
SR 4200	Half	253 ** b	29 ** a	107 b	22 ** b
Sr 4200	Quarter	202 ** b	36 * a	100 b	62 ** b
Brightstar	Full	0 ** a	0 ** a	34 ** a	2 ** a
Brightstar	Half	105 b	35 * a	88 b	19 ** ab
Brightstar	Quarter	144 b	21 ** a	92 b	46 **
Assure	Full	179 ** a	0 ** a	22 ** a	2 ** a
Assure	Half	140 ** a	28 ** a	78 b	13 ** a
Assure	Quarter	192 ** a	49 * a	102 b	49 ** b
Yorktown III	Full	165 * a	8 ** a	24 ** a	3 ** a
Yorktown III	Half	207 ** a	35 * a	96 b	60 ** ab
Yorktown III	Quarter	219 ** a	43 * a	103 b	91 b

Percent less than 100 indicate inhibition and more than 100 stimulation.
* or ** indicate significant difference from the control (100%) at 5% or 1%
for cultivar.
a, b or c indicate significant difference (5%) among full, one-half, or one-quarter
extract concentration for cultivar.

Crabgrass Seed Scarification Procedure. Preliminary germination tests showed that our purchased seedlot of large crabgrass seed had less than 20% germination. By trial and error, we determined that 8 minutes of magnetic stirring in 65% concentrated sulfuric acid, followed by four rinses with cold water and several rinses with deionized water through the seed in a Buchner funnel, air-drying and storing the refrigerator for a few weeks resulted in seed with 70 to 80% germination.

Methods of Soil - Crabgrass Seed Bioassay. Sample plugs (2.5 inch diameter by 2 inches deep) of each ryegrass in the five replications of plots were collected one rep at a time. The plugs were cut off at a 3\8th inch depth. The soil in the plot area was a Captina silt loam (fine-silty, mixed, mesic Typic Fragiudalt). This surface soil was squeezed and teased from the ryegrass roots onto two filter papers in four inch diameter petri dishes. The control soil was from a greenhouse supply of air-dried and stored Captina silt loam. The soil in the dishes was moistened to nearly field capacity. About fifty crabgrass seeds were sprinkled over the soil and dishes covered. The petri dishes were placed in the growth chamber. The day/night temperatures were 30/28 C. Light intensity at the petri dish level was 83 μmol/m^2/sec with a constant 12 hr photo-period. After one week and germinated and ungerminated seeds were counted. All seeds were on the soil surface and thus seen easily.

Results of Soil - Crabgrass Seed Bioassay. This bioassay was conducted with soil collected on 18 July 1996 and 8 May 1997 (Table III.). The germination was greater in the dried (control) soil, but no significant differences occurred among the cultivars in the 1996 or 1997 tests. Of course, this bioassay was not much different than the field plots tests where crabgrass was overseeded into ryegrasses. No differences in ryegrass cultivar affects on crabgrass seed germination were detected.

Powdered Leaf -Shoot Tissue from 12 Ryegrass Cultivars Affect on Crabgrass Seed Germination

This bioassay gave a partial simulation of effects that varying levels clipping debris on the soil surface might have on allelochemical accumulation and inhibition of crabgrass germination. Samples of leaf-shoot tissue collected in June 1996 from the 12 ryegrasses were air-dried, ground to a powder in a Wiley mill, mixed at 0, 500, 1000, 1500 mg rates with 40 g per petri dish of the greenhouse supply of Captina silt loam. The soil - tissue mixes were moistened to uniform dampness of nearly field capacity. The amount of water required increased slightly with increasing powdered tissue rates to compensate for water holding of the tissue. About fifty (+ or - 5) crabgrass seeds were sprinkled over the mix and pressed down lightly. The dishes were covered and placed in the growth chamber at conditions detailed above. One week later all seeds and germinated seeds were counted.

Results of Powdered Tissue - Crabgrass Germination Bioassay. The results are shown in Table IV. Six cultivars, especially Envy, Manhattan II (E) and Yorktown III, inhibited crabgrass germination strongly. The germination percentage decreased step-wise with increasing levels of tissue debris mixed into the soil. The interaction between

Table III. Effects of Surface Soil From Under 12 Ryegrass Cultivars on Crabgrass Seed Germination.

Cultivar	Percent Germination	
	18 July 96	8 May 97
Loretta	32 bc	39
Gator	38 bc	48
Derby	36 bc	44
Derby Supreme	40 bc	49
Envy	30 bc	46
Omega II	32 bc	46
Manhattan II (E)	32 bc	41
Saturn	31 bc	44
SR 4200	31 bc	40
Brightstar	35 bc	41
Assure	32 bc	45
Yorktown III	32 bc	47
Control	54 a	52
Pr > F	0.0001	0.1465
LSD	8.9	NS
C. V.	22	17

cultivars and tissue rate was not significant at the 5% level. The 1500 mg/petri dish rate is roughly 1 ton/A which is less than the clipping yield per year of a ryegrass turf. These results suggest that allelopathy against crabgrass may exist within ryegrass cultivars. inhibited crabgrass germination strongly. The germination percentage decreased stepwise with increasing levels of tissue debris mixed into the soil. The interaction between inhibited crabgrass germination strongly.

Neotyphodium Endophyte Infection Levels Found in the 12 Ryegrass Cultivars

The suppliers of the seedlots of the 12 perennial ryegrass cultivars indicated expected levels of endophyte infection as shown in the left column in Table V. We determined

Table IV.   Effects of Powdered Leaf-Shoot Tissue from 12 Ryegrass Cultivars on Crabgrass Seed Germination.

Cultivar	Mean % Germination
Loretta	44 ab
Gator	38 abc
Derby	37 bc
Derby Supreme	45 a
Envy	19 e
Omega II	27 d
Manhattan II (E)	15 e
Saturn	43 ab
SR 4200	39 abc
Brightstar	33 cd
Assure	28 d
Yorktown III	19 e
Pr > F	0.0001
LSD	7.0
C.V.	27

Rate of Leaf Tissue
mg / 40 g soil

0	44 a
500	35 b
1000	28 c
1500	23 d
Pr > F	0.0001
LSD	4.0
C.V.	27

Cultivars x Tissue Rate

Pr > F	0.0904

endophyte infection following a tissue-print immunoassay (6). In December of 1994, 1995 and 1996 about 25 shoots were randomly selected from all six replications of each cultivar and cut off near the ground. The calculated percentage infection was based on 120 shoots tested for each cultivar each year. The difference between expected and analyzed endophyte infection was quite large for several cultivars. Omega II, Manhattan II (E), Saturn and Assure had much lower levels of infection in the plants than in their respective seed lots. We speculate that excessive time or high temperature during commercial warehouse seedlot storage caused death of endophytes in many seeds. Levels of infection found in 1995 was lower than for 1994 for several cultivars, especially SR 4200. In general, a greater infection frequency was found in 1996 than 1995. For Loretta, Derby, Omega II, Manhattan II (E), Saturn and Assure we speculate that the "stronger" infected plants were gaining in stand percentage over the "weaker" non-infected plants, but such an interpretation does not fit the data for Gator, SR 4200 or Brightstar.

Effects of 99 NTEP Perennial Ryegrass Cultivars on Percent Crabgrass Cover

The 1994 National Turfgrass Evaluation Program Perennial Ryegrass Test was established in mid September 1994. The purpose of obtaining this test was mainly to evaluate 99 cultivars of ryegrasses for effects on crabgrass inhibition in field plots.

Methods for NTEP Ryegrass Test. The cultural program was the same as described above for 12 perennial ryegrass cultivars field plot study. Beginning in mid-March a 21 inch wide strip on the east side of the plots the mowing height was lowered to 0.75 inch and the clippings were removed. Then this strip was spiked several times and seeded to crabgrass at a 1.1 lb/1000 ft^2 rate on April 1, 1995. A visual rating of stand density was made on May 15 on a 1 to 9 scale with 1 being very sparse almost bare ground and 9 being very dense. On July 24 the percent crabgrass cover was estimated visually. We planned to plant the west side of the plots to crabgrass in 1996. But severe winter injury, probably from disease, caused 70 to 90% stand or density reduction on a majority of plots.. Without uniform competition from ryegrasses any crabgrass cover data would have been meaningless.

Results of 99 Ryegrass Cultivars Affects on Crabgrass Cover. As expected, the percent crabgrass cover values had a wide range with abundant statistical overlap among cultivars (Table VI). APM and Top Hat plots had the least crabgrass and DVS NA 9402 and Linn had the most. These four cultivars were chosen for more detailed bioassay evaluations.

The ryegrass density rating data is presented in Table VII. APM was in the middle range (6.0 d-g) of density. Top Hat was in the upper range of density (7.0 a-d). As an indication of variation, the second Top Hat entry was in the mid range of density (6.3 c-f). DVS NA 9402 ranked near the bottom of density (5.3 fgh) and Linn was the least dense. No statistically significant correlation occurred between crabgrass cover and density data.

Table V. Expected vs. Levels of Neotyphodium Endophyte Infection Found by Immunoassay in 12 Perennial Ryegrasses.

	Percent Infection		
	Expected	_Found_	_Found_	_Found_
Cultivar	_1993_	_1994_	_1995_	_1996_
Loretta	0	7	16	74
Gator	0	18	8	18
Derby	5-10	13	20	76
Derby Supreme	40-45	34	8	59
Envy	40	42	13	91
Omega II	76	14	16	52
Manhattan II (E)	50-98	12	15	80
Saturn	80	19	29	45
SR 4200	80-85	94	1	94
Brightstar	90	93	85	92
Assure	94-96	2	22	81
Yorktown III	97	76	45	88
	Pr > F	0.0001	0.0001	0.0001
	LSD	16	26	24
	C.V.	40	78	24

APM, Top Hat, DVS NA 9402 and Linn Extracts Affects on Duckweed Growth

These four cultivars were selected for further study on the basis of having the least and most crabgrass cover among 99 NTEP ryegrasses. The tissue samples were collected from the field plots in mid September 1996. The sampling, preparation of the extracts and duckweed bioassays followed the procedures detailed above.

Results of Duckweed Bioassay with Four Cultivars. Results are presented in Table VIII. Linn and DVS NA 9402 extracts produced highly significant levels of inhibition of duckweed growth at all three extract concentrations. APM and Top Hat extracts produced highly significant inhibition of duckweed growth only at full extract concentrations. This suggests that Linn and DVS NA 9402 had stronger allelochemicals

Table VI. **Percent Crabgrass Cover in 99 NTEP Perennial Ryegrass Cultivars on July 24 Within Low Mown Strips Overseeded with Crabgrass on April 1, 1995.**

Percent*	Ryegrass Cultivars
8 a	APM, Top Hat.
15 ab	Laredo, MVF-4-1, MED 5071, J-1706.
17 abc	Accent, Omni
18 a-d	PST-2M3, WVPB 92-4, Dancer, Riviera II, PST-2R3, Essence, RPBD, Prizm
20 a-e	Calypso II, MB 44, BAR USA 94-II, PSI-E-1, APR 131, WVPB-93-KFK, Achiever, MB 43, Williamsburg, Cutter
22 b-f	SR 4200, PST-2CB, PC-93-1, ISI-R2, Koos-93-3, Express, SR 4010, PST-28M
23 b-g	Elf, BAR Er 5813, LRF-94-C7, ZPS-3DR-94, Precision, LRF-94-CB, Apr 106, MB 45, MB 43, ISI-MHB, Top Hat
25 b-h	LRF-94-MPRH, Figaro, Quickstart, PST-2DLM, PST-GH-94, Brightstar, Assure, DSV NA 9401, Pick 928, PST-2FF
27 b-i	MB 42, APR 124, J-1703, Koos 93-6, Pick PR-84-91, MB-41, Pegasus, PST-2DGR
28 c-i	Nobility, MB 46, Saturn, Imagine, WX3-91, Manhattan III, PST-2ET, SR 4400, ZPS-2NV
30 d-i	APR 066, DLP 1305, LESCO-TWF, Divine, PST-2FE, WVPB-PR-C-2, Night Hawk, TMI-EXFL94, ZPS-PR1, Navajo, Esquire
32 e-i	LFR-94-6B
33 f-j	Pennfine, WX3-2DGR, Edge
35 g-j	Advantage, MB-1-5, Morning Star, Stallion Select, Nine-O-One, Vivid, Ps-D-9
37 hij	ZPS-2ST, Pick Lp 102-92
41 ij	DVS NA 9402
45 j	Linn

Pr > F	0.0001
LSD	13
C.V.	11

* Visual estimates of percent crabgrass cover within ryegrass plots.

378

Table VII. Density Ratings of 99 NTEP Perennial Ryegrass on May 15, 1995

Rating*	Ryegrass Cultivar
8.0 a	LFR-94-B6, MB 46.
7.7 ab	MB 45, Imagine.
7.3 abc	PST-2M3, MB 42, MB 43, MB 41, ISI-MHB.
7.0 a-d	MB-1-5, Pick Lp 102-92, Divine, LRF-94-CB, WX3-93, Top Hat, LRF-94-C7, ZPS-PR1, Pick PR-84-91.
6.7 b-e	ZPS-2ST, MB 44, MED 5071, LRF-94-MPRH, Pick 928, WVPB-PR-C7, CAS-LP23, SR 4200, MB 47, Morning Star, Prizm, Laredo, J-1706, Dancer, PST-2FF, PS-D-9, BAR USA 94-11.
6.3 c-f	Elf, Advantage, PST-2FE, Top Hat, Accent, ZPS-2DR-94, WVPB 92-4, SR 4010, Nobility, Express, Calypso II, Omni, Rivera II, PST-2R3, LESCO-TWF, PST-28M, Precision, WX3-91, Night Hawk, Pegasus, ISI-R2, RPBD, PST-GH-94, PST-GH-94, APR 124.
6.0 d-g	APM, PC-93-1, ZPS-2NV, Essense, PST-2ET, MVF-4-1, ST-2DLM, Edge, J-1703, Cutter, DLP 1305, Vivid, BAR Er Manhattan III, Saturn, Navajo, TMI-EXFLP-94, Williamsburg, APR 066, APR 131.
5.7 e-h	PSI-E-1, SR 4400, Quickstart, Koos 93-6, Nine-O-One, Stallion Select.
5.3 fgh	Koos 93-3, Brightstar, Figaro, PST-2CB, WVPB-93-KFK, DSV NA 9402, APR 106.
5.0 gh	Pinefine, Achiever.
4.7 hi	DSV NA 9401.
3.7 i	Linn

Pr > F 0.0001
LSD 1.18
C.V. 12

* Visual rating of stand density on 1 to 9 scale; 9 = very dense.

than APM and Top Hat. But the poor stand density of Linn and DVS NA 9402 allowed the most crabgrass invasion. Apparently stand density is more important than allelochemicals in suppressing crabgrass.

Table VIII.   Effects of Extracts from APM, Top Hat, DVS NA 9402 and Linn Ryegrasses on Growth of Lemna.

Cultivar	Extract Concentration	Percent of Control
APM	Full	7 ** a
APM	Half	96 b
APM	Quarter	87 * b
Top Hat	Full	39 ** a
Top Hat	Half	96 b
Top Hat	Quarter	87 * b
DVS NA 9402	Full	0 ** a
DVS NA 9402	Half	35 ** b
DVS NA 9402	Quarter	60 ** b
Linn	Full	4 ** a
Linn	Half	41 ** b
Linn	Quarter	60 ** c
	Pr > F	0.0001
	C.V.	19

Percents less than 100 indicate inhibition and more than 100 stimulation.

* or ** indicate significant difference at 5% or 1% from control for cultivar.

a, b or c indicate significant difference at 5% among full, half or quarter extract concentration for cultivar.

Crabgrass Seed Germination in Agar Medium As Affected by Extracts from APM, Top Hat, DVS NA 9402 and Linn Ryegrass Cultivars

The agar - crabgrass seed germination bioassay may be the most direct way of assessing allelopathy of ryegrass cultivar extracts against crabgrass. It was also the most difficult to perform. Micro-organism contamination on agar was a serious problem.

Methods of Agar - Crabgrass Seed Bioassay. The tissue samples were collected from plots in September 1996 by procedures detailed above. The preparation of extracts and scarification of seed followed procedures detailed above. Despite special care to follow sterile procedures, micro-organisms contaminated the agar medium during initial attempts to conduct this bioassay. It proved necessary (and generally successful) to give the seed an ethanol-chlorox surface sterilization treatment to control re-contamination which had occurred after the sulphuric acid seed scarification treatment.

The methods for preparing agar medium were to make a standard agar solution, pour 15 ml of agar and deionized water (for controls) into test tubes, autoclave for 20 minutes and place in 50 C water bath. Next add 15 ml of filter sterilized full, one-half and one-quarter strength extracts (and sterilized water for controls) into test tubes containing agar, mix briefly on vortex stirrer, and pour into sterile petri dishes. Then, using a flamed scoop sprinkle about 50 crabgrass seeds onto solidified agar-extract. The petri dishes were wrapped in parafilm and placed in a growth chamber. After one week, the total number and number of seeds germinated were counted.

Results of Agar-Extract Bioassay with APM, Top Hat, DVS NA 9402 and Linn Ryegrass Cultivars. The extracts of the four ryegrasses reduced crabgrass germination compared to control (Table IX). Top Hat and Linn extracts inhibited crabgrass germination the most. Field plots of Top Hat tied with APM for having the least crabgrass invasion. Top Hat and APM ranked moderate and high in stand density, respectively. Linn plots were least dense and had the most crabgrass. Did this seeming allelopathy of Linn and Top Hat reduce crabgrass invasion? It is difficult to know how to interpret these results.

Comments about Management of This Research Project

The original proposal was for detailed investigation of the 12 ryegrasses. The grant was cut in half and I (John King) considered reducing the number of cultivars to six. I kept hoping some research results would suggest which cultivars to eliminate, but such results didn't occur then. Even now I don't know which six cultivars I should have taken out of the research.

The decision to test full, one-half and one-quarter strength extracts against duckweed growth was scientifically sound. It provided information about the relative strength of allelochemical concentrations. But it tripled the work of duckweed bioassays.

The high variability in duckweed bioassay results in 1994 was very worrisome. In retrospect, the decision to use six and four runs in June 1995 and 1996 duckweed bioassays was an over-reaction. Even with all that effort, the co-efficient of variations remained very high - 63 for Table II data.

The decision to get the 1994 NTEP Perennial Ryegrass Test was good. Some of the best results came from this test.

The downside of these extra efforts was that some important aspects of this research were not done because of lack of time and funds. Despite the difficulties with contamination, the agar- crabgrass seed bioassay is the most direct way to access allelopathic effects from tissue extracts. Early spring and early summer samples should

**Table IX.   Effects of Leaf-Shoot Extracts from APM,
Top Hat, DVS NA 9402 and Linn Ryegrasses in Agar
Medium on Crabgrass Seed Germination.**

Cultivar	Percent Germination Sept 96
Control	54 a
APM	44 b
Top Hat	28 c
DVS NA	39 b
Linn	22 c
Pr > F	0.0001
LSD	9.0
C.V.	16

be studied to determine if seasonal differences in allelochemical strength exist. Soil - seed and powdered tissue - seed bioassays need to be conducted on a seasonal basis also. Any effects on allelopathic responses from nitrogen fertilizer rates need to be determined. The effects of presence or absence of *Neotyphodium* endophyte should be determined. (Our research with Kentucky 31 tall fescue indicated that high and low endophyte level did not affect crabgrass germination, shoot or root length. Article in preparation.) Ultimately, the nature of the allelochemicals need to be determined.

With our present knowledge and with similar further funding, I would select Linn and probably only two other cultivars for intensive laboratory study. The two cultivars would have superb agronomic characteristics, especially stand density. One would have low endophyte and the other high endophyte infection. Some seeds of the high endophyte cultivar would be treated to lower endophyte infection and effects of E+ and E- would be tested. Duckweed and the various crabgrass seed bioassays would be run. Chemical analyzes of the extracts would be done. Overseeding of strips of the NTEP perennial ryegrass cultivars with crabgrass would continue. Larger plots of the three cultivars would provide lab samples and overseeding trials. This approach would enable more thorough testing, but of fewer cultivars.

General Conclusions

Allelopathy may exist in perennial ryegrass cultivars. We have shown this possibility repeatedly in duckweed bioassays and in a few crabgrass seed germination bioassays. But the variability of results within and among the runs of duckweed bioassays was very high.

382

The maceration of leaf-stem tissue may release a vast array of "allelochemicals" into the water extraction brew. The extracts were not analyzed chemically. Naturally the extracts contain many compounds which would be dissipated quickly under field conditions. No correlation was found between endophyte infection level in the 12 ryegrasses and inhibition of duckweed or crabgrass. None of the 100 plus perennial ryegrass cultivars we studied in field plot overseeding trials suppressed crabgrass through competitive and/or allelopathic attributes as well as herbicidal crabgrass controls. Whether plant breeders or bio-technologists can or will concentrate allelopathy into ryegrass cultivars to a highly useful degree depends on their future decisions.

Acknowledgments.
This research was funded partially by a grant from the United States Golf Association Green Section. The 1994 National Turfgrass Evaluation Program Perennial Ryegrass Test was funded partially by a grant from NTEP. Also, the contributions of colleagues and student technicians are gratefully acknowledged.

Literature Citations.
1. Peters, E. J. and A. H .B. Mohammed Zam. **1981**. Allelopathic effects of tall fescue genotypes. *Agronomy J. 73:56-58.*
2. Bob Dilday, John King, Terry Lavy, Dick Oliver and Ron Talbert. Weed Control with Allelopathy. pp. 65-74. July **1997**. *In Alternatives: Accomplishments of the University of Arkansas Alternative Pest Control Center 1989-1995.* Special Report 180. Arkansas Agricultural Experiment Station. Division of Agriculture. University of Arkansas. Fayetteville, AR 72701
3. Hume, D. E. **1993**. Agronomic performance of New Zealand pastures: Implications of Acremonium presence. *Proc. of the Second International Symposium on Acremonium/Grass interactions: Plenary Papers p. 31-38.*
4. Funk, C. R., R. H. White, and J. P. Breen. **1993**. Importance of Acremonium endophytes in turfgrass breeding and management. *Agric. Ecosystems Environ.44:215-232.*
5. Einhellig, F. A., G. R. Leather, and L. L. Hobbs. **1985.** Use of *Lemna minor* L. as a bioassay in allelopathy. *J. of Chem. Ecology 11(1):65-72.*
6. Gwinn, K. D., M. H. Collins-Shepard, and B. B. Reddick. **1992**. Tissue print-immunoblot, an accurate method for the detection of Acremonium coenophialum in tall fescue (Festuca arundinacea L). *Phytopathology 81:747-748.*

Chapter 24

Cultural Control, Risk Assessment, and Environmentally Responsible Management of Scarab Grubs and Cutworms in Turfgrass

D. A. Potter[1], R. C. Williamson[1], K. F. Haynes[1], and A. J. Powell, Jr.[2]

Departments of [1]Entomology and [2]Agronomy, University of Kentucky, Lexington, KY 40546

Cultural tactics that may reduce reliance on insecticides were evaluated against scarab grubs and black cutworms attacking turfgrasses. Withholding irrigation or high mowing during beetle flight, or spring application of aluminum sulfate, reduced infestations of Japanese beetle and masked chafer grubs. Grub densities were not affected by applications of lime or urea, nor by use of a heavy roller. Although tolerance varies, all cool-season turfgrasses, including endophytic cultivars, are susceptible to grubs. Economic thresholds, however, are higher than generally is thought. Feasibility of identifying grub "hot-spots" by monitoring adults with pheromones was demonstrated. Field studies suggested that black cutworm infestations on golf putting greens can be suppressed by clipping management, topdressing, and use of resistant grasses in peripheral areas. Applications and limitations of cultural control for turfgrass insects are discussed.

Managed turfgrasses are attacked by a variety of destructive insects (*1, 2*). Although control of these pests traditionally has relied on broad-spectrum insecticides, this situation is changing. Demand for attractive lawns and uniform playing surfaces for golf and other sports remains high, but societal concerns about potential health risks and environmental hazards of chemicals are creating ever-tighter restrictions on urban pesticide use. Protecting groundwater and wetlands from pollutants is a national initiative (*3*). Agrichemical companies have responded with reduced-risk chemistry, including pyrethroids, chloronicotinyls, and insect growth regulators (*2*). Despite this trend, turfgrass managers continue to face pressures to grow quality turf with fewer chemical inputs. Other emerging problems of insecticides such as resistance by pests, enhanced microbial degradation, and induction of pest resurgences or secondary pest outbreaks underscore the need for effective, nonchemical control options.

Integrated pest management (IPM) relies on a combination of preventive and corrective measures to keep pest densities below levels that would cause unacceptable damage (4). Such measures include site assessment, monitoring, setting reasonable tolerance levels for pests or damage, use of resistant plants, and adjusting cultural practices to foster natural enemies and suppress pest populations (2, 5). Skilled turf managers long have deployed disease-resistant cultivars, and manipulated growing practices to discourage disease development (5). Such tactics are less commonly deployed against turfgrass insects, in part because of difficulty of sampling, uncertain damage thresholds, and absence of reliable non-chemical options (6, 7). We know relatively little about the underlying causes for insect outbreaks, or ways that cultural practices can be modified to reduce their occurrence in lawns or golf courses.

This chapter summarizes recent efforts to evaluate biologically-based methods for risk assessment, host-plant resistance, and cultural control of root-feeding scarab grubs and black cutworms attacking cool-season turfgrasses. Potential applications and limitations of these approaches are discussed. Although our research is emphasized, applicable work from other sources also is reviewed.

Biology and Management of Scarab Grubs

Root-feeding larvae of scarab beetles, commonly referred to as white grubs, are the most destructive insect pests of turfgrasses, especially cool-season grasses, grown in the United States (1, 2, 6). When accompanied by drought or other stress, their damage can be swift and severe. Grub-damaged turf develops irregular dead patches that can be easily lifted from the soil. Moles, skunks, and other vertebrate predators cause further problems by digging or tunneling in infested areas. Our work has focused on native scarab species including the northern and southern masked chafers, *Cyclocephala borealis* and *C. lurida*, respectively, and the green June beetle, *Cotinis nitida*, as well as the Japanese beetle, *Popillia japonica*, an important introduced pest.

Grub control traditionally has relied on short-residual carbamates or organophosphates that are applied in late summer, after eggs have hatched. Improper timing, failure to apply post-treatment irrigation, binding of residues in the highly absorptive thatch layer (8), and accelerated microbial degradation (9) may result in poor control. Although educators urge turf managers to treat selectively, this often is not done. The subterranean habits, patchy distribution, difficulty of monitoring, and destructive potential of white grubs (10), high aesthetic standards, and the view that fence-to-fence treatments provide insurance against damage all tend to encourage nonselective applications (7).

Cultural Control. Cultural practices may influence grub populations by altering the behavior of egg-laying adults, or survival of immature stages. Because eggs and young larvae are unable to survive below critical moisture thresholds, soil moisture probably is the most important factor affecting population dynamics of root-feeding scarabs (11–13). During dry periods, female scarabs may seek out areas with moist soils for oviposition.

We tested this hypothesis in a field study (*14*). Replicated plots of Kentucky bluegrass, *Poa pratensis*, received 0.64 cm of irrigation each morning from late May until September, or natural rainfall only. Even in 1992, a year with above-average rainfall, irrigated turf had higher densities of grubs (Figure 1). The grubs also were heavier and reached their full size more quickly in irrigated turf (*14*). This may explain why irrigated lawns, or moist, non-treated areas of golf courses such as tee banks and rough-fairway borders often harbor dense grub populations.

In another field study, cultural practices were manipulated before or during seasonal flights of *P. japonica* and *Cyclocephala* spp. to study effects on subsequent grub densities in tall fescue turf (*14*). Mowing at an elevated cutting height (18 cm, versus the more customary 7.6 cm), or application of aluminum sulfate (1223 kg/ha) to reduce soil surface pH from ≈ 6.4 to ≈ 5.4 reduced the total biomass of grubs by as much as 55 and 77%, respectively. Both treatments resulted in a tough, more fibrous root system that may have been a poor food resource for grubs. Grub densities were not affected by spring applications of lime, or by aerification of the turf before beetle flights (*14*). Use of a heavy (2247 kg) roller to compact the soil before beetle flights did not affect subsequent grub densities, nor was it effective for remedial control (i.e., crushing to death) of large, third-instar grubs in the fall (*14*).

Nitrogen (N) fertilization seemingly should favor white grubs because most herbivores are limited by dietary N, and turfgrass roots are especially low in nitrogen. Surprisingly, densities of *Cyclocephala* and *P. japonica* were not affected by application of urea at 150 to 300 kg N/ha per year (*14, 15*). Some fertilization practices may have species-specific effects. Green June beetles, for example, reportedly are attracted to sites fertilized with animal manures (*16*). Indeed, we found that application of composted cow manure or activated sewage sludge resulted in higher densities of their grubs in one of two years (*14*).

Management practices that affect turf vigor may mask or accentuate grub damage. For example, irrigation during the fall feeding period, followed by remedial N fertilization, promotes tolerance and recovery of grub-infested turfgrasses (*17*). Unless otherwise stressed, turfgrasses can withstand substantial reduction in root mass from grub feeding without loss of foliage growth (18). Thus, damage thresholds for grubs are much lower in drought-stressed stands than in well-watered turf (19). Scalping, excessive thatch, soil compaction, heavy traffic, disease, or other stresses also would be expected to accentuate grub damage.

Sampling and Risk Assessment. Probably the main reason why turf managers fail to adopt the IPM approach is that monitoring requires too much time. Some soil- or thatch-inhabiting pests (e.g., cutworms) can be detected with a disclosing drench (*2*), but sampling for scarab grubs requires removal of plugs with a golf cup cutter or spade, followed by examination of soil and roots. Eggs and young larvae are especially difficult to detect. Sequential sampling schemes have been developed (*10, 20*), but in practice they are prohibitively time-consuming or destructive, especially for the lawn care industry.

Ability to predict where grub infestations will occur would be beneficial in IPM. Several approaches to risk assessment have been explored. Nyrop et al. (*10*) sampled > 300 home lawns in upstate New York and analyzed correlations between site characteristics and incidence of European chafer, *Rhizotrogus majalis*. High risk sites included front lawns, and lawns that were young, in sunny locations, and had high amounts of Kentucky bluegrass. Whether these predictors apply in other geographic areas with different soils, turfgrasses, and predominant grub species is unknown. Villani (*21*) demonstrated the feasibility of mapping distributions of grubs on golf courses by training maintenance crews to systematically sample each fairway. This was useful for broadly targeting portions of the course with high densities of grubs. For scarabs such as *P. japonica* that feed as adults, the location of favored host plants can be a predictor of grub population densities, provided that adjacent turf areas are suitable for oviposition and survival of immature stages (*22*).

Another study (*23*) tried to predict densities of *Cyclocephala* grubs from numbers of males captured in sticky traps baited with crude extracts of female beetles. Traps were deployed on golf course sites and home lawns during peak beetle flight; nearby turf later was sampled for grubs. Hexane extracts containing sex pheromone were effective in luring males to traps. There was significant, albeit weak correlation ($r^2 = 0.25$) between male captures and grub densities in home lawns. No correlation was found on the golf course, where flight of male beetles between adjacent fairways and roughs confounded the ability to discriminate between high- and low-density patches. Shortage of female extract was a limitation in both tests; only enough virgin females could be collected to provide extract for two nights of sampling.

Trapping throughout the peak of adult flight (about 3 weeks) would provide a better indication of local beetle populations, but to be practical would require a synthetic lure. This approach to risk assessment probably has more potential for *Cyclocephala* and other scarabs that have localized mating flights, than for species such as *P. japonica* in which the adults range widely in search of food.

Identification of the *Cyclocephala* pheromone and production of a commercial lure may be hastened by the discovery that the sex attractant also is present in both male and female grubs (*24*). This greatly extends the period for collecting insects from which pheromone extract can be obtained. We are collaborating with chemists at Cornell University to identify and synthesize this pheromone. Besides potential application in risk assessment, a synthetic lure could be used to monitor beetle flights for timing control actions. Interestingly, *C. borealis* and *C. lurida* are cross-attractive (*24–26*), so one lure would attract both species. The finding that grubs contain volatile chemicals that elicit sexual behavior by males is noteworthy because it represents the first known case of a sex pheromone occurring in a larval insect (*26, 27*).

Damage Thresholds. Usefulness of sampling and risk assessment for white grubs is limited by poorly-defined damage thresholds. Studies with controlled infestation levels in enclosures have shown that thresholds depend on soil moisture, fertility, cutting height, grass and grub species, time of year, and other factors (*17–19, 28, 29*). Much of the industry perceives action thresholds to be ≤ 6 grubs per 0.1 m²; however, we found that nonstressed turf often will tolerate two to three times that density before showing symptoms of damage (*19, 29*). Use of damage thresholds in IPM is complicated by the trend for new-generation soil insecticides (e.g., imidacloprid, halofenozide) to be applied preventively, before the extent of infestation is known.

Host Plant Resistance. Scarab grubs, at least *Cyclocephala* and *P. japonica*, are polyphagous, feeding on roots of all cool-season turfgrasses (*30*) as well as common lawn weeds (*31*). Grubs reared in pots or implanted into field plots of Kentucky bluegrass, creeping bentgrass (*Agrostis stolonifera*), or endophyte-free cultivars of perennial ryegrass (*Lolium perenne*), hard fescue (*Festuca ovina* var. duriuscula L.), or tall fescue showed only subtle differences in growth and survival (*18, 30*). Fecundity of adult *P. japonica* also was similar regardless of which turfgrass they developed in as larvae (*30*). Less is known about relative susceptibility of warm-season turfgrasses, although bermudagrass, *Cynodon dactylon,* is suitable as food for *C. lurida* and *P. japonica* (28). Variation in susceptibility of turfgrasses to grubs is probably affected more by differences in tolerance than by their inherent suitability as food (*18, 28, 30*). Turfgrasses such as tall fescue that maintain summer root growth or initiate growth early in the fall may withstand root-feeding better than other grasses, such as Kentucky bluegrass, that are relatively dormant in late summer.

Infection of perennial ryegrass, tall fescue, and fine-leaved fescues by endophytic fungi (*Neotyphodium* spp.) conveys enhanced resistance to many stem- and foliage-feeding insects (*32*). Although the fungus itself is restricted to above-ground tissues, some endophyte-associated allelochemicals are translocated to roots (*30, 32*).

Research on the effects of *Neotyphodium* endophytes on white grubs is equivocal. Pyrrolizidine and ergot alkaloids incorporated in agar-based medium deterred feeding by *P. japonica* (*33*). Survival and growth of neonate *P. japonica* and growth of *Cyclocephala* were somewhat suppressed in pot-grown, E+ tall fescue (*30*). Fewer *P. japonica* were found in E+ than in endophyte-free (E-) tall fescue field plots in New Jersey in one of two years (*34*). In our field plots, however, survival and growth of implanted *P. japonica* or *Cyclocephala*, and natural grub densities were similar in E+ and E- tall fescue (*15, 30*). Furthermore, neither species discriminated between E+ or E- grass in preference tests (*35*). Female Japanese beetles maturing from grubs that developed in field plots of E+ tall fescue weighed about 10% less than females that developed in E- turf, but lifetime fecundity did not differ between the two groups (*30*).

Endophytes can have subtle interactions with scarab grubs. For example, sublethal stress associated with feeding on endophytic grass may render grubs more susceptible to entomopathogenic nematodes or other pathogens (*36*). Endophytic grasses also may be more tolerant of root-feeding (18).

Biology and Management of Black Cutworms

Golf putting greens are a focal point for environmental concerns because they receive more pesticides per unit area than any other turfgrass sites (*37*). The black cutworm [BCW], *Agrotis ipsilon*, is a major pest of greens, tees, and fairways throughout the United States (*1, 2*). Larvae construct shallow burrows, emerging at night to devour foliage and stems. Feeding by BCW causes sunken pockmarks that reduce uniformity and smoothness of the playing surface. Even greater damage occurs when birds forage for the larvae, gouging the turf surface. Because tolerance for such damage is practically nil, many golf superintendents treat several times each year to control BCW. Relatively little is known about ecology and habits of BCW in turf. Our research sought to determine how cultural practices, especially mowing, aerification, topdressing, and grass cultivar selection, affect BCW infestations on golf courses. More detailed accounts of this work are reported elsewhere (*38–40*).

Oviposition and Removal of Eggs by Mowing. Golf putting greens typically are mowed daily, whereas tees, collars, fairways, and aprons are cut several times per week. Clippings from greens and tees are collected in mowing baskets and discarded, often by strewing them in the surrounding high grass or rough. We studied how such mowing practices affect BCW eggs, and if larvae hatching from eggs on clippings can later reinvade greens.

Replicated plots on creeping bentgrass putting greens were mowed with a walk-behind greens mower at standard cutting heights (3.2 or 4.8 mm). Mated females were confined in open-bottom screened cages for 48 h, and number and location of eggs were determined. Grass blades with eggs were marked with paint. Plots were then mowed to their respective heights, and post-mow egg counts were obtained. Grass clippings were recovered from the mower basket; those with eggs were placed in mesh bags in the rough around putting greens. Eggs were inspected daily for hatch (*38*).

Caged moths laid similar numbers of eggs regardless of cutting height. Nearly all eggs were laid singly, on tips of grass blades. One mowing with a standard, walk-behind greens mower removed an average of 88% and 87% of the eggs at cutting heights of 3.2 and 4.8 mm, respectively (Figure 2). Hatch rates averaged nearly 50%

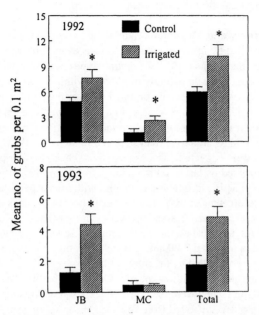

Figure 1. Mean (±SE) density of Japanese beetle (JB) and masked chafer (MC) grubs in Kentucky bluegrass plots that were irrigated or not during beetle flight periods. (Reproduced with permission from ref. 14. Copyright 1996 Entomological Society of America).

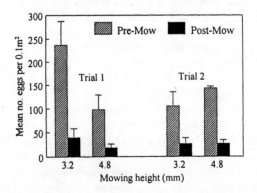

Figure 2. Removal of black cutworm eggs from a creeping bentgrass putting green by mowing. Differences between pre- and post-mow egg counts are significant for both cutting heights ($P<0.05$). (Adapted from ref. 38).

for eggs on mower-harvested clippings that were left in the field, confirming that many eggs survive passage through a greens mower (*38*).

Nocturnal Activity and Movement. Nightly activity and feeding behavior of BCW larvae were studied by inspecting putting greens every 2 h from sunset until sunrise on warm summer nights (*39*). Two distinct types of feeding behavior were observed. Some larvae fed while partially or entirely within a burrow or aerification hole, whereas others were found feeding or crawling over the putting surface while completely outside such a burrow. Samples of larvae engaged in each type of feeding were collected and their instar (age) was determined. Additional putting greens were inspected before dawn to locate larval tracks created in the dew. BCW larvae found at the end of the tracks were collected. Tracks were traced back to their point of origin, and the distances traversed by larvae on the putting surface were determined.

BCW larvae were active throughout the night, with greatest activity between midnight and ≈1 h before sunrise (*39*). Larvae began burrowing back into the turf just before dawn. Heavy bird predation was observed on larvae that remained on putting surfaces at sunrise. As is typical on most golf courses, crews arrived to mow the putting greens between 6:00 and 7:00 am, at which time the small numbers of BCW still on the surface were shredded by the mower blades.

Feeding behavior changed with larval age. Of several hundred BCW observed, >90% of the 3rd and 4th instars were feeding fully exposed on the turf surface, whereas >80% of the older larvae fed from self-made burrows or aerification holes. Larval tracks in the dew averaged 8.8 m (range 5–12 m), indicating that BCW can move considerable distances on putting greens in a single night. Many tracks originated from the peripheral area and terminated on the putting surface (*39*).

Response to Core Aerification and Topdressing. Putting greens typically are core aerified and topdressed several times per year (*41*). Aerification, done with a combustion-powered machine that removes soil cores using hollow tines or spoons, alleviates soil compaction, assists in controlling thatch, and increases air, nutrient, and water infiltration. Topdressing, the incorporation of 100% silica sand or a sand-peat mix into the putting surface, is done to correct irregularities in smoothness, and to control thatch (*41*). Topdressing is sometimes, but not always, associated with core aerification, and on many golf courses is done more frequently.

BCW often will occupy the resulting holes after putting greens are aerified (*1, 2*), but whether BCW are attracted to aerified putting surfaces, resulting in higher densities, was unknown. Silica sand is abrasive to insect cuticle, causing desiccation and mortality of certain soil insects. If BCW are susceptible to such effects, then topdressing might be timed to reduce infestations on putting greens. Response of BCW to combinations of aerification and topdressing therefore was studied (*39*).

Trials were conducted on bentgrass putting greens mowed daily at 4.8 mm. Six combinations of cultural manipulations were tested: aerified versus control, 100% sand topdressing versus control, aerified & topdressing versus control, aerified versus topdressing, aerified & topdressing versus control, aerified & topdressing versus aerified only, and aerified & topdressing versus topdressing only. Rectangular galvanized steel enclosures (91 x 46 x 15 cm) were driven into the turf so that each enclosure overlapped adjacent treatments, half of the enclosed area having received one manipulation and the remaining half another. Twenty 5th instar BCW were released in the center of each enclosure. Preference was determined after 48 h by using a soap solution (2) to flush the larvae. The test was run several times; larval preferences were compared by replicated goodness-of-fit tests (39).

Contrary to expectation, BCW were not attracted to aerified over nonaerified turf (Figure 3). However, about 80% of the larvae recovered from the aerified sides of the plots had exploited an aerification hole as a burrow. In contrast, BCW were somewhat repelled by sand topdressing, whether alone or in combination with aerification. Notably, when both treatments were aerified, larvae avoided the sides of enclosure that had been topdressed (Figure 3).

Host Plant Resistance. Creeping bentgrass is the predominant turfgrass on putting greens, but other grasses often are used in surrounding rough. If suitable as food, these higher-mowed grasses could support reservoir populations of BCW that may invade putting greens. We therefore compared growth, developmental rate, and survival of BCW on common cool-season turfgrasses used on golf courses (40).

BCW were reared to pupation on clippings of 'Penncross' creeping bentgrass, both endophyte-infected (E+) and endophyte-free (E-) cultivars of 'Assure' perennial ryegrass and 'SR8300' tall fescue, and three diverse Kentucky bluegrass cultivars ('Adelphi', 'Kenblue', and 'Midnight'). Grasses were grown in greenhouse flats, and endophyte infection was confirmed as >95% by immunoblot assay. Preferences of BCW among turfgrasses were tested in laboratory assays and paired field plots (40).

All Kentucky bluegrass cultivars were unsuitable as food for BCW. Those larvae were severely stunted; none survived to pupation on Adelphi or Kenblue, and survival on Midnight averaged only 8.8% across three trials. In contrast, E- perennial ryegrass and tall fescue were as suitable as creeping bentgrass (Figure 4). Endophytic tall fescue had no adverse effects on BCW. E+ perennial ryegrass had only slight adverse effects; larvae weighed less at 14 days and required 1-3 days longer to reach pupation than on E- grass. Pupal weights and survival rates, however, did not differ between E+ and E- perennial ryegrass. BCW showed no preference between creeping bentgrass, perennial ryegrass, and tall fescue, but Kentucky bluegrass was avoided.

Another study evaluated a range of bentgrass cultivars for resistance to BCW (42). The 12 cultivars were '18th Green', 'Cato', 'Crenshaw', 'Exeter' (Colonial), 'G-

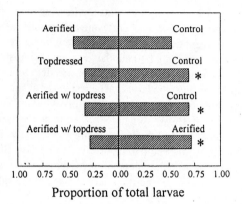

Figure 3. Response of black cutworm larvae to aerification and topdressing, separately or combined, on a creeping bentgrass putting green. Asterisks denote significant preference within paired comparisons (replicated goodness-of-fit tests, $P < 0.05$, 106–173 larvae per comparison; Adapted from ref. 39).

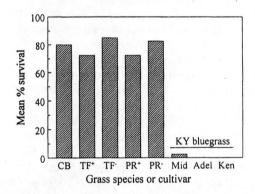

Figure 4. Survival of black cutworms reared on creeping bentgrass (CB), endophyte-infected [+] or endophyte-free [-] tall fescue (TF) or perennial ryegrass, or three diverse cultivars of Kentucky bluegrass: Midnight (Mid), Adelphi (Adel), or Kenblue (Ken). Data are means for two trials, 20 larvae per trial. (Adapted from ref. 40).

2', 'G-6', 'Lopez', 'Penncross', 'Penneagle', 'Providence', 'Seaside', and 'Southshore'. These were planted in replicated field plots as part of a National Turfgrass Evaluation Program trial. Clippings were harvested daily and fed to BCW as above. We found only slight differences in larval growth and pupal weight on the various cultivars (*42*). Large amounts of each cultivar were devoured.

Implications for Management. The finding that mowing removes most BCW eggs may explain why young larvae seldom are observed on putting greens (*38, 42*). Because many eggs survive on grass clippings, disposal of clippings well away from greens may help to prevent reinfestations. We confirmed that large BCW larvae may crawl considerable distances, and that many of those feeding on putting greens may originate from adjacent, higher-mowed turf. Thus, control actions for BCW should also target the reservoir population developing in rough areas. Treating a 8–10 m buffer zone around greens likely would reduce numbers of large larvae reinfesting putting surfaces. This also may reduce how often greens require treatment.

Insecticides for BCW are probably best applied toward evening. This coincides with the nocturnal feeding, may reduce loss of residues from photodegradation and volatilization (*2*), and minimizes exposure to golfers. We found that mid-sized BCW larvae tend to "graze" on the putting surface. Surface-feeding tends not to leave distinct pockmarks, which may be why golf superintendents often fail to notice early symptoms of damage. Pre-dawn mowing, which already is practiced on some golf courses, might provide mechanical control (shredding) of surface-feeding BCW.

Roughs consisting of creeping bentgrass, tall fescue, or perennial ryegrass may harbor higher reservoir populations of BCW than does Kentucky bluegrass, which is relatively resistant. Putting greens surrounded by susceptible grasses may therefore be at greater risk. Our choice tests suggested that BCW crawling through patches of Kentucky bluegrass and encountering a collar or putting green would tend to move into the creeping bentgrass. Endophytes of tall fescue and perennial ryegrass probably will not convey effective resistance to BCW. None of the creeping bentgrass cultivars tested is resistant enough to prevent BCW damage to putting greens.

Contrary to common perception, BCW do not seem to be attracted to core-aerified creeping bentgrass. Large BCW do exploit aerification holes as burrows, which accentuates their feeding damage. Larvae tended to avoid putting surfaces that were freshly topdressed with sand. Timing topdressing to coincide with phenology of BCW might help to reduce infestations on putting greens.

Conclusions

Scientists' perceptions of reasonable IPM systems may be unrealistic from the standpoint of simplicity, reliability, being compatible with the intended use of the turf, and interference with normal maintenance regimes. For example, our experience suggests that turf managers are unlikely to engage in sampling and monitoring because it is prohibitively time-consuming (7). Useful risk assessment methods and cultural controls must be compatible with the agronomic constraints under which turfgrasses are grown. Endophyte-enhanced resistance, for example, can only be implemented when establishing, renovating, or overseeding a turf. Manipulating irrigation or mowing height may be practical for some home lawns, but probably is not feasible for golf fairways. Similarly, Kentucky bluegrass, while resistant to black cutworms, is unsuitable for use on putting greens. It may, however, help to reduce reservoir populations of cutworms if used in surrounding roughs.

Adequacy of cultural controls will be enhanced under management regimes that conserve naturally-occurring beneficial invertebrates. Unless eliminated by broad-spectrum insecticides, predatory and parasitic arthropods normally buffer turfgrasses against pest outbreaks (43). For example, when creeping bentgrass cores upon which black cutworms had oviposited were implanted into golf roughs, ants consumed ≥ 90% of the eggs in a single night (Lopez, R.; Potter, D. A., University of Kentucky, unpublished data). It is encouraging that imidacloprid and halofenozide, two novel insecticides that are gaining wide usage, seem to have relatively little impact on predatory arthropods and earthworms (Kunkel, B. A.; Potter, D. A., University of Kentucky, unpublished data)

Insect populations are rarely uniformly distributed within or between turfgrass sites; rather, certain lawns or portions of golf fairways have recurring problems. Such sites seem to have just the right combination of turf type, soil, and site characteristics to favor particular pest species. Identifying these predisposing factors, and developing a profile of high risk sites (e.g., 10) would enable turf managers to concentrate their monitoring and control actions where they are most needed. One uncertainty is the extent to which such profiles can be extrapolated across geographic zones with different soil types and climate. Will site parameters that predict densities of Japanese beetle grubs on New York golf courses also apply in heavy clay soils of Georgia?

The causes for outbreaks of destructive turfgrass insects have received far less attention from entomologists than has the evaluation of treatments for the symptoms. Insecticides provide varying degrees of control, but the pests soon increase to their former high numbers, necessitating further treatments. Better understanding of how insect pests respond to site characteristics and cultivation may reveal ways that the system itself can be managed to maintain populations at low levels, rather than relying solely on intervention with chemical insecticides.

Acknowledgments

We thank the graduate students and research associates, especially B. A. Crutchfield, A. W. Davidson, D. Held, B. A. Kunkel, R. Lopez, C. G. Patterson (deceased), C. T. Redmond, P. G. Spicer, and D. W. Williams, whose work is reviewed herein. This research was supported by grants from the United States Golf Association, the O.J. Noer Turfgrass Research Foundation, and USDA SRIPM Grant 91-34103-5836. This is paper no. 98-08-128 of the Kentucky Agricultural Experiment Station, Lexington.

395

Literature Cited

1. Tashiro, H. *Turfgrass Insects of the United States and Canada.* Cornell Univ. Press: Ithaca, NY, **1987**.
2. Potter, D. A. *Destructive Turfgrass Insects; Biology, Diagnosis, and Control.* Ann Arbor Press: Chelsea, MI, **1998**.
3. Kenna, M. P. *U.S. Golf Assoc. Green Section Record.* **1995**. *33(1)*, pp. 1–9.
4. Pedigo, L. P. *Entomology and Pest Management.* 2nd ed. Prentice-Hall, NJ, **1996**.
5. Schumann, G. L.; Vittum, P. J.; Elliott, M. L.; Cobb, P. P. *IPM Handbook for Golf Courses.* Ann Arbor Press: Chelsea, MI, **1997**.
6. Potter, D. A.; Braman, S. K. *Annu. Rev. Entomol.* **1991**, *36*, pp. 383–406.
7. Potter, D. A. *Internat. Turfgrass Res. J.* **1993**, *7*, pp. 69–79.
8. Sears, M. K.; Chapman, R. A. *J. Econ. Entomol.* **1979**, *72*, pp. 272–274.
9. Niemczyk, H. D.; Chapman, R. A. *J. Econ. Entomol.* **1987**, *80*, pp. 880–882.
10. Nyrop, J. P.; Villani, M. G.; Grant, J. A. *Environ. Entomol.*, **1995**, *24*, pp. 521–528.
11. Régnière, J.; Rabb, R. L.; Stinner, R. E. *Can. Entomol.* **1979**, *115*, pp. 287–294.
12. Brown, V. K.; Gange, A. C. In *Advances in Ecological Research*; Begon, M.; Fitter, A. H.; Macfadyen, A., Eds.; Academic: New York, NY, 1990, Vol. 20; pp. 1–58
13. Potter, D. A.; Gordon, F. C. *Environ. Entomol.*, **1984**, *13*, pp. 794–799.
14. Potter, D. A.; Powell, A. J.; Spicer, P. G.; Williams, D. W. *J. Econ. Entomol.*, **1996**, *89*, pp. 156–164.
15. Davidson, A. W., Potter, D. A. *J. Econ. Entomol.*, **1995**, *88*, pp. 367–379.
16. Davis, J. J.; Luginbill, P. *N. C. Agric. Exp. Sta. Bull.*, **1921**, No. 242.
17. Crutchfield, B. A.; Potter, D. A.; Powell, A. J. *Crop Sci.*, **1995**, *35*, pp. 1122–1126.
18. Crutchfield, B. A.; Potter, D. A. *J. Econ. Entomol.*, **1995**, *88*, pp. 1380–1387.
19. Potter, D. A. *J. Econ. Entomol.*, **1982**, *75*, pp. 21–24.
20. Ng, Y. S.; Trout, J. R.; Ahmad, S. *J. Econ. Entomol.*, **1983**, *76*, pp. 251–253.
21. Villani, M. G. *Golf Course Manag.*, **1990**, *58(7)*, pp. 14–22.
22. Villani, M.; Nyrop, J.; Dalthorp, D. In *Soil Invertebrates in 1997*; Allsopp, P. G.; Rogers, D. J.; Robertson, L. N., Eds.; Proc. 3rd Brisbane Symp. Soil Invert.; Bureau of Sugar Experiment Stations: Brisbane, Australia; pp. 35–43.
23. Potter, D. A.; Haynes, K. F. *J. Entomol. Sci.*, **1993**, *28*, pp. 205–212.
24. Haynes, K. F.; Potter, D. A.; Collins, J. T. *J. Chem. Ecol.*, **1992**, *18*, pp. 1117–1124.
25. Potter, D. A. *Ann. Entomol. Soc. Am.*, **1980**, *73*, pp. 414–473.
26. Haynes, K. F.; Potter, D. A. *Amer. Entomol.*, **1995**, *41*, pp. 169–175.
27. Haynes, K.F.; Potter, D. A.. *Environ. Entomol.*, **1995**, *24*, pp. 1302–1306.
28. Braman, S. K.; Pendley, A. F. *Internat. Turfgrass Soc. J.*, **1993**, *7*, pp. 370–374.
29. Crutchfield, B. A.; Potter, D. A. *J. Econ. Entomol.*, **1995**, *88*, pp. 1049–1056.
30. Potter, D. A., Patterson, C. G.; Redmond, C. T. *J. Econ. Entomol.*, **1991**, *85*, pp. 900–909.

31. Crutchfield, B. A.; Potter, D. A. *Crop Sci.*, **1995**, *35*, pp. 1681–1684.
32. Siegel, M. R., Latch, G. C. M.; Johnson, M. C. *Annu. Rev. Phytopathol.*, **1987**, *25*, pp. 293–315.
33. Patterson, C. G.; Potter, D. A., *Entomol. Exp. Appl.*, **1991**, *61*, pp. 285–289.
34. Murphy, J. A.; Sun, S.; Betts, L. L. *Environ. Entomol.*, **1993**, *22*, pp. 699–703.
35. Crutchfield, B. A., Potter, D. A. *J. Entomol. Sci.*, **1994**, *29*, pp. 398–406.
36. Grewal, S. K., Grewal, P. S., Gaugler, R. *Entomol. Exp. Appl.*, **1995**, *74*, pp. 219–224.
37. Smith, A. E., Tillotson, W. R. In *Pesticides in Urban Environments*; Racke, K. D.; Leslie, A. R., Eds.; ACS Symposium Series: Vol. 522, Washington, D.C., **1993**; pp. 168–181.
38. Williamson, R. C.; Potter, D. A. *J. Econ. Entomol.*, **1997**, *90*, pp. 590–594.
39. Williamson, R. C.; Potter, D. A. *J. Econ. Entomol.*, **1997**, *90*, pp. 1290–1299.
40. Williamson, R. C.; Potter, D. A. *J. Econ. Entomol.*, **1997**, *90*, pp. 1283–1289.
41. Beard, J. B. *Turf Management for Golf Courses*; Macmillan: New York, NY, **1982**.
42. Williamson, R. C. *Behavior and Ecology of the Black Cutworm in Golf Course Turf;* Ph.D. dissertation, Univ. Kentucky: Lexington, KY, **1996**.
43. Potter, D. A. 1993. In *Pesticides in Urban Environments*; Racke, K. D.; Leslie, A. R., Ed.; ACS Symposium Series: Vol. 522, Washington, D.C., **1993**; pp. 331–343.

Chapter 25

Improved Mole Cricket Management through an Enhanced Understanding of Pest Behavior

R. L. Brandenburg[1], P. T. Hertl[1], and M. G. Villani[2]

[1]Department of Entomology, North Carolina State University,
Box 7613, Raleigh, NC 27695-7613
[2]Department of Entomology, NYSAES, Cornell University, Geneva, NY 14456

An economically and environmentally-sound approach to the management of a serious turfgrass pest to golf courses is effectively developed through an improved understanding of the pests biology and behavior. The subterranean turf pests in the group called mole crickets cause significant turf damage throughout the southeastern United States. The underground habitat of this pest renders this group both difficult and expensive to monitor and control. The development of a sound data base of mole cricket biology, ecology, and behavior is critical for the development an effective management plan. The use of a multi-tactic management approach is essential to provide the desired level of population suppression in an environmentally sound manner. The findings of a broad research effort (1993-1998) directed at mole crickets on golf courses have permitted the integration of new strategies into an effective management program.

Two species of mole crickets, the tawny (*Scapteriscus vicinus*) and its close relative the southern (*Scapteriscus borellii*) are the most troublesome turfgrass pests in the southeastern United States. They can be found throughout the golf course, but are particularly troublesome on tees and greens due to the low tolerance for injury in those sites. Mole crickets are capable of destroying large areas of turf leaving only bare soil unless effective control strategies are applied in a timely manner. The tawny mole cricket feeds almost exclusively on turfgrass roots and is destructive because of its tunneling and feeding. The southern mole cricket is a predator and will even feed on tawny mole crickets of equal or lesser size. Its damage is caused primarily from tunneling.

Basic Pest Biology

Mole crickets range from North Carolina in the East to Texas in the West and there is variability in mole cricket life cycles between the two species (*1*). However, the variation is small and both species have one generation per year throughout most of their range (2). The southern mole cricket may have two generations per year in south Florida. Mole cricket adults oviposit in late April and May in North Carolina. Eggs hatch in June and July and the small nymphs immediately begin feeding. Although these small crickets are very aggressive feeders, the turfgrass damage often goes unnoticed as the warm-season turfgrass is vigorously growing at this time of year and the tunnels are quite small. The nymphs continue to feed and grow through September and October. In the late summer and early fall, the large mole crickets create extensive tunnels and consume considerable amounts of turfgrass. The damage crickets cause is often rated on a '0'-9' scale with a '9' representing severe damage (3). In addition, turfgrass growth slows and the damage may become quite obvious. As soil temperatures cool, large nymph and adult mole crickets overwinter with limited activity in the soil during the winter months. Mole crickets complete development in March and April, mate and oviposition begins. Adult mole crickets die shortly after egg laying is complete.

Traditional Control Strategies

In general, chemical control strategies are most effective when applied to the small, newly-hatched nymphs (1). When crickets are small, however, the damage is not visible making spot treatments difficult. Additionally, egg hatch occurs over a period of several weeks and can vary from year to year. This may result in the insecticide being applied too early and have insufficient residual activity to control the nymphs that emerge at a later date. Conversely, the application may be made too late, after nymphs are larger and more difficult to control and visible damage is already present. The efficacy of control strategies also varies considerably among locations and under different environmental conditions.

Additional frustrations occur when mole crickets are so abundant that even 90% control leaves more crickets than can be tolerated on high quality turfgrass. This results in the need for multiple insecticide applications for adequate control. The current approach to mole cricket management utilizes relatively high initial doses of soil insecticides. Because many areas infested with mole crickets are quite environmentally sensitive (i.e. near water) repeated applications of insecticides are impractical environmentally as well as economically. Therefore, the ability to predict insect development and monitor cricket behavior is important. Two effective means to reduce pesticide load are to use pesticides more efficiently and use alternatives to pesticides. Both approaches require a more thorough understanding of the pest.

Field Biology Studies

Several field, greenhouse, and laboratory studies were conducted over a six year period (1993-1998) to monitor mole cricket egg-laying, egg hatch and nymph development as it relates to soil moisture and soil temperature. Data on mating flights were also collected as an indicator of spring activity. Egg hatch and nymph development were monitored by weekly soil flushes using an irritant solution of infested sites beginning in May of each year. These irritant flushes consisted of a 2% soap solution in water applied to the soil at approximately 8 liters per square meter. The irritant action of the soapy water forces small crickets to the soil surface allowing collection and counting.

Soil temperatures and rainfall were recorded using automated weather systems at each sampling site with readings taken every 15 minutes. Soil moisture data were collected at each site for three soil depths with a soil probe. Soil samples were returned to the laboratory and oven dried and percent moisture calculated. Mating flight activity was monitored using sound traps consisting of commercially available acoustic callers (Eco-Sim, Gainesville, FL, USA) mounted over large funnels to permit collection of the crickets responding to caller. Flight activity and its variations have been well documented for mole crickets in the southeastern United States (4).

Field study data (Hertl, P. T. and R. L. Brandenburg, North Carolina State University, unpublished data) indicate that the initiation of egg laying and egg hatch varies considerably from year to year (Table I), however the termination of egg hatch and development of the population to a damaging age class occurs at about the same calendar date each year. This consistency allows us to target treatment dates with greater confidence. The predictability of small nymph development to an economically damaging level using calendar date rather than soil degree day accumulations can save considerable time and expense.

Table I.   **Degree Day (base 10°C from January 1) Ranges for Egg hatch of the Tawny and Southern Mole Cricket**

Year	First Egg Hatch Tawny	First Egg Hatch Southern	Peak Egg Hatch Tawny	Peak Egg Hatch Southern
1993	---	---	2,300	3,500
1994	2,000	2,400	2,800	3,100
1995	1,800	2,100	2,400	2,900

Soil Moisture Studies

Soil moisture influences the initiation of oviposition thus affecting the timing of egg hatch. Greenhouse studies to determine the effect of soil moisture on oviposition have revealed significant trends that oviposition is delayed under dry soil conditions and accelerated under adequate soil moisture. Soil moisture regimens were established prior to the start of the experiment by combining a weighed portion of distilled water with a weighed amount of oven-dried sand. This procedure was used to produce soil with 2, 7, and 10% moisture based on the dry weight of the soil. Each treatment consisted of 20 PVC oviposition chambers 7.5-cm dia, 15-cm deep (3 inch wide by 6 inch long) similar to those used by Braman (5). Each chamber was filled with a uniform amount of one of the soil mixtures. One randomly selected female southern mole cricket was then confined in each chamber along with a portion of food. Each chamber was then weighed and soil moisture levels were maintained throughout the experiment by weighing the tubes weekly and adding an appropriate amount of water to maintain the starting moisture level. Crickets were fed weekly by adding pre-weighed portions of Mazuri Hi-Ca Cricket Diet (Purina Mills, Inc., Chicago, IL) treated with 0.4 %

potassium sorbate to inhibit fungal growth. Egg-laying was monitored by inspecting the bottom of each enclosure for the presence of eggs. The number of eggs laid in each chamber was recorded daily. Data were analyzed using an arc sine transformation prior to ANOVA and LSD procedures. Significance for mean separation was tested using either LSD (P= 0.05) or LS means procedures.

There were significant differences in both the number of eggs laid per female (Table II), and the percent females laying eggs (Table III). In both cases all three treatment means were significantly different (LSD, P= 0.05) for both measures of oviposition on Day 28. At Day 32 only the 4% treatment means were significantly different from those of the 7% and 10% treatments. The timing of egg-laying was also examined and the means for each treatment show an obvious trend, with mean egg-laying occurring earliest in the high moisture treatment and latest in the low moisture treatment. Mean separation by LSD (P = 0.05) show that on the average, eggs were laid 3.6 days earlier in the 10% treatment than in the 4% treatment, and that this difference is statistically significant (data not presented).

Behavior Studies - Laboratory

Additional studies to monitor cricket behavior have included the use of radiography to monitor mole cricket movement and tunneling in the soil. Laboratory studies using large transparent plastic containers (0.5 m wide X 0.5m high X 0.2 m thick) containing soil were used to monitor mole cricket tunneling and behavior.

Single crickets of each species were placed individually in these containers. A small piece of lead was glued to the prothorax of each cricket to aid in locating its position in the radiograph. The radiographs were taken with a Hewlett-Packard Faxitron (model 43855B, Hewlett-Packard Co., Palo Alto, CA). At specific intervals the boxes were X-rayed and the radiograph provided a permanent record of tunneling behavior. Other studies investigated the behavior response to multiple crickets of the same and different species in single containers. Containers were also X-rayed at specific intervals over several days time to provide documentation of behavior through time.

Table II. **Mean number of eggs laid per female (n = 20, 4 blocks) in three experimental soil moisture treatments in 1997.**

Soil Moisture Treatment	Day 14[a]	Day 21	Day 28	Day 32
10%	1.2 a	5.0 a	11.1 a	11.5 a
7%	0.6 a	2.8 ab	5.8 b	7.8 a
4%	0.4 a	0.8 b	6.3 c	15.2 b

[a]Means in a column followed by the same letter are not significantly different (LSD, P = 0.05. Square root transformation of mean eggs per female prior to ANOVA and LSD).

Table III. **Percent of mole crickets laying eggs in three experimental soil moisture treatments at three dates in 1997 (n = 20, 4 blocks).**

Soil Moisture Treatment	Day 21[a]	Day 28	Day 32
10%	23.8 a	51.3 a	53.8 a
7%	12.5 b	30.0 b	40.0 a
4%	3.8 b	6.3 c	15.2 b

[a]Means in a column followed by the same letter are not significantly different (LSD, P = 0.05; Arcsin transformation of data prior to ANOVA).

Results from the radiograph study revealed consistent traits about mole cricket behavior that may have management implications. Tawny mole crickets typically develop a "Y"-shaped tunnel that provide two routes to the surface (Figure 1). The majority of the root feeding occurs between and around these two "branches" of the tunnel and these routes offer adequate opportunities to escape predators, and as subsequent studies demonstrate, even avoid control measures. In the laboratory these "branches" off the main vertical tunnel were often as short as 5 cm and as long as 15 to 20 cm. In our studies the vertical tunnel then went to the bottom of the 0.5 m high box.

In contrast, southern mole crickets burrowed in a more horizontal fashion seeking prey near the soil surface. When multiple crickets of the tawny species were placed in a box, their behavior remained relatively unchanged. However, when a tawny and a southern mole cricket were placed in a box, the tawny cricket changed its behavior to avoid the southern mole cricket.

Preliminary studies indicate another behavioral characteristic with direct implications for control. Mole cricket behavior was modified when control agents such as insecticides or biological control organisms were applied to the soil surface. The crickets often appeared to avoid contact with the control agent for the duration of the experiments (2 weeks). Many control agents have residual activities of less than 2 weeks and the crickets could thus "escape" the control measure by staying deep in the soil for an extended period of time. This phenomenon may help explain the frequency of 'reverse rate' responses observed in efficacy trials directed toward mole crickets. The higher rates may cause avoidance behavior while lower rates do not induce such behavior and are thus more efficacious (Table IV). This area is under further evaluation to determine under which conditions this behavior occurs most consistently.

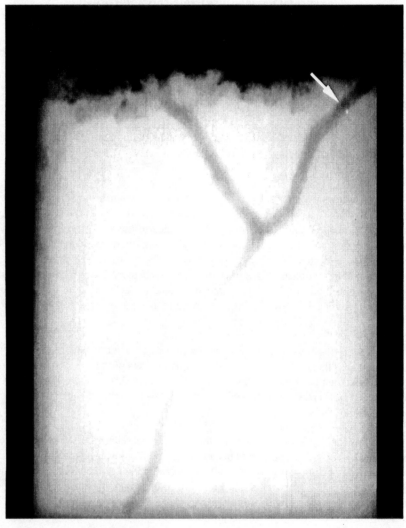

Figure 1. Radiograph showing typical "Y" shaped tunnel structure in soil of the tawny mole cricket.

Table IV.   Subsurface application of granular insecticides for the control of mole crickets nymphs, Fox Squirrel Golf and Country Club, Brunswick County, NC, evaluation of 9 August 1995, (7 DAT), 5 damage grid ratings (0 - 9) per replicate (0 = no damage; 9 = severe damage).

		Mole Cricket damage ratings[a,b]				
Treatment	Rate Kg (AI)/acre	I	II	III	IV	Average
Talstar 0.5 G	0.055	1.4	2.2	6.2	1.8	2.90 a
Talstar 0.2 G	0.11	2.0	1.8	7.8	6.4	4.50 abc
Talstar 0.2 G	0.220	5.6	3.8	8.6	7.4	6.35 bc
Bifenthrin 0.2 G (CAS)	0.055	4.8	7.6	5.2	8.8	6.60 c
Bifenthrin 0.2 G (CAS)	0.11	1.8	3.4	4.6	7.2	4.25 abc
Bifenthrin 0.2 G (CAS)	0.22	2.2	0.4	5.6	4.8	3.25 a
Mocap 10G	11.00	2.2	1.4	6.6	4.4	3.65 ab
Untreated	-----	3.2	5.0	7.4	8.2	5.95 bc

[a]Means followed by the same letter are not significantly different (DNMRT, $P = 0.05$).
[b] Damage ratings (0- 9); 0 = no damage; 9 = severe damage.

Behavior Studies - Field

Field studies during the fall of 1997 and spring of 1998 used melted paraffin poured into the mole cricket tunnels to develop castings of tunnel construction. Paraffin was melted in the field and poured into surface tunnel openings until the tunnel appeared filled. The paraffin was allowed to cool for one hour and then carefully excavated. More than 25 castings were taken and these confirm the "Y" shaped nature of the typical tawny mole cricket tunnel observed in the laboratory radiographs (Figure 2).

Pesticide Application Equipment

Due to the reliance on conventional insecticides to manage mole crickets, application equipment that places the control agent below the surface has shown promising results. (As with any soil insect, contact between the pest and the control agent is critical.) Numerous types of subsurface application equipment are currently available in the United States. These include high pressure liquid injectors, low pressure liquid injectors, and granular formulation slit applicators. While the potential for the various injectors is promising, their value is lessened by the initial high cost and the limited flexibility to use the equipment for anything other than soil insect control. Furthermore some equipment

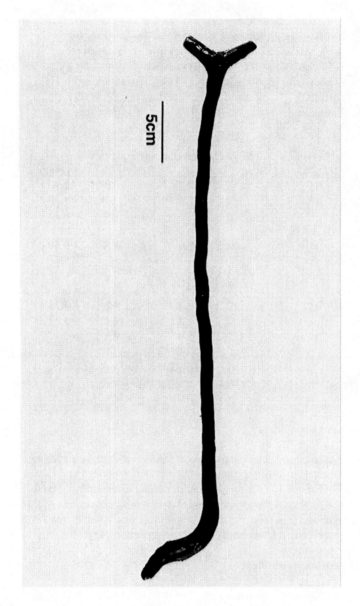

Figure 2. Wax casting from field studies using melted paraffin confirming "Y" shaped tunnel structure observed in laboratory radiographs.

performs poorly when the turfgrass is wet. As a general rule, subsurface applied insecticides performed more effectively as compared to surface applications (Table V). However, these differences are often quite subtle. Additional research is necessary to fully understand the benefits of such application equipment in the application of biological control agents such as entomogenous nematodes. Subsurface application reduces surface residue and drift and possibly minimizes some environmental risk, but does not preclude label requirements for irrigation following the application of many insecticides. Recently, additional turf equipment manufacturers have shown interest in the development of such equipment and this may enhance future research efforts and the potential for the use of such equipment.

Table V.   Comparison of subsurface and surface application of Dursban 2.32G (chlorpyrifos) for mole cricket control on bermudagrass fairways, North Carolina, U.S.A. 1992.

Treatment	Kg (AI) per ha	Damage rating (0-9) (6)[a] 47 days
		Location 1
Dursban 2.32G (chlorpyrifos) subsurface	2.2	0.44a
Untreated	---	6.56b
		Location 2
Dursban 2.32G surface	2.2	4.80a
Untreated	---	6.30a

[a] Means followed by the same letter are not significantly different. (DMRT, P = 0.05).

Biological Control and Cultural Practices

Several species of entomogenous nematodes have been evaluated for control of mole crickets in turfgrass (6, 7). While such control measures are desirable in light of their environmental soundness, they often fail to provide the level of control necessary to maintain the high quality turfgrass expected for golf courses *Steinernema carpocapsae, S. scapterisci,* and *S. riobravis* have been evaluated in numerous field trials. Our trials have focused on *S. scapterisci* and *S. riobravis*. Trials in North Carolina with entomogenous nematodes have never yielded better than a 50% reduction in mole cricket damage and results have been inconsistent (Table VI). Other studies evaluating a commercial formulation of *Beauveria bassiana*, Botanigard ES (Mycotech Corp, Butte, MT) has demonstrated some promise for future development and use in mole cricket and other soil insect pest management (Table VII).

Table VI. Use of the entomogenous nematode, *Steinernema scapterisci*, for adult mole cricket control on bermudagrass fairways, North Carolina, U.S.A. 1993

Treatment	Rate	Damage rating (0-9) (Cobb and Mack 1989) ave. of 20 damage ratings[a]
Steinernema scapterisci	2.54 billion/ha	1.1a
Untreated	----	2.1b

[a]Means followed by the same letter are not significantly different. (DMRT, P = 0.05).

Table VII.   The use of a commercial formulation of Beauveria bassiana for mole cricket control. Bermudagrass fairway, Brunswick Co., NC 1997.

Treatment/ formulation	Rate lb/AI/acre	Application Date	Mean mole cricket damage rating[a]		
			20 Aug	27 Aug	3 Sep
Merit 0.5G + Botanigard ES	0.34Kg+ 0.8 L/ha	27 Jun+ 13 Aug	0.10 a	0.60 ab	1.10 bc
Merit 0.5G + Botanigard ES	0.34Kg+ 3.2L/ha	27 June +	0.85 ab	1.05 bc	1.45 bc
Botanigard ES	3.2L/ha	13 Aug	0.25 ab	1.50 ab	1.30 bc
Merit 0.5G	0.34kg/ha	27 Jun	1.10 b	1.75 c	2.30 cd
Untreated	-----	-----	1.15 ab	1.05 bc	3.05 d

[a]Means within the same column followed by the same letter are not significantly different. (DNMRT, P = 0.05)

Inconsistencies with biological control agents as well as with conventional synthetic pesticides may be related to the avoidance behavior documented in the radiograph studies. The shorter residual activity of many control agents in conjunction with the avoidance behavior of mole crickets may explain the variation in efficacy we observe with some of these products.

IRRIGATION EFFECTS

Studies to monitor the impact of irrigation on the efficacy of mole cricket control agents established on golf course fairways and specific control treatments were replicated four times. Specific treatments included selected insecticide with relatively low water solubility applied under various irrigation regimens including: preirrigation (0.6m prior to treatment), postirrigation (0.6m immediately after application), both pre and postirrigation, and no irrigation. Mole crickets populations were monitored with grid rating sampling in the weeks following treatment. Some study sites revealed trends that are consistent with the concept that irrigation aids in movement of the insecticide down into the soil thus providing better control. However other sites provided data that do not support this hypothesis and indicates many other factors (i.e., soil moisture prior to irrigation) are involved that play a significant role in determining product efficacy.

A better understanding of mole crickets will allow us to use conventional insecticides as well as biological controls much more effectively. This research has unlocked many previously unknown aspects of mole cricket ecology and behavior. The findings of these studies are allowing us to assemble a clearer picture of mole cricket ecology as well as an improved understanding of the factors that affect management strategies.

ACKNOWLEDGMENTS

The authors acknowledge the funding of the United States Golf Association, Turfgrass Council of North Carolina, and the Carolinas Golf Course Superintendents Association in partial support of this research.

LITERATURE CITED

(*1*) Brandenburg, R. L. and M. G. Villani (Eds.). `Handbook of Turfgrass Insect Pests'. (Entomological Society of America: Lanham, MD) 1995.

(*2*) Walker, T. J.. Mole Crickets in Florida. Univ. Fla. Inst. Food Agric. Sci. Bull. 1984. **846**, Gainesville, FL.

(*3*) Cobb, P. P. and T. P. Mack   A rating system for evaluating tawny mole cricket. *Scapteriscus vicinus* Scudder, damage (Orthoptera: Gryllotalpidae). J. Entomol. Sci. 1989. **242**:142-144.

(*4*)Walker, T. J., J. A. Reinert and D. J. Schuster.   Geographical variation in flights of mole crickets, *Scapteriscus* spp. (Orthoptera: Gryllotalpidae). Ann. Entomol. Soc. Am. 1983. **76**:507-517.

(*5*)Braman, S.K.   Progeny production, number of instars, and duration of development of tawny and southern mole crickets (Orthoptera: Gryllotalpidae). J. Entomol. Sci. 1993. 28:327-330.

(*6*)Hudson, W. G., J. H. Frank and J. L. Castner.   Biocontrol of *Scapteriscus acletus* (Orthoptera: Gryllotalpidae) in Florida. Bull. Entomol. Soc. Am. 1988. **345**:192-198.

(*7*)Parkman, J. P., W. G. Hudson and J. H. Frank. Establishment and persistence of *Steinernema scapterisci* (Orthoptera: Gryllotalpidae). J. Entomol. Sci. 1993. **28**:182-190.

Chapter 26

Pasteuria sp. for Biological Control of the Sting Nematode, *Belonolaimus longicaudatus*, in Turfgrass

R. M. Giblin-Davis

University of Florida, Fort Lauderdale Research and Education Center, IFAS, 3205 College Avenue, Fort Lauderdale, FL 33314–7799

A new species of *Pasteuria* (S-1) (obligate, Gram-positive, endoparasitic bacterium) was discovered in Florida that parasitizes the sting nematode, *Belonolaimus longicaudatus*. This bacterium was characterized morphometrically and ultrastructurally along with host attachment studies. A 24-month survey of locations in hybrid bermudagrass plots in southern Florida demonstrated density dependent regulation of sting nematodes in areas infested with *Pasteuria* (S-1). Survey areas that started with low levels of spore encumbrance showed a building trend in encumbrance levels and a corresponding decline in the numbers of sting nematodes. Locations with high spore encumbrance levels cycled and appeared to curtail sting nematode population resurgences, suggesting that *Pasteuria* (S-1) can suppress the sting nematode in a turfgrass ecosystem. A study comparing the effects of inoculating 900 mls of *Pasteuria* (S-1) spore-infested soil (ca. 5,000 endospores/ml) into the center of 1 m^2 turfgrass plots demonstrated that *Pasteuria*-infested soil can be introduced into a turfgrass site with high numbers of sting nematodes to bring about density dependent effects within 12 months.

A major problem in golf courses, athletic fields and lawns in the southern United States is the damage caused to roots by the feeding of phytoparasitic nematodes (*16, 42*). Plant-parasitic nematodes are unsegmented roundworms belonging to the metazoan phylum Nemata. Nematodes are extremely diverse with estimates of about 500,000 species in the world of which only about 10,000 have been described. They are ubiquitous and abundant being found in almost every habitat with enough water to allow for feeding and movement, including marine, freshwater and terrestrial environments (*29*). Most nematodes are free-living and feed on detritus or bacteria. Some are omnivorous and/or predatory. Vertebrate, invertebrate, plant and fungal-parasitic nematode groups have evolved independently several times from free-living nematode ancestors (*6*). These interesting animals are usually colorless, transparent and microscopic in size (250-3,000 μm) (Figure 1C,D), although animal-parasitic forms can be quite large (e.g. *Ascaris lumbricoides* females that inhabit the intestines of humans are up to 35 cm [13.8 in] long and *Placentonema gigantisma* that lives in the placenta of the sperm whale can attain about 8.3 m [27 feet]).

Estimates of nematode numbers in agricultural soils vary from 1 to 10 billion nematodes per acre (*29, 42*). A typical 100 ml soil sample from the top 10 cm of a home lawn or golf course turf in Florida yields 200-2,500 nematodes with about 20-75% being plant-parasitic (bearing a hollow, modified feeding structure called a stylet used for puncturing cell walls and conducting injected digestive or plant growth modifying secretions into the host and then sucking liquid from the host cells) (Figure 1D).

Worldwide crop damage caused by plant-parasitic nematodes is estimated at over $78 billion with about $8 billion worth of damage in the United States (*3*). The actual impact is probably greater because of the difficulty of proper diagnosis (*3*). Nematodes are considered to be one of the most important invertebrate problems in golf course turf (*27*) and cost estimates of invertebrate pesticide usage on Florida's 1300 golf courses are more than three times that of any of the other U.S. regions surveyed (*27*). About $4 million dollars would have been spent for nematode control in 1996 assuming that 10% of the pesticide budget for an average Florida golf course is used for nematode suppression.

One of the most damaging nematodes in the warm-season turfgrass ecosystem in sandy soils is the ectoparasitic sting nematode, *Belonolaimus longicaudatus* (Figure 1) (*19, 42*). There are many other species of plant-parasitic nematodes that can cause root damage (*16, 41, 42*). However, based upon pathogenicity estimates, the sting nematode is the most damaging phytonematode to turf roots in the state of Florida with estimates of 10 (*16*) to 35 (*42*) or 40 (*8*) sting nematodes per 100 mls of soil causing symptoms on many turf types. Thirty to fifty percent dry root weight reductions occur in controlled inoculation studies with sting nematode on commonly cultivated hybrid bermudagrasses that are used for golf course greens such as 'Tifgreen' and 'Tifdwarf' (*23*) and on certain diploid St. Augustinegrasses used for lawns (Figure 1B) (*7, 8, 19*). A reduction in the biomass and health of host roots can cause a significant decrease in host transpiration (*19*) and nutrient uptake leading to chronic water and nutrient stress especially in plants grown in sandy soils with low cation exchange and water holding capacity. This leads to the typical above ground symptoms of patchily distributed chlorotic and underperforming turf (Figure 1A).

The sting nematode has a wide host range (eurytrophic) and is a major pest of a variety of grasses, vegetables and perennial crops. It is most damaging to seedlings and young plants (*31*). It is a relatively large plant parasite (ca. 2,000 μm long) (Figure 1C) and goes through its life cycle in about 28 days at 27° C (*35*). The sting nematode does best in soils with > 80% sand (*31, 35*). *Belonolaimus longicaudatus* and closely related species are documented pests in the sandy soils of the Coastal Plains from Florida north to Virginia and along the Gulf Coast into Texas. They also occur in Arkansas, Kansas, Oklahoma, Missouri, and Nebraska (*10*). *Belonolaimus*

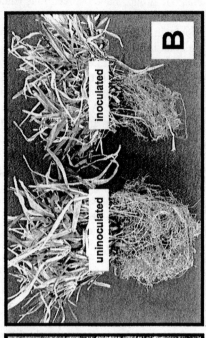

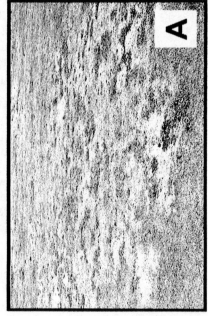

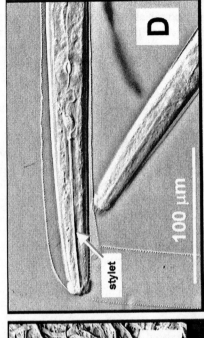

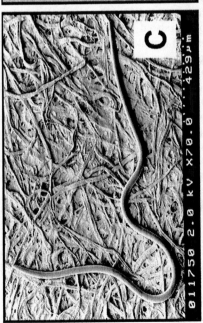

Figure 1. Symptoms and signs of the sting nematode, *Belonolaimus longicaudatus*. A. 'Tifgreen II' hybrid bermudagrass fairway with over 100 sting nematodes per 100 mls of soil showing characteristic chlorosis and weed encroachment. B. Root mass differences between a potted FX-313 St. Augustinegrass control (uninoculated; left) and one inoculated with 50 sting nematodes (right) 168 days after inoculation. C. Scanning electron micrograph of a female sting nematode. (courtesy Dr. Bill Wergin, U.S.D.A.) D. Light micrograph of two sting nematodes showing stylet.

longicaudatus has also been recently introduced into southern California where it is causing problems in golf course greens (*30*).

Current management of phytoparasitic nematodes for perennial crops such as turfgrass is largely dependent upon postplant application of restricted use organophosphate pesticides, such as fenamiphos (ethyl 3-methyl-4[methylthio]phenyl[1-methylethyl]phosphoramidate) (*27*). Fenamiphos is nematostatic at the concentrations achieved in the field and usually requires multiple applications for short-lived (< 4 weeks) suppression of phytoparasitic nematode populations (*20, 21, 22, 24, 43*). Chronic exposure of nematodes and the soil microflora to sublethal doses of nematicides can encourage microbial decomposition of the parent molecule into less toxic or non-toxic metabolites (*28)* lowering the dose and exposure time below efficacious levels for sting nematode control (*24*). Also, cultural practices in the turf environment can increase the probability that water soluble chemicals, such as the metabolites of fenamiphos (fenamiphos sulfoxide and fenamiphos sulfone) will not persist long enough in the rhizosphere to be efficacious (*13, 24, 38*). There are environmental risks in using pesticides like fenamiphos as well. The low LD_{50} values of fenamiphos for non-target mammals, birds and fish and the possibilty for misapplication or groundwater contamination can create human and wildlife safety problems (*33*). Because of the foregoing problems with efficacy and safety, alternative management strategies for plant-parasitic nematodes in turfgrass are greatly needed.

Biological control using host specific (stenotrophic) members of the obligate, Gram-positive, nematode and cladoceran endoparasitic bacterial genus *Pasteuria*, may provide an alternative or supplement to traditional chemical control. These endospore-forming bacteria, most closely related to *Alicyclobacillus* spp. (formerly *Bacillus*) (*17)* attach to, and infest the host nematode's pseudocoelom via the cuticle (Figure 2). The parasitized nematode becomes incapable of reproduction and eventually is filled with sporulating endospores of the bacterium, that are released into the environment upon host disintegration (*4, 36*). Some forms of these bacteria attack juveniles and do not sporulate until the nematode becomes an adult female, e.g. *P. penetrans* sensu strictu in root-knot nematodes (*Meloidogyne* spp.) producing up to 2.5 X 10^6 spores (*4, 36*). Other species, such as *Pasteuria thornei* can attack and sporulate in juvenile or adult nematodes and produce only several hundred spores (*39*). The potential assets of *Pasteuria* as biological control agents of turfgrass nematodes are; 1) their apparent ability to persist for long periods of time (> 1 year), 2) apparent host specificity, 3) compatibility with pesticides, and 4) lack of environmental risk to humans and other non-target organisms (*15*). Spores of *Pasteuria* are resistant to desiccation and exposure to nematicides and have been reported adhering to, or infesting, 205 species of nematodes from 51 countries worldwide (*40*). Only a few species of the *P. penetrans* group are well characterized, however, and little is known about the ecology of the group in native or managed soil ecosystems (*2, 11, 25*).

Preliminary Survey and Greenhouse Studies with *Pasteuria* in Florida Turfgrass

Survey work done from 1985-1989 demonstrated that isolates of *Pasteuria* were widely distributed in bermudagrass fairway turf in southern Florida (*18, 25*). Five morphometrically distinct isolates of the bacteria were observed on five species of

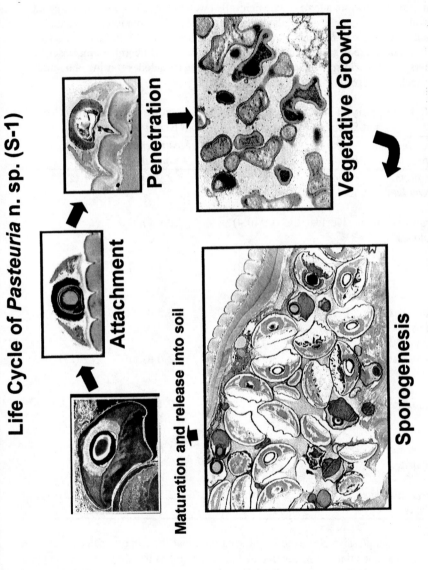

Figure 2. Transmission electron micrographs illustrating the life cycle of *Pasteuria* n. sp. (S-1).

plant-parasitic nematodes, including the sting nematode (Table I)(*15, 18, 25*). A one year greenhouse study demonstrated that soil, infested with a large-spored isolate of *Pasteuria* (6.1 μm endospore diameter) from sting nematode (Table I), that was heat-treated at 47° C for 72 hours to kill the nematodes but leave the bacteria viable, was suppressive to large populations of the sting nematode on potted 'Tifgreen II' bermudagrass grown in a temperature controlled water bath (27° C) when compared with soil that was autoclaved to kill all organisms (*18*). Soil containing this isolate was not suppressive to *B. longicaudatus* in the first 6 months (*18, 25*) but caused a significant decrease in sting nematodes after one year with concomitant increases in numbers of *Pasteuria* sp.-encumbered and filled sting nematodes starting at 6 months after inoculation (*18*).

Table I. Summary of Measurements of Mature Endospores from Different Species and Selected Isolates of *Pasteuria* from Nematode Hosts.

Nematode Host Species	Pasteuria Isolate	Mean Endospore Diameter (S.D.)	Mean Central Body Diameter (S.D.)	Ref.
Belonolaimus longicaudatus	Florida (S-1) n. sp.	6.1 (0.4)	2.9 (0.2)	(*25*)
Helicotylenchus microlobus	Florida	3.9 (0.4)	2.0 (0.2)	(*25*)
Heterodera spp.	*Pasteuria nishizawae*	5.3 (0.3)	2.1 (0.2)	(*37*)
Hoplolaimus galeatus	Florida (L-1)	7.3 (0.4)	3.5 (0.2)	(*25*)
Hoplolaimus galeatus	Florida (LS-1)	3.9 (0.3)	2.0 (0.2)	(*25*)
Meloidogyne spp.	*Pasteuria penetrans*	4.5 (0.3)	2.1 (0.2)	(*25, 36*)
Pratylenchus brachyurus	*Pasteuria thornei*	3.5 (0.2)	1.6 (0.1)	(*39*)
Tylenchorhynchus annulatus	Florida	4.6 (0.3)	2.5 (0.2)	(*25*)

This suppressive soil experiment suggested that the sting nematode isolate of *Pasteuria* had potential for inoculative biological control of the sting nematode in golf course greens and other turf or perennial cropping situations where small amounts of soil infested with the bacteria or *Pasteuria* encumbered nematodes could be used for inoculation into a site devoid of it. We hypothesized that high populations of sting nematodes in these perennial ecosystems could be used as the replicating and dispersal mechanism for a (currently) unculturable (*5*) obligate parasite. The United States Golf Association (USGA) funded a research project (1994-1998) to; 1) determine if this *Pasteuria* isolate was a new species, 2) show if it had potential for density dependent regulation of the sting nematode in natural populations in turfgrass, and 3) see if it could be successfully manipulated for biological control in the managed turfgrass ecosystem. This chapter summarizes the findings of that project.

A New Species of *Pasteuria*

The biology and taxonomy of *Pasteuria* have been obscure and difficult to study because of the small size and highly host-specific and obligate-parasitic nature of these interesting bacteria (*17, 36*). *In vitro* culture attempts have been largely unsuccessful owing to a lack of knowledge about the bacteria and the physiology of their nematode hosts (*5, 34*). Without reliable *in vitro* culture methods, traditional procedures for characterization of new bacterial species have not been available (*36*) forcing a reliance upon morphometrics, ultrastructural comparisons of mature endospores and development with scanning electron microscopy (SEM) and transmission electron microscopy (TEM), and host attachment and specificity trials. Recent work with PCR amplification and sequencing of 16S rDNA from *Pasteuria ramosa* and *P. penetrans* is allowing for phylogenetic hypotheses about the placement of the genus within the bacterial kingdom and providing independent data for generating hypotheses about species relationships (*1, 17*). Currently, there are four named species of *Pasteuria*; one from water fleas (cladocerans; *P. ramosa*) (*17*) and three from nematodes; *P. penetrans* from root-knot nematodes *Meloidogyne* spp. (*36*), *P. thornei* from lesion nematodes *Pratylenchus* spp. (*39*), and *P. nishizawae* from cyst nematodes *Heterodera* and *Globodera* (*37*).

   Pasteuria n. sp. (S-1) from the sting nematode from Ft. Lauderdale, Florida is considered a new species based upon ultrastructure, morphometrics, development, and host attachment studies (*26*). We are currently generating 16S rDNA sequence data for independent comparisons with other isolates and nominal species (Anderson et al. Unpubl. Data). Figure 3 depicts a schematic drawing of a transverse-section of a mature endospore of the new species with labelled morphological features to help with the following discussion. SEM work demonstrates that the external morphology of attached endospores of *Pasteuria* n. sp. (S-1) is significantly different from any of the named species. In SEM micrographs, some of the peripheral fibers of the endospore of *Pasteuria* n. sp. (S-1) protrude around the exposed spherical outer coat of the spore creating a crenate border which gives the endospore the appearance of a fried egg with a scalloped ring around the yolk (Figure 4B). Spores described from all of the other described nematodes appear like a "fried egg" without a scalloped border, except a recent isolate from *Hoplolaimus galeatus* from Peru (*12*). The sporangium and endospore diameters of *Pasteuria* n. sp. (S-1) were on the average at least 1.0 and 0.5 cm wider than these respective measurements for the other described species of *Pasteuria* or other host isolates of *Pasteuria* from southern Florida fairways (Table I).

   In TEM micrographs, the epicortical wall of *Pasteuria* n. sp. (S-1) surrounds the cortex in a sublateral band and the basal cortical wall thins to expose the inner endospore, similar to *P. thornei* but different from the other two species. The spore

Terms for Mature Endospore of *Pasteuria* n. sp. (S-1)

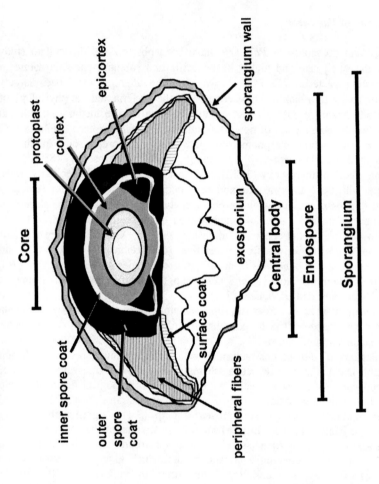

Figure 3. Schematic drawing of a mature endospore of *Pasteuria* n. sp. (S-1) illustrating morphological terms and terminology used for measurements.

pore diameter, measured from TEM micrographs, is larger than any other described species of *Pasteuria*. The outer coat wall thickness at its thickest point is 1/6 the diameter of the central body for *Pasteuria* n. sp. (S-1) compared with 1/8-1/30 for the other described species of *Pasteuria*. TEM comparisons of mature endospores of *Pasteuria* n. sp. (S-1) from sting from different geographical regions and host plants in Florida (Ft. Lauderdale from turfgrass, Orlando from turfgrass, and Hastings from potato) show consistency in the ultrastructure and morphometrics (Giblin-Davis and Williams unpubl. Data). This is consistent with a Linnean species concept or definitional species which is the best we can do for these bacteria until we have independent data from gene sequences or other biochemical or physiological data.

A brief description of the life cycle of *Pasteuria* n. sp. (S-1) based upon light microscopy (LM) and TEM follows; after attachment of a mature endospore to the cuticle of the host, penetration ensues via a germ tube through the cuticle into the pseudocoelom of the sting nematode (Figure 2). Spore attachment is hypothesized to be mediated via glycopeptide ligand receptors (spore adhesins) in the microfibrillary layer of the spore coat that recognizes and attaches to corresponding lectin ligands on the host nematode cuticle (*14, 32*).

All stages from second-stage juveniles (J2) through adults were observed with attached endospores of *Pasteuria* n. sp. (S-1) on the cuticle and J3 through adults were observed with internal infections of vegetative and sporulating *Pasteuria* n. sp. (S-1). Counts of spore-filled cadavers showed that juveniles (J3-J4; N = 5), males (N = 5), and females (N = 5) contained averages of 4,700, 1,483, and 3,633 spores/nematode, respectively (Hewlett and Giblin-Davis unpubl. Data). A mycelial microcolony is formed which increases in size in the pseudocoelom. Mycelial filaments are divided by septa and possess double-layered cell walls (Figure 2). Endospores are produced endogenously and the formation sequence (sporogenesis) for *Pasteuria* n. sp. (S-1) is similar to the other described species of *Pasteuria* and species of *Bacillus* (*9*).

Stages II-VII of sporogenesis were observed in *Pasteuria* n. sp. (S-1) with TEM. The forespore septum is formed during stage II. The forespore is fully engulfed and peripheral fibers appear during stage III. In stage IV, the cortex is formed, the sporangial wall appears, there is enlargement of the peripheral fibers, and there is dorso-ventral flattening of the endospore. Formation of the spore coats begins in stage V; the epicortex appears first and surrounds most of the cortex, thinning near the pore. As stage V progresses, the epicortex appears to migrate equatorially. The exosporium appears in stage VI, as the spore coats are still forming. The outer spore coat appears dorsally as a cap and descends around the cortex as the inner spore coat is formed. Concordant with the development of the outer and inner spore coats, the epicortical region continues to recede into a species characteristic sublateral band and the cytoplasm bounded by the exosporium is lessened. Spore maturation occurs during stage VII (Figures 2 and 4). The outer spore coat continues to thicken dorsally and laterally as the cytoplasm in between the spore and the exosporium disappears. Also, a microfibrillar layer (spore coat) appears on the dorsal and ventral surfaces of the peripheral fibers (Figures 2 and 4).

Host attachment studies have demonstrated that *Pasteuria* n. sp. (S-1) endospores only attach to different geographical and host isolates of *Belonolaimus*

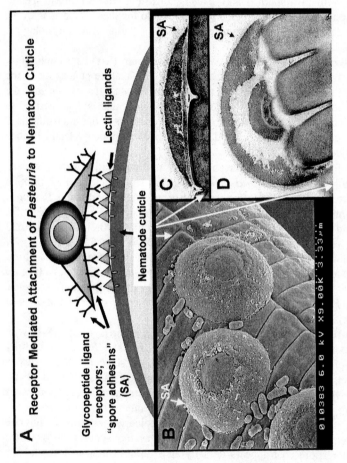

Figure 4.   Attachment of endospores of *Pasteuria* to the cuticle of nematodes. A. Schematic drawing of hypothetical receptor mediated attachment process for nematode-associated *Pasteuria*. B. Scanning electron micrograph of *Pasteuria* n. sp. (S-1) attached to the cuticle of a sting nematode. C and D. Transmission electron micrographs through the spore adhesin (SA) coat and peripheral fibers of *Pasteuria* n. sp. (S-1) attached to sting nematode cuticle.

longicaudatus (*26*) and *B. euthychilus*. This is consistent with field work that we have done in southern Florida greens and fairways where we only see the *Pasteuria* n. sp. (S-1) attaching and completing its life cycle in sting nematodes, even when there are many other species of nematodes present in the same sample (i.e. *Hoplolaimus galeatus*, *Tylenchorhynchus annulatus*, *Meloidogyne* spp., *Helicotylenchus microlobus*, *Hemicriconemoides annulatus*, *Criconemella ornata*, *Trichodorus proximus*, and several freeliving nematode species) (*18, 25*).

Laboratory Time-course Study of Sting Nematodes with or without *Pasteuria* n. sp. (S-1)

A laboratory pot assay was conducted to study the population dynamics of the sting nematode and compare the ability of *Pasteuria* to suppress the establishment of *B. longicaudatus* on FX-313 St. Augustinegrass [*Stenotaphrum secundatum*] using a previously described bioassay (*19*). Nematodes were extracted from soil using centrifugal flotation and attached spores of *Pasteuria* n. sp. (S-1) were visualized after staining with crystal violet stain. Treatments involved a harvest factor (harvested 42, 84, 126, 210, 308, and 392 days after inoculation) and a *Pasteuria* encumbrance factor. There were four treatments; 1) no sting nematodes with no bacteria, 2) sting nematodes (99 $\pm$ 10) with no bacteria, 3) sting nematodes (99 $\pm$ 10) + 10 sting nematodes encumbered with 8 $\pm$ 6 spores of *Pasteuria* n. sp. (S-1), and 4) sting nematodes (99 $\pm$ 10) + 25 sting nematodes encumbered with 8 $\pm$ 6 spores of *Pasteuria* n. sp. (S-1). Spore encumbered *B. longicaudatus* were harvested from a Ft. Lauderdale, FL, field site with *Pasteuria* n. sp. (S-1). The resulting 20 combinations were arranged in a randomized complete block design with 9 replications for time periods 42, 84, and 126 days and 6 replications for the 210, 308, and 392 day harvest times.

The hypothesis was that inoculated sting nematodes that were encumbered with spores of *Pasteuria* n. sp. (S-1) would eventually succumb during sporogenesis, releasing bacteria that would negatively affect the healthy population. Unfortunately, the results did not confirm this hypothesis. Population dynamics of the healthy sting nematodes were significantly increased by the addition of "encumbered" nematodes (high *Pasteuria* treatment was highest at 84 days, the low *Pasteuria* treatment was highest at 126 days, and the no *Pasteuria* treatment did not peak until 168 days) suggesting that spore encumbrance is not a good indicator of spore production potential. Root dry weights for the different treatments confirmed that root loss was greatest in the treatments receiving the most nematodes. Although spore encumbered sting nematodes were recovered throughout most of the 390 day study the levels were never greater than 1% from treatments receiving spores. This suggests that inoculative release of "encumbered" nematodes is unacceptable for establishment and population suppression work.

Seasonal Depth Survey of Hybrid Bermudagrass Sites with Different Levels of Sting Nematode and *Pasteuria* n. sp. (S-1)

In 1995, a monthly survey was begun of six different sites of fairway-managed hybrid bermudagrass at the Ft. Lauderdale Research and Education Center where

Pasteuria n. sp. (S-1) occurs naturally at different levels to monitor its suppressive effects on sting nematodes at three different soil depths (0-10 cm, 10-20 cm and 20-40 cm). This study was completed 24 months later in April, 1997. Weekly maximum and minimum soil temperatures were recorded at 5, 15, 25, and 40 cm depths at a central on-site location using Fisher Digital internal/external thermometers with external sensors. About 30 sites around the station were pre-sampled for sting nematodes and *Pasteuria* (S-1) n. sp. for final selection of the six locations used. Nematode and bacterial sampling and root dry weights were collected as described above.

Greater than 98% of all roots were recovered from the top 10 cm of soil which is consistent with previous studies (*23*). Maximum and minimum soil temperatures were most extreme at the 5 cm soil depth and the level of temperature fluctuation was the least variable at the 40 cm soil depth. After 24 months of sampling, sting nematode significantly suppressed root dry weight (P < 0.01) in a density dependent manner (y = -0.0003x + 0.2105; r = 0.207; x = sting nematode number, y = root dry weight). Also, *Pasteuria* n. sp. (S-1) suppressed sting nematode in each of the six survey locations. Regression analyses of density dependence of *Pasteuria* n. sp. (S-1) parasitism of sting nematodes (juvenile, male and female nematodes combined) demonstrated that the suppression was density dependent at each of the six locations surveyed for depths 1 and 2, and all three depths combined but not always at depth 3. Plots of the percentage of sting nematodes encumbered with *Pasteuria* n. sp. (S-1) and sting nematode counts versus time showed that survey locations that started with low levels of spore encumbrance slowly built in encumbrance levels with a corresponding decline in the numbers of sting nematodes over time. Locations starting with high spore encumbrance levels cycled and appeared to suppress sting nematode population resurgence, suggesting that *Pasteuria* n. sp. (S-1) might help produce suppressive soil for the sting nematode in a turfgrass ecosystem.

Effects of *Pasteuria* (S-1) Infested Soil on Sting Nematode in a Bermudagrass Green

Soil that was infested with S-1 *Pasteuria* was collected from a bermudagrass fairway area on 29 February 1996. Soil was mixed uniformly, cleaned of roots and rock with a #10 U.S.A. Standard Testing Sieve (2.00 mm openings), sampled for sting nematode and spore counts, and divided in half. One half of the soil was autoclaved for 90 min. and dried in a drying oven at 45-46° C for 48 hours to kill all *Pasteuria*. The other half of the soil was only dried in the drying oven at 45-46° C for 48 hours to kill nematodes but leave the *Pasteuria* alive. An area of 'Tifdwarf' bermudagrass (certified from Tifton, Georgia stock; obtained from Rapid-Turf) that had been established in 20 X 25 ft plots since 20 November 1992 was pre-sampled for nematodes and discovered to have high counts (about 200 sting nematodes per 100 mls of soil) without *Pasteuria* n. sp. (S-1). This area was marked in a grid of 1 m² plots with 15 cm borders and pre-sampled and stratified for sting nematode counts. Twenty sting nematodes were stained and examined from each of the twenty plots and unfortunately one nematode was found with one attached spore of *Pasteuria* n. sp. (S-1). This suggests that there may have been very low background contamination of the plots with *Pasteuria* n. sp. (S-1). The study was continued in spite of this find because of the amount of time and energy that had already been expended to set up the experiment. Plots with equal sting counts were paired and inoculated with 900 g of soil (autoclave treatment versus dried soil). There were 10 replications for each treatment. Inoculations were done on 28 March 1996 by removing a 15 cm core from the center of each plot and triming 900 g of soil from the bottom of it. The treated soil was added into the inoculation hole and covered with the trimmed core.

The test plots were monitored about every 6 months for 18 months at the center of the plot, at 25 cm from the center, and at 50 cm from the center. In the first 6 months, *Pasteuria* encumbrance increased significantly at the point of inoculation (center) and 25 cm from the center of each plot receiving soil that was not autoclaved (live *Pasteuria* n. sp. [S-1]) (Table II). Contamination of the control plots (autoclaved) with *Pasteuria* n. sp. (S-1) was evident at 6 months after inoculation and built-up through 18 months (Table II). The contamination could have been due to a build-up of the very small but detectable level of *Pasteuria* n. sp. (S-1) in the plots at the start of the experiment (Table II). Also, *Pasteuria* n. sp. (S-1)-infested nematodes from treated areas could have spread the bacteria or equipment might have helped spread spores. At 13 and 18 months after treatment, there were significantly less sting nematodes in the *Pasteuria* n. sp. (S-1)-treated plots than in the autoclaved (control) treatment at the center and at 25 cm from inoculation (Table III). This suggests that the significant increase in the proportion of sting nematodes encumbered with *Pasteuria* n. sp. (S-1) observed at 6 months after treatment led to a decline in the populations of the sting nematodes by 13 months after treatment. At 18 months after inoculation, the mean sting nematode counts were different but less disparate for both treatments because of an apparent delayed epizootic of *Pasteuria* n. sp. (S-1) caused by contamination of the control plots.

TABLE II. Percentage encumbrance of *Belonolaimus longicaudatus* with *Pasteuria* n. sp. (S-1) in a soil inoculation study on a 'Tifdwarf' hybrid bermudagrass green.

Treatment	START 0 - 50 cm	6 months			13 months			18 months		
		0 cm	25 cm	50 cm	0 cm	25 cm	50 cm	0 cm	25 cm	50 cm
Autoclave	0	8b	12b	10 b	26b	51b	33b	72	61	70b
Heat-treat	0.5	91a	75a	35a	85a	84a	77a	67	100	100a

Means in column with different letters are significantly different with a Student's t-test at P < 0.05.

These data are consistent with previous greenhouse work (*18*) and with the 2-year field survey (see above) where density dependent regulation of sting nematodes caused by parasitism of *Pasteuria* n. sp. (S-1) appeared to occur. These data are very encouraging because they suggest that a relatively small amount of *Pasteuria* n. sp. (S-1)-infested soil can be inoculated into a USGA site with high numbers of sting nematodes to bring about density dependent regulation within a relatively short period of time. It is important to emphasize the necessity of inoculation of *Pasteuria*

422

n. sp. (S-1) into sites with high numbers of sting nematode because this obligate parasitic bacterium requires large numbers of hosts to build-up sufficiently to cause an epizootic.

TABLE III. Counts of *Belonolaimus longicaudatus* per 100 mls of soil from a *Pasteuria* n. sp. (S-1) soil inoculation study on a 'Tifdwarf' hybrid bermudagrass green.

Treatment	START 0 - 50 cm	6 months 0 cm	25 cm	50 cm	13 months 0 cm	25 cm	50 cm	18 months 0 cm	25 cm	50 cm
Autoclave	296	100	183	139	143a	123a	109a	14a	9a	8a
Heat-treat	294	88	103	129	7b	33b	87b	0.3b	0.4b	0.3b

Means in column with different letters are significantly different with a Student's t-test at P < 0.05.

In addition to the destruction of sting nematode hosts after sporogenesis of *Pasteuria* n. sp. (S-1), the precipitous decline in sting nematodes may have been caused by differences in the ability of spore encumbered nematodes to move relative to unencumbered nematodes. Sting nematode is relatively large and requires coarse sand soils (> 80% sand) to survive. An increase in clay or silt to > 20% prevents this nematode from prospering (*31*). *Pasteuria* n. sp. (S-1) is a large spore (Table I) that could physically hinder the movement of these nematodes after attachment preventing the amphimictic nematodes from reaching each other for mating and/or for moving to feeding sites on plant roots. It is possible that the receptor mediated attachment of surface adhesins on spores of *Pasteuria* to the cuticles of plant-parasitic nematodes (Figure 4) may be potentially useful for the development of strategies to selectively bind to and tie up sting nematodes or as selective targeting tools for attaching toxins to the nematode for biorational control strategies.

Acknowledgments

Special thanks to Don W. Dickson, John L. Cisar, Barbara J. Center, Donna Williams, Tom Hewlett, Z. X. Chen, Henry Aldrich, James F. Preston, William Wergin, Janete Brito, Ole Becker, Sadia Bekal, Monica Elliott, Jennifer Anderson and Marcus Prevatte for help, discussions and ideas used in this project.

References

1.   Anderson, J. M.; Maruniak, J. E.; Preston, J. F.; Dickson, D. W. *J. Nematol.* **1998**, 30, (in press).
2.   Atibalentja, N.; Noel, G. R.; Liao, T. F.; Gertner, G. Z. *J. Nematol.* **1998**, 30, 81-92.
3.   Barker, K. R. *Plant & Soil Nematodes: Societal Impact and Focus for the Future;* Report by the Committee on National Needs and Priorities in Nematology. Society of Nematologists: Lakeland, FL, 1994; 11 pp.
4.   Bird, A. F.; Bird, J. *The Structure of Nematodes.* Academic Press, Inc.; San Diego, CA, 1991; Chapter 12, pp. 316.
5.   Bishop, A. H.; Ellar, D. J. *Biocontrol Science Technology.* **1991**, 1, 101-114.

6. Blaxter, M. L.; De Ley, P.; Garey, J. R.; Liu, L. X.; Scheldeman; Vierstraete; Vanfleteren, J. R.; Mackey, L Y.; Dorris, M; Frisse, L. M.; Vida, J. T.; Thomas, W. K. *Nature.* **1998**, 392, 71-75.
7. Busey, P.; Giblin-Davis, R. M.; Center, B. J. *Crop Science.* **1993**, 33, 1066-1070.
8. Busey, P.; Giblin-Davis, R. M.; Riger, C. W.; Zaenker, E. I. *Suppl. J. Nematol.* **1991**, 23, 604-610.
9. Chen, Z. X.; Dickson, D. W.; Freitas, L. G.; Preston, J. F. *Phytopathology.* **1997**, 87, 273-283.
10. Cherry, T.; Szalanski, A. L.; Todd, T. C.; Powers, T. O. *J. Nematol.* **1997**, 29, 23-29.
11. Cianco, A. *Phytopathology.* **1995**, 85, 144-149.
12. Cianco, A.; Vega Farfan, V.; Carbonell Torres, E.; Grasso, G. *J. Nematol.* **1998**, 30, 206-210.
13. Cisar, J. L.; Snyder, G. H. *Intl. Turfgrass Soc. Res. J.* **1993**, 7, 971-977.
14. Davies, K. G.; Danks, C. *Nematologica.* **1993**, 39, 53-64.
15. Dickson, D. W.; Oostendorp, M.; Giblin-Davis, R. M.; Mitchell, D. J. In *Pest Management in the Subtropics. Biological Control - A Florida Perspective.* Rosen, D.; Bennett, F. D.; Capinera, J. L., Eds. Intercept Ltd., Andover, U.K., 1994, Chapter 26, pp 575-601.
16. Dunn, R. A.; Noling, J. W. *1995 Florida Nematode Control Guide;* Series SP-54; University of Florida Cooperative Extension Service: Gainesville, FL, 1995; Chapter 2, 170 pp.
17. Ebert, D.; Rainey, P.; Embley, T. M.; Scholz, D. *Phil. Trans. R. Soc. Lond. B.* **1996**, 351, 1689-1701.
18. Giblin-Davis, R. M. *Proc. Fl. State Hort. Soc.* **1990**, 103, 349-351.
19. Giblin-Davis, R. M.; Busey, P.; Center, B. J. . *J. Nematol.* **1992**, 24, 432-437.
20. Giblin-Davis, R. M.; Cisar, J. L.; Bilz, F. G. *Suppl. J. Nematol.* **1988**, 20, 46-49.
21. Giblin-Davis, R. M.; Cisar, J. L.; Bilz, F. G. *Nematropica.* **1988**, 18, 117-127.
22. Giblin-Davis, R. M.; Cisar, J. L.; Bilz, F. G.; Williams, K. E. *Nematropica.* **1991**, 21, 59-69.
23. Giblin-Davis, R. M.; Cisar, J. L.; Bilz, F. G.; Williams, K. E. *Suppl. J. Nematol.* **1992**, 24, 749-756.
24. Giblin-Davis, R. M.; Cisar, J. L.; Snyder, G. H.; Elliott, C. L. *Intl. Turfgrass Soc. Res. J.* **1993**, 7, 390-397.
25. Giblin-Davis, R. M.; McDaniel, L. L.; Bilz, F. G. *Suppl. J. Nematol.* **1990**, 22, 750-762.
26. Giblin-Davis, R. M.; Williams, D.; Hewlett, T. E.; Dickson, D. W. *J. Nematol.* **1995**, 27, 500.
27. Golf Course Superintendents Association of America. *1996 Golf Course Superintendents Report;* Golf Course Superintendents Assoc. Am.: Lawrence, KS, 1996, pp. 161.
28. Johnson, A. W. *J. Nematol.* **1998**, 30, 40-44.
29. Marquardt, W. C.; Demaree, R. S., Jr. *Parasitology;* Macmillan Publishing Co.: New York, NY, 1985; Chapter 30, pp 636.
30. Mundo-Ocampo, M.; Becker, J. O.; Baldwin, J. G. *Plant Disease.* **1994**, 78, 529.
31. Perry, V. G.; Rhoades, H. L. In *Nematology in the Southern United States;* Riggs, R. D., Ed.; Southern Cooperative Series Bulletin 276; Florida Agricultural Experiment Station: Gainesville, FL, 1982; Chapter 13, pp 144-149.
32. Persidis, A.; Lay, J. G.; Manousis, T.; Bishop, A. H.; Ellar, D. J. *J. Cell Science.* **1991**, 100, 613-622.
33. Peterson, D.; Winterlin, W.; Costello, L. R. *California Agriculture.* **1986**, 46, 26-27.

424

34. Riese, R. W.; Hackett, K. J.; Sayre, R. M.; Huettel, R. N. *J. Nematol.* **1988,** 20, 657.
35. Robbins, R. T.; Barker, K. *J. Nematol.* **1974,** 6, 1-6.
36. Sayre, R. M.; Starr, M. P. *Proc. Helminth. Soc. Wash.* **1985,** 52, 149-165.
37. Sayre, R. M.; Wergin, W. P.; Schmidt, J. M.; Starr, M. P. *Res. Microbiol.* **1991,** 142, 551-564.
38. Snyder, G. H.; Cisar, J. L. *Intl. Turfgrass Soc. Res. J.* **1993,** 7, 978-983.
39. Starr, M. P.; Sayre, R. M. *Ann. Inst. Pasteur/Microbiol.* **1988,** 139, 11-31.
40. Sturhan, D. *Nematologica.* **1988,** 34, 350-356.
41. Todd, T. C.; Tisserat, N. A. *Plant Disease.* **1990,** 74, 660-663.
42. Todd, T. C.; Tisserat, N. A. *Golf Course Management* **1993,** 38.
43. Veech, J. A.; Dickson, D. W. *Vistas on Nematology: A Commemoration of the Twenty-fifth Anniversary of the Society of Nematologists;* Society of Nematologists: Hyattsville, MD, 1987; Chapter 61, 509 pp.

INDEXES

Author Index

Subject Index

442

Bestsellers from ACS Books

The ACS Style Guide: A Manual for Authors and Editors (2nd Edition)
Edited by Janet S. Dodd
470 pp; clothbound ISBN 0–8412–3461–2; paperback ISBN 0–8412–3462–0

Writing the Laboratory Notebook
By Howard M. Kanare
145 pp; clothbound ISBN 0–8412–0906–5; paperback ISBN 0–8412–0933–2

Career Transitions for Chemists
By Dorothy P. Rodmann, Donald D. Bly, Frederick H. Owens, and Anne-Claire Anderson
240 pp; clothbound ISBN 0–8412–3052–8; paperback ISBN 0–8412–3038–2

Chemical Activities (student and teacher editions)
By Christie L. Borgford and Lee R. Summerlin
330 pp; spiralbound ISBN 0–8412–1417–4; teacher edition, ISBN 0–8412–1416–6

Chemical Demonstrations: A Sourcebook for Teachers, Volumes 1 and 2, Second Edition
Volume 1 by Lee R. Summerlin and James L. Ealy, Jr.
198 pp; spiralbound ISBN 0–8412–1481–6
Volume 2 by Lee R. Summerlin, Christie L. Borgford, and Julie B. Ealy
234 pp; spiralbound ISBN 0–8412–1535–9

The Internet: A Guide for Chemists
Edited by Steven M. Bachrach
360 pp; clothbound ISBN 0–8412–3223–7; paperback ISBN 0–8412–3224–5

Laboratory Waste Management: A Guidebook
ACS Task Force on Laboratory Waste Management
250 pp; clothbound ISBN 0–8412–2735–7; paperback ISBN 0–8412–2849–3

Reagent Chemicals, Eighth Edition
700 pp; clothbound ISBN 0–8412–2502–8

Good Laboratory Practice Standards: Applications for Field and Laboratory Studies
Edited by Willa Y. Garner, Maureen S. Barge, and James P. Ussary
571 pp; clothbound ISBN 0–8412–2192–8

For further information contact:
Order Department
Oxford University Press
2001 Evans Road
Cary, NC 27513
Phone: 1-800-445-9714 or 919-677-0977

Highlights from ACS Books